1995

Regulation of Atmospheric CO$_2$ and O$_2$ by Photosynthetic Carbon Metabolism

Regulation of Atmospheric CO$_2$ and O$_2$ by Photosynthetic Carbon Metabolism

Edited by

N. E. Tolbert

Jack Preiss

New York   Oxford

Oxford University Press

1994

Oxford University Press

Oxford   New York   Toronto
Delhi   Bombay   Calcutta   Madras   Karachi
Kuala Lumpur   Singapore   Hong Kong   Tokyo
Nairobi   Dar es Salaam   Cape Town
Melbourne   Auckland   Madrid

and associated companies in
Berlin   Ibadan

Copyright © 1994 by Oxford University Press, Inc.

Published by Oxford University Press, Inc.,
200 Madison Avenue, New York, New York 10016

Oxford is a registered trademark of Oxford University Press

Library of Congress Cataloging-in-Publication Data
Regulation of atmospheric CO_2 and O_2 by photosynthetic carbon metabolism
edited by N. E. Tolbert, Jack Preiss.
p. cm.   Includes bibliographical references and index.
ISBN 0-19-507932-9
1. Atmospheric carbon dioxide.
2. Photosynthesis.
3. Plants, Effect of atmospheric carbon dioxide on.
I. Tolbert, N. E. (N. Edward), 1919–
II. Preiss, Jack, 1932–
QC879.8.R44   1994
551.5'11—dc20   93-4724

9 8 7 6 5 4 3 2 1

Printed in the United States of America
on acid-free paper

Contents

CONTRIBUTORS

Joseph A. Berry
Carnegie Institution of Washington
Department of Plant Biology
290 Panama Street
Stanford, California 94305-4101

Clanton C. Black
Biochemistry Department
The School of Chemical Sciences
Boyd Graduate Studies Research Center
University of Georgia
Athens, Georgia 30602

G. James Collatz
Carnegie Institution of Washington
Department of Plant Biology
290 Panama Street
Stanford, California 94305-4101

William R. Emanuel
Environmental Sciences Division
Oak Ridge National Laboratory
Oak Ridge, Tennessee 37831-6335

Victoria J. Fabry
Department of Biology
California State University
San Marcos, CA 92096

Marilyn D. Fogel
Carnegie Institution of Washington
Geophysical Department
5251 Broadbranch Road
Washington, D.C. 20015-1305

Hideya Fukuzawa
Laboratory of Plant Molecular Biology
Department of Agricultural Chemistry
Kyoto University
Kyoto 606-01, Japan

Joanna Gemel
The Interdisciplinary Plant Group
University of Missouri
Columbia, Missouri 65211

Larry Giles
Botany Department
Duke University
Durham, North Carolina 27706

Robert D. Guy
Department of Forest Science
University of British Columbia
Vancouver, British Columbia
V6T 1W5, Canada

Osmund Holm-Hansen
Marine Research Division
Scripps Institution of Oceanography
University of California
La Jolla, California 92093-0202

A. M. Johnston
Department of Biological Sciences
University of Dundee
Dundee DD1 4HN, Scotland

Anthony W. King
Environmental Sciences Division
Oak Ridge National Laboratory
Oak Ridge, Tennessee 37831-6335)

Dan Lubin
California Space Institute
Scripps Institution of Oceanography
University of California
La Jolla, California 92093-0221

Jan A. Miernyk
The Interdisciplinary Plant Group
University of Missouri
Columbia, Missouri 65211

Shigetoh Miyachi
Marine Biotechnology Institute Co. Ltd.
Head Office, 2-35-10 Hongo
Bunkyo-ku, Tokyo 113, Japan

John Morrison
VG Isogas Ltd.
Aston Way
Middlewich, Cheshire
CW10 0HT England

William L. Ogren
Agricultural Research Service
U.S. Department of Agriculture
1102 S. Goodwin Avenue
Urbana, Illinois 61801

Donald R. Ort
Photosynthesis Research Unit
U.S. Department of Agriculture
Agricultural Research Service &
   Departments of Plant Biology and
   Agronomy
University of Illinois
Urbana, Illinois 61801

Barry Osmond
Research School of Biological Sciences
Australian National University
GPO 475
Canberra Act 2601, Australia

Wilfred M. Post
Environmental Sciences Division
Oak Ridge National Laboratory
Oak Ridge, Tennessee 37831-6335

Jack Preiss
Department of Biochemistry
Michigan State University
East Lansing, Michigan 48824

Douglas D. Randall
The Interdisciplinary Plant Group
University of Missouri
Columbia, Missouri 65211

J. A. Raven
Department of Biological Sciences
University of Dundee
Dundee DD1 4HN, Scotland

T. ap Rees
Botany School
University of Cambridge
Downing Street
Cambridge CB2 3EA
United Kingdom

S. Steven Sikes
The Mineralization Center
LSB 124
Department of Biological Sciences
University of South Alabama
Mobile, Alabama 36688

Helmut Stabenau
University of Oldenburg
Fachbereich Biologie
Postfach 2503
D-2900 Oldenburg, Germany

Irwin P. Ting
Department of Botany and
Plant Sciences
University of California
Riverside, California 92521

N. E. Tolbert
Department of Biochemistry
Michigan State University
East Lansing, Michigan 48824

James C. G. Walker
Space Physics Research Laboratory
Department of Atmospheric, Oceanic
    and Space Sciences and Department of
    Geological Sciences
The University of Michigan
Ann Arbor, Michigan 48109

Dan Yakir
Department of Environmental
    Sciences and Energy
    Research
The Weizmann Institute
Rehovot 76100, Israel

Regulation of Atmospheric CO$_2$ and O$_2$ by Photosynthetic Carbon Metabolism

Overview by Editors

N. E. TOLBERT and JACK PREISS

We organized this symposium to discuss the role of photosynthesis in regulating atmospheric CO_2 and O_2. We therefore asked the contributors not to concentrate on the effect of increasing atmospheric CO_2 on plants but, rather, to deal with the effect of plants on the atmospheric CO_2. Unfortunately, there has been little research specifically dealing with photosynthetic regulation of atmospheric CO_2 beyond the concept of a biological global carbon cycle whereby photosynthesis removes the CO_2. Consequently, the contributors reverted in part to the effect of increasing atmospheric CO_2 on plants and photosynthetic productivity, for which there are data and much interest. These two aspects of the subject of photosynthesis and increasing atmospheric CO_2 have in common the same basic processes for CO_2 fixation and O_2 evolution.

The chapters of this volume have been organized around the topic of regulation of atmospheric CO_2 and O_2. Because they represent the proceedings of a symposium developed to recognize the transition of Professor Nathan Edward Tolbert to Emeritus status, it is he who opened the symposium with an overview on this topic based on his work and that of our selection of speakers. There are two contributions from outside the subject area of photosynthesis: environmentalists W. R. Emanual et al. provide data on the magnitude of currently increasing atmospheric CO_2, and J. C. G. Walker gives a geologist's viewpoint about long-term changes in global carbon pools. The rest of this book is concerned with short-term changes in atmospheric CO_2. Impressive recent data on ultraviolet-B (UV-B) inhibition of photosynthetic CO_2 reduction are provided, in the chapter by O. Holm-Hansen, since such a reduction from decreasing ozone should compound our concern about controlling or limiting the increase in atmospheric CO_2.

Tolbert reviews the self-regulating role of the carboxylase versus the oxygenase activity of Rubisco which controls net photosynthetic CO_2 reduction. Symposium discussions of the enzymatic properties of Rubisco are not included in this volume, but numerous references to Rubisco and other aspects of plant biochemistry are found in the 16 volume series *The Biochemistry of Plants*, edited by P. Stumpf and E. Conn (Academic Press, 1980). On photosynthetic carbon metabolism, J. Preiss describes the C_3 reductive photosynthetic carbon cycle which, with higher atmospheric CO_2, leads to increases in sucrose and starch synthesis. These increases are in general beneficial to the plant, as sucrose and starch synthesis represent the first steps in the reduction of atmospheric CO_2 for storage of carbon as plant biomass. The regulatory role of the C_2 oxidative photosynthetic carbon cycle to waste excess photosynthetic energy through photorespiration is detailed in three chapters by Tolbert, Ogren, and Randall. Calculations of the energy drain due to photorespiration vary and are complicated by numerous reactions and by variations in gross photosynthetic and photorespiratory rates due to physiological changes, particularly temperature, CO_2 concentration, water use efficiency, and light intensity. Ogren based his calculations of energy loss from the reactions of the C_2 cycle, whereas Tolbert also included regeneration of the ribulose bisphosphate from phosphoglycerate and refixation of the CO_2 as part of the total expense of phororespiration. In his chapter, A. P. Rees notes that the energy and CO_2 loss from dark respiration are additional factors in reducing net plant biomass and net CO_2 removal from the air. A further complication, as discussed by Randall, deals with photorespiratory inhibition of dark respiration. Those who want to simplify this topic can reason that in the light the cell has adequate energy (ATP and NADPH) from photosynthesis and thus needs less dark respiration. Down-regulation of mitochondrial dark respiration may also be controlled by photorespiration through high intercellular pools of ATP and reduced pyridine nucleotides.

Traditionally, photosynthetic research has been divided between (1) light reactions plus electron transport with ATP and NADPH synthesis, and (2) the reactions of photosynthetic carbon metabolism (C_2, C_3, and C_4 cycles). It is natural that global regulation of atmospheric CO_2 would deal primarily with photosynthetic carbon metabolism, since, globally, light intensity is overall in excess. Of course, the chloroplastic light reactions can also limit photosynthesis, and Dr. Ort explores their regulation in controlling the final atmospheric CO_2 and O_2 levels.

To increase the efficiency of photosynthetic CO_2 reduction from low limiting levels of CO_2, photosynthetic tissues have, in addition to their C_3 and C_2 cycles, CO_2-concentrating processes—namely, the C_4 cycle in C_4 plants and the algal dissolved inorganic carbon (DIC) concentrating mechanisms or pumps. These CO_2 pumps could lower the atmospheric CO_2 well below the equilibrium level achieved with only C_3 plants. In the past, these CO_2 pumps have been investigated in the hope of increasing plant and algal growth. C. C. Black and I. Ting have discussed the C_4 cycle in C_4 and Crassulacean acid metabolism (CAM) plants. J. A. Raven with A. M. Johnson, C. C. Black, and N. E. Tolbert have speculated on the overall potential of using these CO_2 concentrating processes to reduce atmospheric CO_2. Major components in the algae DIC pump as well as in plants are carbonic anhydrases that facilitate conversion of HCO_3^- to CO_2. S. Miyachi discusses recent research on this enzyme. C. S. Sikes explores the limited information available on the association of the algal CO_2 pumps with carbonate deposition, which has long been a major CO_2 sink. Marine biologists particularly need to expand their knowledge about the role of photosynthesis in carbonate formation and whether it can be increased as a CO_2 sink for increasing atmospheric CO_2. An alternative would be to increase the number of C_4 plants to lower the atmospheric CO_2, as described by C. C. Black. However, C. Somerville (unpublished) has concluded that the phenotypic changes would be too numerous for this molecular modification to be feasible, and N. E. Tolbert, in his overview, points out that at higher CO_2 levels C_4 plants actually become less efficient at utilizing CO_2 than C_3 plants.

The concepts of oxygen inhibition of photosynthesis and photosynthetic CO_2 compensation points are well known and have been used in the past to differentiate between C_3 plants with a high (40 ppm) CO_2 compensation point and C_4 plants with a low (1–2 ppm) CO_2 compensation point. These same principles apply to the atmospheric CO_2 equilibrium level, which conceptually amounts to a higher CO_2 compensation point from additional CO_2 input by the global carbon cycle. J. Berry discusses the CO_2 equilibrium, but otherwise there are few older and no recent reviews on the CO_2 compensation point. Even though the established phenomenon of O_2 inhibition of photosynthesis is used in this volume to explain photorespiration and the CO_2 compensation point, the concept of O_2 compensation points has not been used to evaluate the atmospheric O_2 level (see the introductory chapter by N. E. Tolbert). B. Osmond's leadership in this field has been demonstrated by reviews on the regulatory role of photorespiration and in his chapter in this book, on O_2 effects. We hope that this book leads to further discussion about why the atmospheric O_2 is at 21% and whether it can be altered by changes in atmospheric CO_2 over long time periods.

One cannot help but be encouraged by the fact that the steps for photosynthetic CO_2 reduction to carbohydrate by C_3 plants proceed equally well from about 40 to well over 1000 ppm CO_2 (or even to 1–2% CO_2), although the physiological effects and predicted environmental consequences should seriously affect overall plant growth. In this statement we have reverted to the effect of higher CO_2 on plants, and we have to force ourselves to restate the data to recognize that from 1000 down to 40 ppm CO_2, the rate of photosynthetic CO_2 removal from the atmosphere by C_3 plants is slowed down because more of the photosynthetic energy is wasted on photorespiration. As a consequence, the lower limit of atmospheric CO_2 has been predetermined by the competing properties of photo-synthetic CO_2 fixation and photorespiratory waste of excess photosynthetic energy. In this range of 1000 to 40 ppm CO_2, a global rate-limiting factor for growth of plants is the CO_2 concentration. Many other components (temperature, nutrient, moisture, and light) also affect the rate of photosynthesis at any given moment and place. Even though these other factors are seldom optimum, restrictions from low CO_2 reduce net photosynthesis even further. Photosynthetic carbon metabolism, as the sum of CO_2 reduction and photorespiration, regulates the net rate of photosynthesis and the global carbon cycle on a short time basis.

Because increasing atmospheric CO_2 increases photosynthesis, it is itself considered beneficial to plants and, thus, the attitude in agriculture is of less concern. Plants and algae evolved in high CO_2 and pulled the atmospheric CO_2 down to an equilibrium level of 180 to 285 ppm CO_2—which has probably existed for millions of years. If (or, better, as) the atmospheric CO_2 suddenly doubles in a geological moment, the photosynthetic process in the dominant C_3 plants can use the higher CO_2 to produce more carbohydrates. However, physiological growth processes within the plants will adapt slowly to the higher carbohydrate level and overall plant growth will not increase proportionally to the CO_2 increase. In this book we have discussed how plants established that low CO_2 equilibrium level. If increasing atmospheric CO_2 were to cease, plants and algae should slowly reestablish the past low CO_2 level, which we now consider to be normal or safe. If we can explain these processes in molecular terms, might we be able to expedite the return to normal lower CO_2 levels by photosynthesis? The reader will not find answers to this question in this volume specializing in biochemical photosynthesis, but rather the subject needs consideration by a broader spectrum of research disciplines. It is hoped that geologists, plant geophysiologists, environmentalists, ecologists, and researchers in other disciplines concerned about the significance of increasing atmospheric CO_2 can obtain sufficient insights into the biochemistry of the photosynthetic CO_2 and O_2 atmospheric equilibria, so that we can all communicate better toward our common goal.

The chapters on regulation of atmospheric CO_2 reflect the many disagreements or divergence of views on this topic. The disparities are understandable and reflect the need for further work. There was agreement that atmospheric levels of CO_2 are increasing and that the increasing CO_2 will increase the net rate of CO_2 fixation by plants; however, the extent of this increase is uncertain, since other factors limiting plant growth will reduce the theoretically attainable increase in photosynthetic rate. Geologists studying the long-term global carbon cycle may conclude that short-term changes in photosynthesis are of little consequence; however, photosynthetic plants and other forms of biological life exist only on a short-term basis. That both research disciplines—geology and botany—influence each other is still not integrated into an understanding of the overall life support system on this planet. Investigators of photosynthetic carbon dioxide fixation have not integrated the C_2, C_3, and C_4 carbon cycles and algal CO_2 concentrating processes into one system for regulating atmospheric CO_2. There are many facets of this regulation, and it is hoped that the theme of this book will help extend the concept. But even within individual topics controversy remains. While Tolbert, Ogren, Randall, Osmond, and Berry concur on the steps in photorespiration and its reduction of net photosynthesis, they present such diverse details that investigators from other disciplines will hardly be able to arrive at a simple overview. Although we have marveled at the efficiency of CO_2 removal by C_4 plants, no consensus has been reached on whether these plants can be used effectively to lower the higher atmospheric CO_2 in the future. C. C. Black believes they will, but N. E. Tolbert points out why they may not, and C. Somerville said that converting C_3 plants into C_4 plants is not practical. Finally, the role of $CaCO_3$ deposition by marine photosynthetic systems to remove excess CO_2 is so poorly understood that everyone can find something with which to disagree or, at least, something requiring further investigation.

Different authors have used different phrases and acronyms to specify a particular term. These terms were not edited for consistency between chapters since the reader will find all of them in the literature. A partial list of the most common variant phrases and acronyms follows:

1. CO_2 concentrations in air may be expressed as 0.03% or 300 ppm or ppmv or 300 $\mu l \cdot L^{-1}$ or 300 μbar.
2. The atmospheric O_2 concentration is 20.976%, but 21% or 20 mbar is generally used.

3. Net CO_2 assimilation = net CO_2 fixation or photosynthesis = A_n.
4. Reductive photosynthetic carbon cycle = PCR cycle = C_3 cycle = Calvin cycle.
5. Oxidative photosynthetic carbon cycle = PCO cycle = C_2 cycle.
6. Ribulose-1,5-bisphosphate = ribulose-P_2 = RuP_2 = RuBP = RuDP.
7. 3-Phosphoglycerate = 3-PGA = PGA.
8. Rubisco = ribulose-1,5-bisphosphate carboxylase/oxygenase = RuBP carboxylase/oxygenase. Rubisco was previously called RUDP carboxylase or dismutase or protein fraction I.
9. C_3 plants have only the C_3 and C_2 cycles of photosynthetic carbon metabolism. C_3 plants include most of the temperate crop plants and all trees.
10. C_4 plants have, in addition to the C_3 cycle a C_4 cycle for concentrating atmospheric CO_2. C_4 plants include in general some tropical grasses, such as corn and sorghum.
11. DIC = dissolved inorganic carbon ($CO_2 + HCO_3^- + CO_3^{2-}$) = Ci.

Financial support for the symposium and in part for the publication of this book came from the Michigan State Alumni Association and the Department of Biochemistry at Michigan State University.

1

Role of Photosynthesis and Photorespiration in Regulating Atmospheric CO_2 and O_2

N. E. TOLBERT

Remember the turtle, he only makes progress when his neck is out.

WINSTON CHURCHILL

The increasing levels of atmospheric CO_2 are of international concern to many research disciplines from geology and ecology to the social sciences. In few of these discussions are the plant and algal scientists and investigators of photosynthesis well represented. Yet major factors in controlling atmospheric CO_2 and O_2 include photosynthetic CO_2 fixation and photorespiratory CO_2 release by plants and algae. In this volume, investigators of photosynthetic carbon metabolism, and a few others who study atmospheric CO_2, consider the role of plants and algae in regulating CO_2 and O_2 levels. Plant scientists have held numerous conferences on how increases in atmospheric CO_2 levels will increase the growth of plants and algae and their nutrient requirement, and alter their composition and ecological distribution (24). The present volume considers another aspect of this problem: how atmospheric CO_2 and O_2 equilibria are, in turn, regulated by plants and their pathways for photosynthetic CO_2 fixation. The discussions are based on the biochemistry of photosynthetic carbon metabolism. Only peripheral attention is given to the environment and paleobotany. The often neglected area of marine photosynthesis is included because of its potential for lowering atmospheric CO_2.

Almost everything plant biochemists write on the regulation of atmospheric CO_2 by photosynthesis is limited by omission of findings from related specialties. One conclusion to be drawn from this volume is the need for more interdisciplinary conferences on this complex subject. Conclusions drawn solely from knowledge of biochemical processes cannot simply be extrapolated and applied to whole plants and to the environment. Moreover, consensus is needed in as many basic principles in photosynthetic carbon metabolism as is possible, so that these may be communicated to researchers working in other areas of environmental investigation. For instance, plant scientists must focus on the reactions of photosynthesis and photorespiration that are most involved in maintaining the atmospheric CO_2 equilibrium. We need to explain photorespiration and plant CO_2

and O_2 compensation points, even to their own associates. We need to give priority to those topics needing further investigation—for example, the properties and potential modification of Rubisco. As experts on photosynthetic carbon metabolism, we have a responsibility to consider the feasibility of doing research on the regulation of the atmospheric CO_2 by plants and algae. Plant scientists are best qualified to judge certain hypotheses—for example, modification of Rubisco, incorporation of enzymes into more plants for concentrating CO_2, or use of biological calcification deposits as sinks for excess CO_2. It would be useful to investigate the feasibility and effectiveness of implementing other suggestions for reducing atmospheric CO_2, such as the planting of additional trees. In this volume we restrict ourselves mainly to research on Rubisco and photosynthetic CO_2 fixation and photorespiration, which are introduced in this chapter. This volume does not explore the latest advances in photosynthetic carbon metabolism, except for new results pertinent to the theme of regulation of atmospheric CO_2, such as UV inhibition of CO_2 fixation.

Biological processes are blanketed by regulation for controlling growth and development. This book deals with biological regulation of atmospheric CO_2 and O_2 by photosynthetic carbon metabolism. The complex regulation of photosynthetic electron transport, the C_3 cycle, and starch and sucrose synthesis are reviewed briefly to provide an overall understanding of photosynthetic control. Regulation of gross CO_2 fixation of photorespiration is emphasized because it appears to be most directly associated with atmospheric composition. The key enzyme, Rubisco, is highly regulated, but because most of these factors affect carboxylase and oxygenase activities similarly, they may not directly alter the atmospheric composition. Rather, regulation is achieved through the ratio of CO_2 to O_2 levels in the air. Biological regulation of photosynthetic carbon metabolism by the whole plant is yet an even more complex area of investigation than is the biochemistry. A complete review should include both physiological and ecological aspects such as gas diffusion, water use efficiency, and C_3 and C_4 plant distribution relative to differential effects between photosynthesis and photorespiration.

INCREASING ATMOSPHERIC CO_2

From analyses of glacier ice cores we know that preindustrial levels of atmospheric CO_2 for the past 160,000 years fluctuated between a low of 180 ppm in cold, glacier periods to a high around 285 ppm during warm, interglacial times (6). These changes in atmospheric CO_2 are consistent with the predicted greenhouse effects: low temperatures when the atmospheric CO_2 was low, and higher temperatures with higher CO_2. It is not known which was cause and which was effect, and whether plants were involved in causing these changes. After the industrial era started, atmospheric CO_2 was reported at 290 ppm in 1890, just 100 years ago (Fig. 1.1). When I was a graduate student after World War II, it was 303 ppm, and since 1958, when it had risen to 315 ppm, the CO_2 level has been continuously recorded (19). It is now about 370 ppm. During the research career of our present graduate students, the CO_2 level should rise to around 500 ppm. I hope that they will be able to do more than just talk about it. If one uses 250 ppm as the average

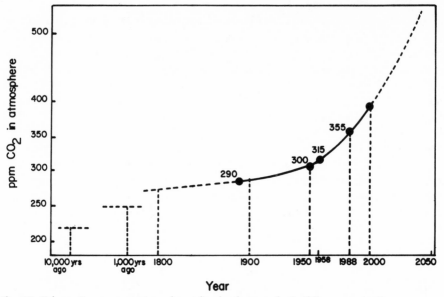

Fig. 1.1. Schematic representation of past levels of atmospheric CO_2 concentration.

preindustrial CO_2 concentration, the CO_2 level has now risen about 120 ppm or 48%, and in the next century it will have doubled to 500 ppm. Thus 500 ppm CO_2 (not 600 ppm as if starting in 1945), represents a doubling of the atmospheric CO_2 from previous normal values. As far as we know, such a high level of CO_2 has not occurred since an atmosphere consisting of 180–285 ppm CO_2 and 21% O_2 was established over 100 Myr ago. Changes in the environment and plant growth are expected and/or are already occurring as a result of this higher CO_2 level. But this volume deals primarily with our specialty—the photosynthetic carbon cycles as related to the global carbon cycles.

RUBISCO AND THE C_3 PHOTOSYNTHETIC AND C_2 PHOTORESPIRATORY CARBON CYCLES

All plants and algae have the same bifunctional enzyme for photosynthetic carbon metabolism (Fig. 1.2): ribulose bisphosphate carboxylase/oxygenase, or *Rubisco*. The carboxylase reaction initiates gross CO_2 fixation by the C_3 reductive photosynthetic carbon cycle, and the oxygenase reaction initiates the C_2 oxidative photosynthetic carbon cycle for energy and CO_2 loss by photorespiration. The two cycles coexist as competitive reactions for the two gaseous substrates, CO_2 and O_2 (18, 24, 28, 29, 35, 36). Together they are photosynthetic carbon metabolism, and together they control the rate for net photosynthesis. Photorespiration occurs only in the light because it is a part of photosynthetic carbon metabolism.

The reactions of the C_3 reductive photosynthetic carbon cycle (4) are discussed in Chapter 5. The key enzymes are activated in the light by increases in pH and Mg^{2+} concentration in the chloroplast stroma and by specific thioredoxins (except

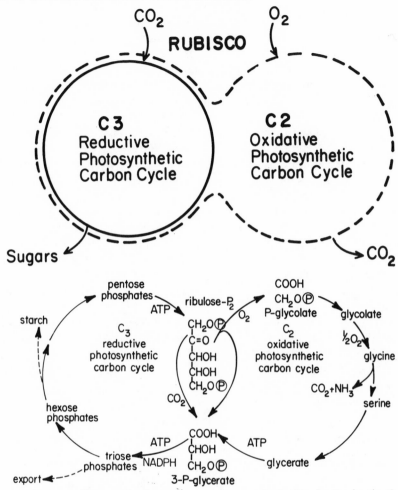

Fig. 1.2. Schemes to illustrate simultaneous gross photosynthetic CO_2 fixation by the C_3 cycle and photorespiration by the C_2 cycle. The most simple scheme (like a bicycle) is at the top; an intermediate scheme is at the bottom. See Figure 1.3 for the cycle that includes most of the chemical reactions.

for Rubisco). The overall rate is controlled by the availability of CO_2, as the high $K_m(CO_2)$ for Rubisco in air is over $26\,\mu M$. The substrate for the carboxylase reaction is CO_2, not bicarbonate. It has been estimated that free CO_2 is less than $5\,\mu M$ and is derived from CO_2 diffusion into the chloroplast stroma from low levels in air. In the chloroplasts at pH 8.3, most of the dissolved inorganic carbon (DIC) is present as bicarbonate. Thus the carboxylase reaction might not exceed 2% to 10% of V_{max} without increasing atmospheric CO_2 or utilizing CO_2-concentrating processes. In an apparent effort in evolution to counter this limitation up to half of the soluble protein in the chloroplast is now Rubisco, yet much of it seems to be in an inactive form when quickly extracted from the leaf. The two Rubisco reactions dominate all considerations of photosynthetic carbon regulation.

The $K_m(O_2)$ for the oxygenase activity of Rubisco is about 400 μM, which is available in aqueous solutions from air; in the chloroplast, the O_2 level could be even higher, as O_2-saturated water contains 1300 μM O_2 at 20°C. It is not clear how the carboxylase reaction in a C_3 plant in air or in a C_3 macroalgae has a specificity factor about four times faster for the carboxylase activity than for the oxygenase activity. Research must continue on the microstructure of the active site of Rubisco.

Today, for C_3 plants in sunlight, photorespiration (as CO_2 release) has been estimated to be 25 to 50% the rate of CO_2 fixation (44). Many past experiments have proven that altering the levels of atmospheric CO_2 or O_2, substrates of Rubisco, directly affects the carboxylase/oxygenase activity in vivo. Changing the CO_2/O_2 ratio by increasing CO_2 or lowering O_2 increases net photosynthesis, whereas changing the ratio in the opposite direction, by increasing O_2 or lowering CO_2, increases photorespiration while decreasing net photosynthesis. As atmospheric O_2 increases from 2% to 21%, O_2 inhibition of photosynthesis by C_3 plants rises to about 25%—a phenomenon first described for algae by Warburg (cited in Ref. 41). Plant growth changes proportionally. When CO_2 exchanges from the C_3 and C_2 cycles are equal, the plant is at its CO_2 compensation point, which is around 40–60 ppm CO_2 at 20°C for C_3 plants in 21% O_2. At the compensation point, net photosynthesis is zero and no growth occurs. In the past, the CO_2 compensation point has been used to evaluate how plant growth is controlled by the atmospheric O_2 and CO_2 levels. How the CO_2 compensation point limits photosynthetic CO_2 removal and thus controls the atmospheric composition has been considered in much less detail.

PHOTORESPIRATION

The two products of the Rubisco oxygenase reaction are phosphoglycolate and phosphoglycerate. Two molecules of phosphoglycolate are converted, via the C_2 cycle, to one CO_2 and another phosphoglycerate (Fig. 1.3) (18, 30, 31, 37, 38). A unique property of this C_2 cycle in plants and multicellular algae is compartmentation of part of the reactions, including catalase, glycolate oxidase, and glyoxylate aminotransferases in the peroxisomes. As a result of this compartmentation, two glycolates form two glycines, rather than an uncontrolled peroxidation of glyoxylate to CO_2 or glyoxylate reaction with –SH or –NH_2 entities. Two glycines are then converted to one serine, one CO_2, one NH_3, and one NADH in the mitochondria. This system thus conserved CO_2 by limiting its formation during photorespiration to one carbon out of four from two phosphoglycolates.

Fig. 1.3. The C_2 cycle or the oxidative photosynthetic carbon cycle. Sequential parts of the C_2 cycle are (a) the irreversible glycolate pathway from the oxygenase activity of Rubisco for P-glycolate synthesis to glycine and serine [top center, and middle], (b) mitochrondrial pathway for glycine and serine metabolism [bottom], (c) the photorespiratory nitrogen cycle for refixing the ammonia into glutamate [top, right], (d) the C_3 cycle for refixing the CO_2 [top left], (e) the reversible glycerate pathway between serine and 3-P-glycerate from both the Rubisco oxygenase and the glycerate pathway to ribulose-1,5-bisphosphate [middle, and top

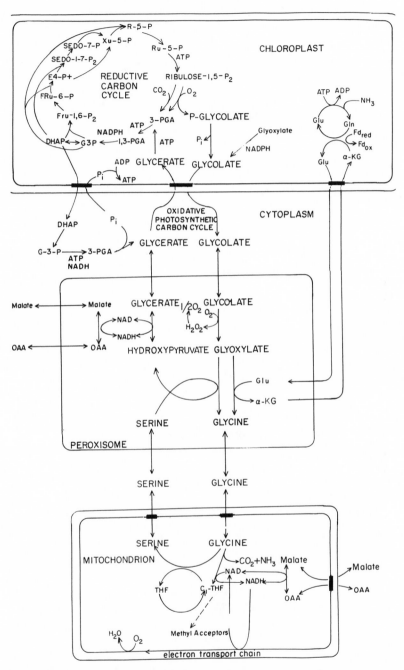

left]. An additional component to produce enough glycerate for glycolate exchange from the chloroplast involves (f) production of glycerate from the triose-P shuttle between the chloroplast and cytosol [top left] (40). In the C_2 cycle, as presented here, O_2, CO_2, and NH_3 uptake and refixation are equal (see Table 1.1). There is no net synthesis of carbohydrates or protein.

Table 1.1. Energy Consumption per Six CO_2 Turned Over During Photorespiration.[a]

12 Ru-1,5-P_2 $\xrightarrow{12\,O_2}$ 12 P-Glycolate + 12 P-Glycerate

12 Glycolate $\xrightarrow{6\,O_2}$ 12 Glyoxylate

12 Glycine $\longrightarrow$ 6 Serine + 6 CO_2 + 6 NH_3 + 6 NADH

6 NH_3 + 6 Glutamate $\xrightarrow{6\,ATP}$ 6 Gln + 6 αKg $\xrightarrow{6\,FdH_2\,(NADPH)}$ 12 Glu

6 Hydroxypyruvate + 6 NADH $\longrightarrow$ 6 Glycerate

6 Glycerate $\xrightarrow{6\,ATP}$ 6 P-Glycerate

18 P-Glycerate $\xrightarrow[\text{18 NADPH}]{\text{18 ATP}}$ 18 Triose-P

12 Ru-5-P $\xrightarrow{12\,ATP}$ 12 Ru-1,5-P_2

Subtotal: 24 NADPH + 42 ATP + 18 O_2 + 6 CO_2 from the C_2 cycle

Refixing 6 CO_2 $\longrightarrow$ 12 NADPH + 18 ATP + $C_6H_{12}O_6$

Total: 36 NADPH + 60 ATP used, with zero net change in O_2, CO_2, and NH_3

[a] *Note:* Only reactions of the C_2 cycle involving CO_2, O_2, NADPH, or ATP are shown.

The C_2 cycle uses or wastes immense amounts of photosynthetic assimilatory capacity per molecule of CO_2 turnover (Table 1.1) (30, 31, 37–9). In considering the total energy loss by the C_2 cycle, one must combine all facets of photorespiration, which include oxidation of ribulose bisphosphate to CO_2, regeneration of the ribulose bisphosphate from 3-phosphoglycerate, and refixation of the CO_2 and ammonia. Thus photorespiration uses the reactions of the C_2 cycle to generate CO_2, plus those of the C_3 cycle to regenerate ribulose bisphosphate and to refix the CO_2. At the CO_2 compensation point, all photosynthetic assimilatory power is used by photorespiration; there is no net CO_2 fixation. Refixation of the CO_2 does not result in net CO_2 fixation, so its energy drain is charged to photorespiration.

Photosynthetic reduction of 6 CO_2 to a hexose by the C_3 cycle requires 12 NADPH and 18 ATP (4). Calculations of the amount of photosynthetic energy wasted per turnover of six CO_2 by photorespiration in the light are based on the reactions of the C_2 cycle shown in Fig. 1.3 for photorespiration. The energy-utilizing steps that occur during photorespiratory release and refixation of 6 CO_2 and 6 NH_3, listed in Table 1.1, add up to 36 NADPH and 60 ATP. After O_2 uptake during photorespiration is subtracted from O_2 evolved during photosynthetic generation of the wasted NADPH, it is apparent that no net O_2 exchange occurs during overall photorespiration. Because 1 O_2 is evolved per 2 NADPH, a total of 18 O_2 are evolved and 18 O_2 are taken up in the first two equations. This calculation assumes that NADH formed in the mitochondria during glycine oxidation, or its equivalent, is shuttled to the peroxisomes for hydroxypyruvate reduction to glycerate.

The carbon atoms are balanced as follows: Twelve molecules of Ru-1,5-P_2 contain 60 carbons which start the cycle to form 18 triose-P and 6 CO_2—or 60 carbons. Refixing the 6 CO_2 into the carbohydrate pools of the C_3 cycle provides a total of 12 Ru-5-P with 60 carbons to maintain the cycle. The net result is only the loss of the energy.

Glycolate excretion by the chloroplast is accomplished through an exchange with a glycerate import. These calculations do not include the generation of an additional 6 glycerates in the cytosol to facilitate the excretion of 12 glycolates from the chloroplast (see Fig. 1.3). From 2 glycolates excreted by the chloroplast, only 1 glycerate is regenerated. Additional glycerate in the cytosol may be provided from the triose phosphate shuttle (see a more detailed discussion of this process under the glycerate cycle, in Ref. 38). Phosphorylation of the 6 additional glycerates for the complete C_2 cycle when evolving 6 CO_2 would require wasting 6 additional ATP. However, this cyclization of glycerate would also transfer 6 ATP and 6 reduced pyridine nucleotides from the chloroplast to the cytosol (Fig. 1.3). Experiments to evaluate whether this apparent wheel actually occurs among the photosynthetic carbon cycles have not been performed.

There is no net change in CO_2 or NH_3 during photorespiration, as the CO_2 and NH_3 evolved are refixed. Additional net CO_2 reduction occurs only by the C_3 cycle, if there is sufficient CO_2. During mitochondrial respiration, net CO_2 evolution and O_2 uptake occur. The idea that photorespiration evolves CO_2 in the light is misleading since an equivalent amount of CO_2 must be refixed. When the light is turned off, the momentary burst of CO_2 release is indicative of the depletion of the C_2 cycle intermediates in the dark without CO_2 refixation, but underestimates the internal turnover of CO_2.

For each CO_2 turnover during photorespiration, the amount of photosynthetic energy wasted exceeds the amount of energy that is needed for net CO_2 reduction to carbohydrates or that is generated per CO_2 during glycolysis and the tricarboxylic acid cycle.

Based on the use of photosynthetic energy, the turnover of 6 CO_2 by the C_2 cycle during photorespiration consumes 3 times (36 NADPH and 60 ATP) as much energy as the reduction of 6 CO_2 by the C_3 cycle (12 NADPH and 18 ATP) (Table 1.1). Although the amount of CO_2 fixation has been about 3 times more than the amount of CO_2 generated by photorespiration, energy consumption per CO_2 turnover is 3 times more by photorespiration than by CO_2 fixation. Consequently, in full sunlight, with 250–300 ppm CO_2 and 21% O_2 in the atmosphere, as in the past, the C_2 cycle wastes about as much energy as is used for net photosynthesis. The C_2 cycle represents a consumption of much photosynthetic energy, which cannot be used for net CO_2 fixation. Because CO_2 is insufficient at the global CO_2 equilibrium, excess global photosynthetic light energy has been wasted on photorespiration. This evaluation on the basis of carbon metabolism does not take into account the loss or waste of light energy also by processes during photosynthetic electron transport (Ref. 2, see also Chapter 6).

The large energy loss by photorespiration is accomplished with minimal CO_2

loss, as carbon flows all the way around the C_2 and C_3 cycles, drawing off energy. Thus CO_2 is conserved, yet the photosynthetic electron transport reactions are kept operational. When photorespiration utilizes or wastes excess NADPH and ATP from photosynthetic electron transport, this part of photosynthetic carbon metabolism itself serves as a "Hill" process or as an alternate photosynthetic electron acceptor—alternate to net CO_2 reduction. Although O_2 evolution by PS II continues, as in a Hill reaction, no net O_2 is evolved, as the O_2 produced is balanced by O_2 uptake by the C_2 cycle. Thus photorespiration does not alter the photosynthetic ratio of net O_2 evolved relative to net CO_2 fixed, which remains 1:1.

The combined photosynthetic and photorespiratory processes explain how plants have survived and can grow faster as atmospheric CO_2 increases from 180 to about 1000 ppm CO_2 in 21% O_2. With increasing atmospheric CO_2, the energy wasted by photorespiration at low CO_2 is used to reduce CO_2 at a faster rate, and photorespiration decreases. Experimentally the atmospheric CO_2 needs to increase to around 1000 ppm before photorespiration is reduced >90% and other factors such as global light intensity become overall limiting (12, 23). These predictions explain higher rates of net photosynthesis with increasing atmospheric CO_2 without increases in sunlight intensity. Similar considerations have been used to explain how photorespiration protects the plant in high light and low CO_2 from phototoxicity, by wasting excess photosynthetic energy (31). Thus the photosynthetic carbon system, based on the properties of Rubisco, can accommodate fluctuations in either CO_2 levels or light intensities by distributing the energy between net CO_2 reduction and photorespiration.

A consequence of photorespiration is to severely limit the rate and amount of CO_2 that plants can remove from the air as long as the CO_2 concentration is low and limiting. In the light the rates of photosynthesis and photorespiration greatly exceed the rate of dark respiration. For a C_3 plant in air and sunlight, photorespiration based on CO_2 exchange can be as much as 50% the rate of CO_2 fixation, which is the equivalent of 1 CO_2 formed for every 2 CO_2 reduced to carbohydrates. To simplify these calculations of photosynthetic rates, dark respiration is not often included, because it is around 5% the rate of photosynthetic CO_2 fixation, and in the light it may be further reduced (Ref. 7, see also Chapter 9). The total amount of mitochondrial dark respiration at night is, of course, important in the total energy balance and in regulating plant growth (see Chapter 10).

ATMOSPHERIC AMMONIA AND THE PHOTORESPIRATORY NITROGEN CYCLE

During photorespiration equal amounts of CO_2 and NH_3 are evolved during glycine oxidation, and both are refixed. Atmospheric NH_3 at very low levels is another gas closely regulated by plants and algae through their C_2 photorespiratory carbon cycle. Both CO_2 and NH_3 have been reduced to low equilibrium limits by plant and algal photosynthesis. Many references for details of photorespiratory nitrogen cycle are found (see reviews in Refs. 18, 20, 28, 38). The CO_2 and NH_3 already in the cells are preferentially refixed over the atmospheric gases, but there is nevertheless a CO_2 pressure for CO_2 release from the leaf. The NH_3 released inside the cell is nearly all refixed by the photorespiratory nitrogen cycle. In plants,

NH_3 is toxic to electron transport, but fortunately its refixation is made very efficient by abundant glutamine synthetase and glutamate synthase reactions with their low K_m values. The abundance of these enzymes in a C_3 leaf far exceeds that necessary for protein synthesis. Nevertheless, there is a very small NH_3 pressure in a leaf, for measurable exchange with the atmosphere. Because atmospheric NH_3 is so low, whether the photorespiratory nitrogen cycle regulates atmospheric NH_3 has not been evaluated. According to the known biochemical reactions, the plant and algal photorespiratory nitrogen cycle has a great capacity to remove NH_3 from the atmosphere. Ammonia or urea (converted in the leaf to NH_3), applied directly to leaves, has been used as a nitrogen fertilizer (43).

HEAT EXCHANGE DURING PHOTOSYNTHESIS AND PHOTORESPIRATION

Light energy is used with a certain biological efficiency to reduce CO_2 to carbohydrate or plant material, and dark respiration, fermentation, or burning of the plant material releases the same amount of CO_2 and energy. In the past, photosynthetic energy was used, with limited efficiency, to reduce CO_2 to carbohydrate, and this energy was not released as heat until the carbohydrate was burned back to CO_2. About an equal amount of that photosynthetic energy was consumed by photorespiration and immediately wasted or converted back to heat. Although the amount of light energy stored in plant biomass has been a minor component in regulating global temperatures, it merits recognition. If the atmospheric CO_2 were increased to 1000 ppm CO_2, which would nearly suppress photorespiration, more of the photosynthetic energy would be used to reduce CO_2 to carbohydrate and none would be immediately reconverted to heat by photo-respiration. If the newly formed plant material were stored as wood or as algal organic carbon deposits, the storage, as long as it lasted, would prevent the release of that CO_2 and energy. It seems unlikely that the difference between photosynthetic CO_2 reduction and photorespiratory reconversion of photosynthetic energy back to heat exerts any signficant effect on temperature, even though the mass of plants and algae is great. Whereas the competition between photosynthesis and photo-respiration exerts a relatively large change on the very small CO_2 levels, changes in net heat during photorespiration may be relatively minuscule compared to the total light energy reaching the globe. Global temperature modelers should include plant transpiration, reflectance, and other processes of heat loss or retention by plants whose population may increase with more CO_2.

The total net amount of sun energy captured by the planet is large relative to how much of this energy is trapped photosynthetically. Products of photosynthesis represent 0.1–1.0% of the energy of sunlight falling on a field in the course of a year (24). The lowest (0.1%) efficiency of biomass production has been estimated for C_3 plants, such as wheat and soybeans, and the highest (0.95%) efficiency has been reported for sugar cane, a C_4 plant. Relative to the total global heat balance, these low values are reduced further by biologically nonproductive areas of the globe. Thus it would seem that photosynthetic and photorespiration may not have a significant direct effect on global heat exchange, but are involved indirectly by regulating atmospheric CO_2.

DISSOLVED INORGANIC CARBON (DIC)-CONCENTRATING PROCESSES

Rubisco has no known catalytic binding site for CO_2 or O_2 (25); therefore, the idea of modifying its constituent amino acids in order to change the molecular configuration at the "active site" becomes a complex problem. Increases in Rubisco specificity factor (30) (see Chapter 7), or in the ratio of the carboxylase/ oxygenase activity, have evolved in nature from cyanobacteria to C_3 plants. Whether further improvements, such as an increase V_c/V_o, can be produced in C_3 plants by molecular biologists is debatable. Further increases would also be of environmental concern, as an increase in the efficiency of the carboxylase could ultimately lower the atmospheric CO_2 level too much. Certainly, an understanding, at the molecular level, as to how the active site of Rubisco provides some specificity for CO_2 over O_2 utilization is a critical problem that has not yet been solved. To date, attempted modifications of the amino acids of Rubisco have resulted in a lower specificity factor because of a relatively unchanged oxygenase activity. Changes in the efficiency ratio for utilizing CO_2 versus O_2 by Rubisco isolated from photosynthetic bacteria or C_4 plants have been reported to be in the direction of a higher $K_m(CO_2)$ and a higher V_{max}—which is acceptable for these tissues that have CO_2-concentrating processes (30). It also seems unlikely that net CO_2 fixation can be improved by further increases in the amount of Rubisco present, as it is already the major protein in the leaf. Today the ubiquitous C_3 plants, which have no known CO_2-concentrating process, have the lowest $K_m(CO_2)$ for Rubisco and the best efficiency ratio for CO_2 versus O_2. For further CO_2 efficiency, nature has increased the effective carboxylase/oxygenase ratio in vivo by adding reactions for concentrating CO_2 at the site of Rubisco in the chloroplast, thus increasing the carboxylase rate and suppressing the oxygenase reaction by competition from higher CO_2.

There are two different CO_2-concentrating processes: (a) the C_4 cycle in C_4 and CAM (i.e., Crassulacean acid metabolism) plants and macro algae (Ref. 10; see Chapters 11 and 12) and (b) the dissolved inorganic carbon (DIC)-concentrating processes in unicellular algae (1, 8, 27). But in these plants it is Rubisco and the C_3 cycle that ultimately perform CO_2 reduction to carbohydrate. Processes that concentrate CO_2 or HCO_3^- are much more CO_2-efficient than is the Rubisco system alone. Phosphoenoylpyruvate carboxylase of the C_4 cycle fixes bicarbonate (the predominant form of DIC at pH 8.3 in the chloroplast), and its $K_m(HCO_3^-)$ is low. With bicarbonate as the DIC substrate, O_2 does not compete for the carboxylase reaction in a C_4 plant. Thus 21% atmospheric oxygen does not severely inhibit photosynthetic CO_2 fixation by a C_4 plant.

There are probably two unicellular algal DIC pumps, one for CO_2 and one for bicarbonate (Fig. 1.4). As in a C_4 plant the algal DIC-concentrating processes increase (~ 1000 fold) the concentration of DIC in the cell and chloroplast to favor the carboxylase reaction. All DIC-concentrating processes, however, require extra photosynthetic energy beyond that of the C_3 cycle, but since the higher level of internal CO_2 reduces photorespiration, part of the energy that might have been wasted by photorespiration is utilized instead by the DIC-concentrating processes. Increasing atmospheric CO_2 around the plant is a better way to increase net photosynthesis, as there is no energy drain for a DIC pump. Increasing

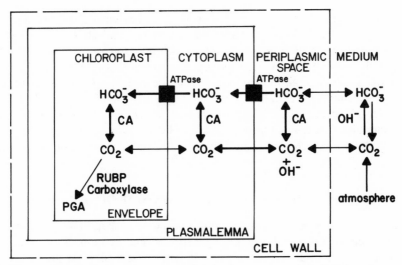

Fig. 1.4. A working model for concentrating dissolved inorganic carbon (DIC) by micro green algae (29, 35). For other models, see Refs. 8 and 27.

atmospheric CO_2 is being used successfully by researchers for greenhouse growth of plants (11), but when planet Earth is the greenhouse, the environmental consequences limit utilization of the higher CO_2 levels. Until recently, increases in atmospheric CO_2 have generally been beneficial to plants because the amount of net photosynthesis has increased nearly proportionately (24).

Most, if not all, algae can concentrate DIC up to 1–10 mM from only a mere micromolar (μM) level in the medium, provided that they have been grown photoautotropically in low atmospheric levels of CO_2 (27). For algae such as the *Chlorophyceae* and *Charaphyceae*, >1 μM external CO_2 is required in order for the DIC pump to attain $\frac{1}{2} V_{max}$ during photosynthetic O_2 evolution. When these algae are grown at high (1–5%) CO_2, the DIC pumps are suppressed, and $\frac{1}{2} V_{max}$ for photosynthetic O_2 evolution due only to CO_2 diffusion requires a CO_2 level similar to that for isolated Rubisco. The increased internal CO_2 from DIC pumps stimulates CO_2 fixation and decreases phosphoglycolate biosynthesis and thus lowers the CO_2 compensation point. For C_4 plants and for algae having DIC pumps, the atmospheric compensation point is about 1–2 ppm CO_2; without the DIC pump the compensation point is about 40 ppm CO_2 (29, 39). Thus the unicellular algal DIC-concentrating mechanisms and the C_4 cycle in plants or macro algae accomplish the same end—namely, to increase net photosynthesis and decrease photorespiration by altering the CO_2/O_2 ratio in the chloroplast at the site of Rubisco. Plant scientists might evaluate which CO_2-concentrating system would be most feasible to bioengineer future plants to modify atmospheric CO_2 levels.

Understanding the DIC-concentrating processes in algae has been and is currently an active research area with freshwater, unicellular, green, and blue-green algae (1, 8, 27). There are three ways in which DIC may enter or exit the algal cell or chloroplast (Fig. 1.4): by CO_2 diffusion; by a CO_2 pump between pH 4 to

8; or by HCO_3^- transport between pH 8 to 11. External carbonic anhydrase, often a component of the CO_2 pump, rapidly converts external HCO_3^- to more CO_2 for import into the cell, and inhibitors of carbonic anhydrase inhibit the CO_2 pump (1). HCO_3^- import by *Scenedesmus* cells at pH 9 is inhibited by vanadate (35)—indicating that a plasmalemma-type ATPase may be involved in that HCO_3^- transport mechanism. Uptake of DIC by isolated intact chloroplasts from air-adapted *Chlamydomonas* or *Dunaliella* cells (13) is also inhibited by vandate; thus even during active CO_2 uptake, an ATPase, which uses photosynthetic energy, may be involved at the chloroplast envelope to concentrate DIC.

MARINE PHOTOSYNTHESIS AND CALCIUM CARBONATE DEPOSITION

The DIC-concentrating processes among marine algae need considerably more research. Ionic HCO_3^- uptake should be a specific, energy-requiring process, as it is in freshwater algae. Research on the DIC pumps in freshwater algae has been active now for two decades; yet we still do not know the biochemical steps involved. A major research effort will be required in order for marine biology to catch up.

Calcium carbonate depositions and biotic reef formation in antiquity and today occur in association with the photosynthetic activity of calcareous algae. Although calcification has long been described by marine biologists and geologists (see Chapter 15), its connection with photosynthesis has not been explained in molecular terms, nor has its role as a CO_2 sink as yet been adequately considered in regard to increasing atmospheric CO_2. $CaCO_3$ deposition by algae varies widely. It includes amorphous or crystalline deposits outside the cell, in or on the cell surface, and inside the cell. Reviews (e.g., Ref. 5) cite a photosynthetic carbon balance of two HCO_3^- from the medium, with one CO_2 taken up and reduced to organic carbon by the algae, and one carbonate (CO_3^{2-}) by-product that is deposited as a calcium or magnesium salt. Both the reduced carbon as phytoplankton and carbonate salts sink into long-life pools. When many algae, both marine and freshwate, accumulate DIC, they alkalinize the media, and the hydroxy ion (OH^-) in turn can be neutralized by dissolving more CO_2 (carbonic acid) or by the conversion of HCO_3^- to CO_3^{2-}, as shown in Figure 1.5.

Current biochemical investigations of the DIC concentrating processes, which have focused particularly on freshwater unicellular green and blue-green algae, need to be further expanded to include marine algae of all kinds. Photosynthetically active DIC accumulation and calcification seem to be directly related processes. Every HCO_3^- converted to organic carbon during photosynthesis by marine algae with excess HCO_3^- is potentially accompanied by deposition of a second HCO_3^- as $CaCO_3$. Thus marine photosynthesis may be filling $CaCO_3$ sinks with excess CO_2. I hasten to add that marine biologists and geologists tell us how extremely complex this simple process can be. Algal blooms of *Emiliania huxleyi*, which have internal $CaCO_3$ deposits, have been monitored by satellite. About 40% of the CO_2 currently being put into the atmosphere is not accounted for by photosynthetic refixation by plants and algae (see Chapter 15). Could these unknown sinks be in part algal $CaCO_3$ deposits, or is the carbonate cycle in the ocean saturated or

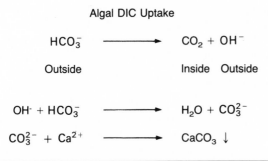

$$\text{Algal DIC Uptake}$$

$$HCO_3^- \longrightarrow CO_2 + OH^-$$

Outside   Inside Outside

$$OH^- + HCO_3^- \longrightarrow H_2O + CO_3^{2-}$$

$$CO_3^{2-} + Ca^{2+} \longrightarrow CaCO_3 \downarrow$$

Photosynthesis

DIC Pump

$$2\ HCO_3^- \longrightarrow CO_2 + CO_3^{2-} + H_2O$$

Outside   Inside,   Deposited as salts
                   as reduced C

Fig. 1.5. Equations for conversion of HCO_3^- by calcareous algae to carbonates during photosynthetic uptake of CO_2 by the algal DIC pump for CO_2 fixation by photosynthesis.

balanced? Lowering atmospheric CO_2 by the global carbon cycle for land plants (see the following section) may not be feasible, because the photosynthate is continually being recycled. To lower atmospheric CO_2, the excess carbon needs to be reburied, like the coal, gas, and oil from which it comes, into an inaccessible form or location. Is increased formation of $CaCO_3$ deposits in the oceans by algae a possible depository? What are the natural limitations to $CaCO_3$ deposition, such as saturation of the ocean carbonate cycle, algal nutrition, temperature, pH, and light? If photosynthetic options are to be investigated for controlling atmospheric CO_2, marine organic and carbonate deposition should be considered. If we were faced with the opposite situation—too low an atmospheric CO_2—science fiction could consider converting the massive $CaCO_3$ deposits of the world, such as the white cliffs of Dover, back to CO_2 by biological acidification.

THE C$_2$ CYCLE IN ALGAE

Algae have DIC concentrating processes that facilitate CO_2 and CHO_3^- uptake from water, where DIC diffusion is slow. These DIC pumps also suppress photorespiration by the high internal DIC levels. There are other differences in photorespiration between algae and plants, but the oxygenase activity of Rubisco in the algae is similar to that in plants. Most microalgae do not have peroxisomes nor glycolate oxidase (36), although *Mougeotia* and *Spirogyra* and other multicellullar Charaphycean algae have leaf-type peroxisomes (34, 40; see Chapter 8). In most unicellular algae, which contain direct DIC-concentrating processes, there are no leaf-type peroxisomes, and the leaf-type peroxisomal enzymes for reactions of the C_2 cycle are located instead in the algal mitochondria. In algal mitochondria there is a glycolate or D-lactate dehydrogenase, whose measurable activity is too low for the indicated rate of the C_2 cycle. The evolutionary significance of a change from algae, with their CO_2 or HCO_3^- concentrating processes but no peroxisomes, to plants, with or without the C_4 cycle for concentrating CO_2, but always with the wasteful peroxisomal C_2 photorespiratory cycle, is not understood; in fact, the change should have been unnecessary or deleterious to the evolving plants. The change seems to have occurred while the atmospheric O_2 was increasing from 2% to 21%, and may have ultimately served some purpose for limiting photosynthesis and regulating the atmospheric CO_2 level. The change seems to have occurred during the evolution of multicellular Charaphycean algae over 500 Myr in this one line of algae that led to higher plants. Data for this line of reasoning are reviewed in the proceedings from another symposium (34).

INHIBITION OF PHOTOSYNTHETIC CO_2 FIXATION BY UV-B

UV inhibition of photosynthetic CO_2 fixation would enhance the increase in atmospheric CO_2. Holm-Hansen (Chapter 3) and others (32, 33) have reported a great decrease in CO_2 fixation during photosynthesis by the phytoplankton under the ozone-deficient hole over the Antartic Ocean. Our research years ago indicated that low dosages of UV for short times did not directly inhibit electron transport or the reactions of the C_3 cycle for CO_2 fixation (45). How, then, does low chronic UV cause a decrease in photosynthesis? Small dosages of UV-B of wavelength around 280–310 nm block the formation of the DIC-concentrating process in *Chlamydomonas* and in the saltwater alga *Dunaliella* without otherwise inhibiting CO_2 fixation (14). UV light of 280 nm has been reported to inhibit the induction of the extracellular carbonic anhydrase of *Chlorella*, although the enzyme activity was not affected by this UV wavelength (9). Thus one site for UV sensitivity may be in the formation of DIC pumps. The protein D1 of photosystem II, which transfers electrons into the plastoquinone pool, is another UV-sensitive site (15). Further biochemical investigations of UV inhibition of photosynthesis are needed.

THE GLOBAL CARBON CYCLE AND THE CO_2 COMPENSATION POINT

In simplistic terms the global carbon cycle entails net CO_2 fixation by plants and the conversion of an equal amount of reduced organic carbon back to CO_2 by respiration or long-term decomposition. So that an atmospheric CO_2 equilibrium can be discussed, this complex global carbon cycle is broken down in Figure 1.6, with net photosynthesis divided into gross CO_2 fixation (measured in the absence of O_2 or presence of excess CO_2) and CO_2 release from photorespiration, although much of the CO_2 is quickly refixed. Experimentally, the CO_2 compensation point for a plant is the CO_2 equilibrium level in the air of the greenhouse or bell jar with 21% oxygen at 20°C at which there is no net change in atmospheric CO_2. At the CO_2 compensation point, the CO_2 released by photorespiration is all being

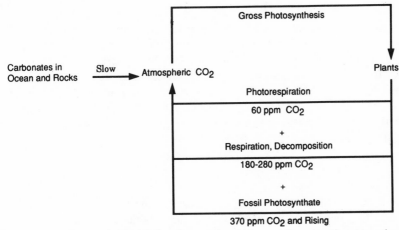

Fig. 1.6. Components of the global carbon cycle which contributes to the atmospheric equilibrium level. With only photorespiration, the fast CO_2 turnover at the CO_2 compensation point would equilibrate at 40–60 ppm CO_2. Other natural sources of CO_2 increased the CO_2 equilibrium to around 250 ppm CO_2, the level that existed in the past. Today the burning of fossil photosynthate provides an additional source of CO_2 which has raised the equilibrium to around 370 ppm CO_2.

refixed; this is the equilibrium shown in Figure 1.3 and Table 1.1. There is no net CO_2 fixation. This is the minimal, downside limit for the air CO_2 level in the presence of C_3 plants without other sources of CO_2. For C_3 plants the reported CO_2 compensation points in full sun light range from 40 to 65 ppm CO_2 (depending on how the tests were done) in 21% O_2 and at 20°C. At 30°C the CO_2 compensation point is over 110 ppm CO_2; at around 37°C, the compensation point is about 300 ppm CO_2 in 21% O_2. As the temperature rises CO_2 solubility in water decreases faster than O_2, so that the ratio of CO_2/O_2 concentration decreases (22). Above 37°C, C_3 plants grow very slowly or not at all because of near zero net photosynthesis from accelerated photorespiration. This temperature (37°C) has been thought of as a temperature compensation point for C_3 plants for current atmospheric CO_2 and O_2 levels. At the compensation points, plants can live, and all the photosynthetic processes are turning over, but net photosynthesis is zero. For C_4 plants or algae with DIC pumps, the CO_2 compensation points are much lower, because the DIC pumps have concentrated the external CO_2.

In the large closed system of planet Earth, other sources of CO_2 in addition to that from photorespiration—from respiration, decomposition, the ocean, and rocks—have shifted the atmospheric equilibrium, as controlled by plants, in the past, to values of 180–280 ppm. In a sense, the atmospheric CO_2 equilibrium is a global CO_2 compensation point. Competing Rubisco carboxylase and oxygenase reactions have CO_2 turnover times of a second, which is much more rapid than CO_2 release from the other sources, with turnover times of years or centuries. Total atmospheric CO_2 turnover by photosynthesis is but a few years. In the global carbon cycle, the photosynthetic/photorespiratory part of the cycle has a very rapid turnover, whereas the other sources of CO_2 have a very slow turnover. The other sources of CO_2 determine the total CO_2 equilibrium concentration. Because the atmospheric CO_2 equilibrium is higher than the plant compensation point from photosynthesis and photorespiration, net photosynthesis exists. In this century, an additional source of atmospheric CO_2 has quicly occurred from the burning of fossil photosynthate to CO_2. As a consequence, the atmospheric CO_2 equilibrium has risen to 370 ppm. Consequently, photosynthetic CO_2 fixation has increased and CO_2 release by photorespiration has decreased. At atmospheric levels of 1000 ppm CO_2, photorespiration would be decreased by about 90% of that at 250–300 ppm CO_2 (12, 23).

C_4 PLANTS

If planet Earth could suddenly be populated with more C_4 plants or new plants with algal DIC-concentrating processes, the atmospheric CO_2 level should drop to a lower equilibrium level, representing a compensation equilibrium from a plant system highly efficient for removing CO_2. The component for photorespiratory CO_2 release in Figure 1.6 would essentially disappear and the equilibrium concentration for CO_2 fixation would be lowered by prior bicarbonate fixation by phosphoenoylpyruvate carboxylase. Within the canopy of a corn (C_4 plant) field, the CO_2 level is lower than in the air above it (24).

In the long term, attempting to regulate atmospheric CO_2 by growing more C_3 plants appears futile, because the global CO_2 equilibrium would not be changed, as the new growth would only be recycled in the global carbon cycle. Although carbon recycling would also occur from an increase in C_4 plants, the global CO_2 equilibrium should be lower because of a much more CO_2-efficient, O_2-insensitive, CO_2-concentrating component. This hypothesis has not been tested in macro plant growth facilities. However, experiments with closed plant growth chambers containing both C_3 and C_4 plants together in the light showed the death of the C_3 plants in a few days due to continuous photorespiratory loss of CO_2, but sustained growth of the C_4 plant (42). From photorespiratory oxidation of the C_3 plant, the low CO_2 equilibrium level in the closed growth chamber quickly fell below the CO_2 compensation point of the C_3 plant (40 ppm) but remained above that of the C_4 plant (2 ppm).

In the future (with over 350 to 400 ppm CO_2), increasing atmospheric CO_2 will favor growth of the C_3 plant (Fig. 1.7), whereas in the past, with only 250–300 ppm CO_2, growth of C_4 plants was favored over C_3 plants. Atmospheric CO_2 today at 370 ppm is at this crossover point, and from now on higher CO_2 will favor the growth of C_3 plants over C_4 plants. In the past, with lower CO_2, C_4 plants were photosynthetically more CO_2-efficient than C_3 plants. In these theoretical discussions, all other factors affecting photosynthesis and C_3 versus C_4 plant growth are considered to be the same. Above 350–400 ppm CO_2, photosynthesis by C_4 plants will increase relatively less rapidly than that by C_3 plants, because of the obligatory use by C_4 plants of their constitutive, energy-requiring C_4 cycle. Ecologically, this CO_2 change by itself might in the long term favor a decrease in the presently limited population of C_4 plants, which in the past, as part of the plant population, have kept the atmosphere equilibrated at a lower CO_2 level.

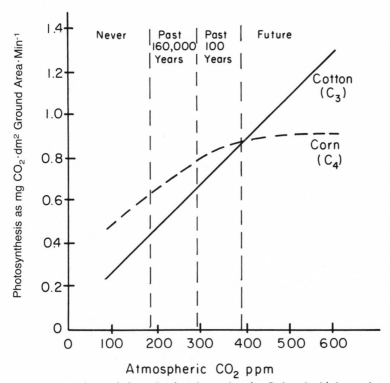

Fig. 1.7. Comparison of growth for a C_3 plant (cotton) and a C_4 (corn) with increasing atmospheric CO_2 (3). These data were from long-term growth experiments, and the figure represents only a part of the results. Superimposed upon the data are the time periods over which the various levels of CO_2 have existed: 160,000 Myr ago to the year 1890, with CO_2 at 180–280 ppm (6); and the past 100 years, with CO_2 rising from 280 to 380 ppm (19). The atmospheric CO_2 probably "never" dropped below about 180 ppm CO_2, owing to the down-side stop from the competition between CO_2 reduction and photorespiration at low CO_2 levels. The decreasing rates are from data in experimental growth chambers. In the future if levels of CO_2 approach 1000 ppm, photorespiration will be nearly repressed; at that point, other factors become limiting, and a further increase in net photosynthesis with even more CO_2 will become nil.

A limitation for using C_4 plants in order to control further the CO_2 level is that they contain so many morpholgical and biochemical differences that simple molecular modification does not seem feasible. In addition, no C_4 plants are known to fix N_2. We do not know why this incompatibility of "C_4ism" and N_2 fixation exists. Nitrogen becomes an ever more limiting nutrient deficiency as more CO_2 is available to increase plant growth. Fortunately, the increased photosynthate available with higher CO_2 is used by nitrogen-fixing C_3 plants, such as soybeans, for a longer period and higher rate of N_2 fixation (16). Thus, at CO_2 levels above 370–400 ppm, superior photosynthetic CO_2 efficiency by C_4 plants no longer exists. In spite of their lower CO_2 compensation point, the idea of reducing atmospheric CO_2 by more C_4 plants is probably not feasible.

In addition, it is not known whether the atmospheric CO_2 equilibrium shifts

Table 1.2. Ratio of C_3 to C_4 Plants: an Environmental Teeter-Totter.

Glacial Periods: Low CO_2 and Temperature	Interglacial Periods up to 1850: Intermediate CO_2 and Temperature	The Future After 2000: Higher CO_2 and Temperature
Low temperature favors C_3 plants.	High temperature favors C_4 plants.	Higher temperature favors C_4 plants.
But there is higher CO_2 fixation by C_4 plants relative to C_3 plants in the tropics.	*But* there is more total CO_2 fixation by C_3 plants relative to C_4 plants.	*But* CO_2 fixation by C_3 plants becomes more efficient.

or is shifted by the ratio of C_3 or C_4 plants. From current biochemical information, we can hypothesize that CO_2 concentration, temperature, and water regulate the ratio of C_3 to C_4 plants, and in turn the ratio of C_3 to C_4 plants has influenced the atmospheric CO_2 level and the temperature in some complex balance. A teeter-totter in the population between C_3 and C_4 plants is summarized in Table 1.2. In the glacier periods, when the atmospheric CO_2 level was as low as 180 ppm CO_2, the cold temperature would have limited the area for C_4 plant populations. In most C_4 phosphenoylpyruvate carboxylase and pyruvate, phosphate dikinase for phosphoenoylphospate synthesis are cold labile. C_4 corn plants might be modified by molecular biology to be more cold tolerant, but in the past corn plants have not grown much north of the 45th paraolel. On the other hand, at 180 ppm CO_2, C_4 plants in the tropical areas would have been greatly favored photosynthetically, as compared to C_3 plants with insufficient internal DIC to suppress photorespiration. In the interglacial periods, at 250–280 ppm CO_2, the higher temperature would have expanded the growth habitat for C_4 plants, which in turn effectively held down the higher CO_2 level. On the other hand, now at near 400 ppm, the net rate of photosynthesis per unit ground area by C_4 plants is near that for C_3 plants (Fig. 1.7), and at still higher CO_2, C_4 plants are at a photosynthetic disadvantage. At 500–600 ppm CO_2, the C_4 plants are less photosynthetically efficient that C_3 plants. Environmentally, we have entered a new era of higher CO_2 and high O_2 which, in prolonged periods, ought to influence the ecological distribution of C_3 and C_4 plants. We have no records to predict how this will affect the global environment.

EVOLUTIONARY DEVELOPMENT OF RUBISCO

Current dogma is that early anaerobic organisms (still found in the Dead Sea) photosynthetically fixed CO_2 by reversing the tricarboxylic acid cycle with iron sulfur peptides (ferredoxins) with their high reducing potentials gained from photosystem I. These reduced substances were so unstable in the presence of O_2 when photosystem II evolved that Rubisco and the C_3 reductive carbon cycle evolved to utilize NADPH and ATP. Today some oxidation by molecular O_2 of the reduced iron sulfur peptides from photosystem I for forming H_2O_2 remains a point of photosynthetic inefficiency (the Mehler reaction) (2). It is assumed that Rubisco, which does not directly use a reducing agent, evolved at a time of very high CO_2 as a carboxylase with a high $K_m(CO_2)$, which was not then a disadvantage. Oxygenic (aerobic) photosynthesis began about 2800 Myr ago, after a period of anaerobic photosynthesis from about 3500 Myr ago (17). For nearly 1000 Myr, the atmospheric O_2 remained low, because the new O_2 reacted with reducing components of the Earth's surface, such as Fe^{2+} and sulfur. Today, surviving forms of the anaerobic and aerobic photosynthetic prokaryotes, as well as the eukaryotic single-cell algae, use Rubisco for their primary carboxylation reaction in photosynthesis. Rubisco from these organisms has both carboxylase and oxygenase activity, similar to that in higher plants, except that the $K_m(CO_2)$ is higher. The oxygenase activity of Rubisco should not have been significant while algae and plants evolved. Atmospheric O_2 was low, CO_2 was high relatively until the last 100 Myr, and these algae (that remain today) contain DIC-concentrating mechanisms. These conditions should have severely suppressed the oxygenase activity of Rubisco. The Rubisco oxygenase activity becomes large only with 21% O_2 and with low CO_2, as in our current atmosphere. It is likely that aerobic conditions inside prokaryote cells evolving O_2 could have resulted in some Rubisco oxygenase activity, which essentially represented a limited system for glycolate catabolism for the synthesis of essential glycine, serine, and the methyl donor (34, 40). Thus it is likely that, in general, during early evolution of algae and plants, alteration or elimination of the oxygenase activity of Rubisco was never needed or severely challenged. If true, how was the oxygenase activity of Rubisco, which would be used today, 2600 Myr later, so fortuously incorporated into the system? Because of the molecular mechanism of the carboxylase reaction, the oxygenase activity of Rubisco may have started out as an inadvertent and unavoidable consequence or side reaction from the presence of O_2 (24). Regardless, the atmospheric composition is now *very dependent* on the oxygenase reaction of Rubisco—a reaction that may have not been necessary during the first 2600 Myr of evolution of algae and plants.

The carboxylase activity or Rubisco, which evolved at a time of high CO_2 and low O_2, has an extremely slow rate (1 mole$\cdot$s$^{-1}\cdot$promoter^{-1}) for an enzymatic reaction. Enzymatic rates are generaly a thousand times faster. The slow rate is apparently due to the absence of a CO_2 binding-site cofactor. If Rubisco had evolved with a CO_2 binding cofactor, such as biotin or a heme, or if it had used HCO_3^- as does the C_4 cycle enzyme phosphoenoylphosphate carboxylase, it should have had a faster turnover and a much lower $K_m(DIC)$, and it would not be an oxygenase. The first time that such an improvement might have occurred in the past, the atmospheric CO_2 would probably have been depleted even further—with dire consequences. The unfavorable catalytic properties of the carboxylase activity of Rubisco prevents C_3 plants from removing all of the atmospheric CO_2; at best, plants can only approach their CO_2 compensation point.

O_2 CONCENTRATION AND O_2 COMPENSATION POINT

During aerobic photosynthesis one O_2 is produced for every CO_2 fixed. This process requires 2 NADPH and 3 ATP, which are generated during the transport of four electrons through photosystems II to I. In the global carbon cycle, the organic products are metabolized back to CO_2 with the same amount of O_2 uptake, so that today the total CO_2 and O_2 in the air should remain nearly balanced. In the past, when the organic carbon formed by photosynthesis was buried or removed from the global carbon cycle, the atmospheric O_2 level increased. This is readily demonstrated during growth of algae or plants in a closed system with a constant supply of CO_2; CO_2 is a converted into plant mass, and O_2 accumulates. Today with more photosynthetic CO_2 fixation, a proportional increase in the amount of O_2 evolution also occurs on a short-term basis. However, the increase in reduced organic products, when they are later reoxidized, would result in a proportional increase in O_2 uptake. As a consequence, the atmospheric O_2 at equilibrium should not increase more than does CO_2 fixation on a short-term basis, or not at all on a long-term basis.

The question of why the atmosphere equilibrated at 21% O_2 in the presence of 0.025% CO_2 is not clear, nor has it been experimentally considered. One might speculate that when the high atmospheric CO_2 was buried as fossil photosynthate, O_2 accumulated because it was not used to oxidize the buried organic carbon or to further oxidize the Earth's crust. By the time the atmospheric CO_2 was reduced to 0.025%, the atmospheric O_2 just happened to have risen to 21%, and after that the global carbon and oxygen concentration remained unchanged, because for every CO_2 reduced by photosynthesis one O_2 is evolved and for every carbon reoxidized in respiration one O_2 is used. This concept is not satisfactory to explain why the atmospheric O_2 equilibrated at 21% and not some other level. Instead I would add the hypothesis that atmospheric CO_2 and O_2 both are ultimately equilibrated by photosynthesis to specific levels based on the kinetic properties of Rubisco. Just as the CO_2 compensation point provides down-side restriction on how much CO_2 the carboxylase action of Rubisco can remove from the air, the O_2 compensation point, derived from O_2 inhibition of photosynthesis by the oxygenase activity of Rubisco, sets up-side limits on the amount of atmospheric O_2. The final CO_2 and O_2 equilibria should be the same for all C_3 plants. In addition to the C_3 plants, the ecological balance with C_4 plants and algae, with their lower CO_2 compensation points and higher O_2 compensation points (less O_2 inhibition), has contributed to the final greenhouse equilibria of 0.025% CO_2 and 21% O_2.

O_2 inhibition of photosynthetic CO_2 fixation is well established (41). As an approximation for C_3 plants, O_2 inhibits photosynthetic CO_2 fixation only a few percent at 2–5% atmospheric O_2, but there is 25% inhibition at 21% O_2, about 50% at 40–60% O_2, and around 90% inhibition at 99% O_2 (12, 22, 41). In general, the O_2 compensation points have not been established, but the approximate value of 100% O_2 against 0.025% CO_2 at 20°C for a C_3 plant is used in Table 1.3. At near 100% O_2, net photosynthesis would be nil. At higher temperatures, CO_2 solubility decreases relative more than O_2 solubility decreases so that a similar inhibition of CO_2 fixation would occur at lower levels of O_2 (22). As indicated for the CO_2 compensation point at 37°C, 21% O_2 nearly completely inhibits CO_2 fixation by C_3 plants in 250–300 ppm CO_2. In a small closed chamber at 20°C, with a C_3 plant supplied with constant high levels of 300 ppm CO_2, the O_2 concentration rises to nearly 100% (the O_2 compensation point with 300 ppm CO_2)

Table 1.3. Theoretical Changes in Short-Term Photosynthetic Rates by C_3 Plants at 20°C with Changing Atmospheric CO_2 and O_2.

Atmospheric		Gross Photosynthesis (CO_2 Fixation and O_2 Evolution	Photorespiration (CO_2 Evolution and O_2 Uptake)	Net Photosynthesis (CO_2 Fixation and O_2 Evolution
CO_2 (ppm)	O_2 (%)			
			mg $CO_2 \cdot h^{-1} \cdot dm^{-2}$	
10	21	5	15	−10
60	21	20	20	0[a]
250	21	70	35	35
1000	21	100	5	95
250	~2	70	5	65
250	21	70	35	35
250	100	70	70	0[a]
1000	~2	100	0	100
1000	21	100	5	95
1000	100	100	20	80

[a] The *compensation point*, or zero net CO_2 exchange, is defined as the point where the rate of CO_2 fixation by the C_3 cycle equals the rate of CO_2 production by the C_2 cycle and dark respiration. The compensation point of about 40–60 ppm CO_2 with 21% O_2 in the light has been verified by photosynthetic rates and growth of plants in closed systems (21). The compensation point attained with increasing O_2 has been deduced from measurements of O_2 inhibition of net CO_2 fixation, which approaches zero with over 90% O_2 (12, 23).

before CO_2 fixation is almost completely stopped. Just as the atmospheric CO_2 equilibrium at 21% O_2 is higher than the CO_2 compensation point with 21% CO_2, the O_2 equilibrium at 21% O_2 is lower than the O_2 compensation point.

Insight into what holds the amount of O_2 in the atmosphere at about 21% with 250–300 ppm CO_2 may lie in the above principles of an O_2 compensation point for plants. With an atmosphere of about 300 ppm CO_2 and about 99% O_2, net photosynthesis at 20°C by C_3 plants would be near zero owing to runaway photorespiration at an O_2 compensation point. This extreme would exist in the chloroplast, where O_2 is evolved, were it not for O_2 diffusion outwards to 21% O_2. In the opposite direction, limitation on CO_2 diffusion into the cell is a major restriction on net CO_2 fixation. CO_2 and O_2 diffusion into and out of the chloroplast and plant cells to a given concentration outside establishes the CO_2 and O_2 compensation points. These equilibria in turn may establish atmospheric equilibrium points for both O_2 and CO_2, which are set by the properties of Rubisco and the rate of CO_2 and O_2 exchange between the chloroplasts and the atmosphere. At 21% O_2 in the air, a low enough O_2 level in the chloroplast is established for net carboxylase activity of Rubisco with over 60 ppm CO_2.

As shown in the top part of Table 1.3, with 21% atmospheric O_2 the CO_2 compensation point at 20°C is at 60 ppm CO_2. With higher CO_2 levels up to 1000 ppm CO_2, net photosynthetic CO_2 fixation and O_2 evolution would increase equally, while photorespiration decreases. There should be an insignificant change in the 21% atmospheric O_2 concentration over this small change in CO_2 concentration. The consequence of increasing atmospheric O_2, when the atmospheric CO_2

is kept low at 250 ppm, is shown in the middle of Table 1.3. If the atmospheric CO_2 becomes higher, the O_2 compensation level for Rubisco should also be higher. At 1000 ppm CO_2 and 20°C, there ought not to be much photorespiration if the atmospheric O_2 remains near 21% O_2. However, at temperatures higher than 30°C, photorespiration may again become significant even with 1000 ppm CO_2, but I do not know whether this has been experimentally tested. Parts of Table 1.3, based on the compensation point are obviously speculative, and need to be further extrapolated to atmospheric CO_2 and O_2 equilibria by experimental data.

Any scheme to bury reduced carbon from photosynthesis to lower CO_2 levels would increase the atmospheric O_2 proportionally. However, the magnitude of such changes would probably be insignificant relative to the amount of global O_2. A CO_2 sink, such as $CaCO_3$ deposits from marine photosynthesis, ought not to alter atmospheric O_2 levels because there is a one-to-one correspondence between the number of CO_2 molecules reduced to organic carbon and the number of O_2 molecules generated, and both enter their respective cycles. The second HCO_3^- deposited as $CaCO_3$ requires only ATP from cyclic photosynthetic phosphorylation for the CO_2 pump; photolysis of water with O_2 release by photosystem II is not needed. Thus $CaCO_3$ deposits would be (have been) a photosynthetically feasible way to remove excess CO_2 without altering the atmsopheric O_2 levels.

Today we can make a good case for no significant increase in the very large atmospheric O_2 pool over several decades or centuries with doubling or tripling of the trivial atmospheric CO_2, just as long as the excess photosynthate is not buried or removed from the global carbon cycle. Any excess of O_2 from fixing the increasing CO_2 is probably being used to oxidize the old photosynthate. $^{16}O/^{18}O$ measurements have demonstrated that the atmospheric O_2 is turning over probably in quantities as large as those of CO_2 (see Chapter 17). The difference in the CO_2 and O_2 pool sizes represents the relatively fast turnover of atmospheric CO_2 (requiring only a few years) versus that of the vast O_2 pool, requiring centuries to turn over. Nevertheless, with increasing atmospheric CO_2, a higher O_2 compensation equilibrium theoretically would occur over the long term, particularly if the temperature increased as well. The extent of these increases needs to be evaluated by considering all of the other environmental factors controlling photosynthesis and the CO_2 and O_2 global cycles.

GAIA

Gaia, or Mother Earth, is named after the Greek Earth goddess. Gaia is the concept of a self-regulation entity—a living Earth in which the composition of the atmosphere and of the oceans is biologically controlled. These concepts have been elaborated by Lovelock in his book *GAIA* (26). What has been described as the global carbon cycle (as in this chapter) is attributed by Lovelock to Gaia, as a feedback system to optimize and maintain the environment. Because of Gaia or biological life, the atmosphere of a life-bearing planet should be different from that of a dead planet. Lovelock writes that "the presence of fossils shows that the Earth's climate has changed very little." The Gaia concept is that "the entire range of living matter on Earth ... constitutes a single living entity, capable of manipulating the Earth's atmosphere." Over the past 2800 Myr, Rubisco, found in all plants and algae and with its dual activity for both photosynthesis and photorespiration, fulfilled this regulatory concept. By removing and replacing CO_2 in the atmosphere, Rubisco has been the photosynthetic mechanism for controlling and maintaining the atmospheric composition and the CO_2 greenhouse. Lovelock speculated that "the atmosphere is not merely a biological product, but more probably a biological construction—a living system designed to maintain a chosen environment." The ratio of carboxylase to oxygenase activities of Rubisco during photosynthetic carbon metabolism constitutes this biological entity to maintain the environment. How these activities developed during evolution needs to be considered if we would try to modify our atmosphere in the future by photosynthesis. Rubisco of the primitive, surviving photosynthetic bacteria or algae has both carboxylase and oxygenase activities, although in the originally high CO_2 and low O_2 atmospheres the oxygenase activity was probably non-functional in photorespiratory control. Only relatively recently has the vital role of Rubisco oxygenase activity in regulating net photosynthesis and the composition of the atmosphere become critical. Has this just been fortuitous? Does Gaia have an ethereal component, or is the current "essential" biological environment just an evolutionary consequence of photosynthesis? Much further thought and research are needed to evaluate how much Rubisco controls the atmospheric gases to regulate the greenhouse and to maintain optimal conditions for life. It seems that plants and algae at current temperatures have a built-in photosynthetic carbon system to maintain atmospheric CO_2 between 180 and 1000 ppm CO_2 and 21% O_2. At higher CO_2 levels, other plant growth factors become overall limiting. The balancing and regulating components are contained in the properties of one enzyme, Rubisco: carboxylase for CO_2 fixation by the C_3 cycle, and oxygenase as required for the photorespiratory C_2 cycle to dissipate variable amounts of excess energy.

REFERENCES

1. Aizawa, K., and Miyachi, S. 1986. Carbonic anhydrase and CO_2 concentrating mechanisms in microalgae and cyanobacteria. *FEMS Microbiol. Rev.* **39**: 215–33.

2. Asada, K., and Takahashi, M. 1987. Production and scavening of active oxygen in photosynthesis. In *Photoinhibition*, ed. D. J. Kyle, C. B. Osmond, and C. J. Arntzen, pp. 227–87. Elsevier Sci. Publ. BV, New York.

3. Baker, D. N., and Lambert, J. R. 1979. In *Report of the AAAS–DOE Workshop on Environmental and Societal Consequences of a Possible CO_2-Induced Climate Change.* Annapolis, Md.

4. Bassham, J. A., and Calvin, M. 1957. *The Path of Carbon in Photosynthesis.* Prentice-Hall, Englewood-Cliffs, N.J.

5. Borowitzka, A. 1987. Calcification in algae: Mechanism and the role of metabolism. *CRC Crit. Rev. Plant Sci.* **6**: 1–45.

6. Branola, J. M., Raynaud, D., Korotkevich, Y. S., and Lorius, C. 1987. Vostok ice core provides 160,000-year record of atmospheric CO_2. *Nature* **329**: 408–14.

7. Budde, R. J. A., and Randall, D. D. 1990. Pea leaf mitochondrial pyruvate dehydrogenase complex is inactivated in vivo in a light-dependent manner. *Proc. Natl. Acad. Sci. USA* **87**: 673–76.

8. Coleman, B., and Turpin, D. H., organizers. 1991. Second International Symposium on Inorganic Carbon Utilization by Aquatic Photosynthetic Organisms. *Can. J. Bot.* **69**: 907–1160.

9. Dioniso, M. L., Tsuzuki, M., and Miyachi, S. 1989. Blue light induction of carbonic anhydrase activity in *Chlamydomonas reinhardtii*. *Plant Cell Physiol.* **30**: 215–19.

10. Edwards, G. E., and Huber, S. C. 1981. The C_4 pathway. In *The Biochemistry of Plants*, ed. P. Stumpf and E. Conn, Vol. 8: *A Comprehensive Treatise*, ed. M. D. Hatch and N. K. Boardman, pp. 327–81. Academic Press, New York.

11. Enoch, H. Z., and Kimball, B. A. 1986. Carbon dioxide enrichment of greenhouse crops, Vols. I and II. CRC Press.

12. Forrester, M. L., Krotkov, G., and Nelson, C. D. 1966. Effect of oxygen on photosynthesis, photorespiration and respiration in detached leaves. I. Soybean. *Plant Physiol.* **41**: 422–27.

13. Goyal, A., and Tolbert, N. E. 1989. Uptake of inorganic carbon by isolated chloroplasts from air-adapted *Dunaliella*. *Plant Physiol.* **9**: 1264–9.

14. Goyal, A., and Tolbert, N. E. 1992. Blue light induction of the dissolved carbon concentrating mechanism in *Chlamydomonas*. *Plant Physiol.* **99**: 5–609.

15. Greenberg, B. M., Gaba, V., Canaani, O., Malkin, S., Mattoo, A. K., and Edelman, M. 1989. Separate photosensitizers mediate degradatkon of the 32-kDa photosystem II reaction center protein in the visible and UV spectral regions. *Proc. Natl. Acad. Sci. USA* **86**: 6617–20.

16. Hardy, R. W. F., and Havelka, U. D. 1975. Photosynthate as a major factor limiting N_2 fixation by field grown legumes with emphasis on soybeans. In *Symbiotic Nitrogen Fixation in Plants*, ed. R. S. Nutman. pp. 421–39. Cambridge: Cambridge University Press.

17. Hayes, J. K. 1987. Development of the Earth's atmosphere. *Encycl. Brit.* **14**: 305–12.

18. Husic, D. W., Husic, H. D., and Tolbert, N. E. 1987. The oxidative photosynthetic carbon cycle or C_2 cycle. *Crit. Rev. Plant Sci.* **5**: 45–100.

19. Keeling, C. D. 1958–86. Atmospheric CO_2 concentration—Mauna Loa Observatory. Hawaii. Carbon Dioxide Information Center, Oak Ridge National Laboratory, Oak Ridge, Tenn.

20. Keys, A. J., Bird, I. F., Cornelius, M. J., Lea, P. J., Wallsgrove, R. M., and Miffin, B. J. 1978. Photorespiratory nitrogen cycle. *Nature (London)* **275**: 741–42.

21. Krenzer, E. G., Moss, D. N., and Crookston, R. K. 1975. CO_2 compensation points of flowering plants. *Plant Physiol.* **56**: 194–209.

22. Ku, S.-B., and Edwards, G. E. 1978. Oxygen inhibition of photosynthesis. III. Temperature dependence of quantum yield and its relation to O_2/CO_2 solubility ratio. *Plants* **140**: 1–6.

23. Ku, S.-B., Edwards, G. E., and Tanner, C. B. 1977. Effects of inhibition of photosynthesis and transpiration in *Salanum tuberosum*. *Plant Physiol.* **59**: 868–72.

24. Lemon, E. R., ed. 1983, *CO_2 and Plants: The Response of Plants to Rising Levels of Atmospheric Carbon Dioxide*. Westview Press, Boulder, Co,

25. Lorimer, G. H. 1981. The carboxylation and oxygenation of ribulose 1,5-bis-phosphate: The primary event in photosynthesis and photorespiration. *Annu. Rev. Plant Physiol.* **32**: 349–83.

26. Lovelock, J. E. 1979. *GAIA*, Oxford University Press, New York.

27. Lucas, W. J., and Berry, J. A. eds, 1985, Inorganic carbon uptake by aquatic photosynthetic organisms. Waverly Press, Baltimore, Md.

28. Mifflin, B. J., and Lea, P. J. 1980. Ammonia assimilation. In *The Biochemistry of Plants*, Vol. 5, ed. B. J. Mifflin, pp. 169–202. Academic Press, New York.

29. Moroney, J. V., Husic, H. D., and Tolbert, N. E. 1986. CO_2 and HCO_3^- accumulation by microalgae. In *Regulation of Chloroplast Differentiation*, ed. G. Akoyunoglou and H. Senger. Plant Biology Series, Vol. 2, pp. 715–24. Alan R. Liss, New York.

30. Ogren, W. L. 1984. Photorespiration: Pathways, regulation, and modification. *Annu. Rev. Plant Physiol.* **35**: 415–47.

31. Osmond, C. B. 1981. Photorespiration and photoinhibition: Some implications for the energetics of photosynthesis. *Biochem. Biophysica Acta* **638**: 77–98.

32. Smith, R. C., Baker, K. S., Holm-Hansen, O., and Olson, R. 1980. Photoinhibition of photosynthesis in natural waters. *Photochem. Photobiol.* **31**: 585–92.

33. Smith, R. C., Prezelin, B. B., Baher, K. S., Bidigare, R. B., Boucher, N. P., Coley, T., Karentz, D., MacIntyre, S., Mallick, H. A., Menzies, D., Ondrusek, M., Wan, Z., and Waters, K. J. 1992. Ozone depletion: Ultraviolet radiation and phytoplankton biology in Antarctic waters. *Science* **255**: 952–9.

34. Stabenau, H. 1992. *Phylogenetic Changes in Peroxisomes of Algae—Phylogeny of Plant Peroxisomes.* pp. 1–442. University of Oldenburg Press, Oldenburg, Germany,

35. Thielman, J., Tolbert, N. E., Goyal, A., and Senger, H. 1990. Differentiating between two systems for concentrating inorganic carbon during photosynthesis by *Scenedesmus*. *Plant Physiol.* **92**: 622–9.

36. Tolbert, N. E. 1974. Photorespiration by algae. In *Algal Physiology and Biochemistry*, ed. W. D. P. Stewart, pp. 474–504. Blackwell Scientific Publications Ltd., Oxford, U.K.

37. Tolbert, N. E. 1980. Photorespiration. In *The Biochemistry of Plants*, ed. P. Stumpf and E. Conn, Vol. 2: *Metabolism and Respiration*, ed. D. D. Davies, pp. 488–525. Academic Press, New York.

38. Tolbert, N. E. 1983. The oxidative photosynthetic carbon cycle. In *Current Topics in Plant Biochemistry and Physiology*, ed. D. D. Randall et al., Vol. I. pp. 63–77. University of Missouri, Columbus, Mo.

39. Tolbert, N. E., Husic, H. D., Husic, D. W., Moroney, J. V., and Wilson, B. J. 1985. Relationship of glycolate excretion to the DIC pool in microalgae. In *Organic Carbon Uptake by Aquatic Photosynthetic Organisms*, ed. W. J. Lucas and J. A. Berry, pp. 211–23. Amer. Sci. Plant Physiol., Rockville, Md.

40. Tolbert, N. E. 1992. Comparison of peroxisomes and the C_2 oxidative photosynthetic carbon cycle in leaves and algae. In *Phylogenetic Changes in Peroxisomes of Algae—Phylogeny of Plant Peroxisomes*, ed. H. Stabenau, pp. 204–32, University of Oldenburg Press, Oldenburg, Germany.

41. Turner, J. S., and Brittain, E. G. 1962. Oxygen as a factor in photosynthesis. *Biol. Rev.* **37**: 130–70.

42. Widholm, J. M., and Ogren, W. L. 1969. Photorespiratory-induced senescence of plants under conditions of low carbon dioxide. *Proc. Natl. Acad. Sci. USA* **63**: 668–75.

43. Wittwer, S. H., and Taubner, F. G. 1959. Foliar absorption of mineral nutrients. *Annu. Rev. Plant Physiol.* **10**: 13–32.

44. Zelitch, I. 1975. Improving the efficiency of photosynthesis. *Science* **188**: 626–33.

45. Zill, L. P., and N. E. Tolbert. 1958. The effect of ionizing and ultraviolet radiations on photosynthesis. *Arch. Biochem. Biophys.* **76**: 196–203.

I
THE ENVIRONMENT AND GEOCHEMISTRY

2

Changes in Atmospheric CO$_2$ Concentration and the Global Carbon Cycle

WILLIAM R. EMANUEL, ANTHONY W. KING, and
WILFRED M. POST

A global network is currently measuring unprecedented increases in atmospheric CO$_2$ concentration (24), tracking the response of a fundamental biogeochemical system—the Earth's carbon cycle—to fossil fuel emissions and other human activities as well as to natural events. We need to predict the consequences of further fossil fuel use, but we now only partially understand the Earth's systems that control atmospheric CO$_2$. Our accounting of the current situation is incomplete, and our synthesis of past events that led to these conditions is inconsistent (5, 16, 37).

Fossil fuel use almost certainly caused most of the observed CO$_2$ increase, but the accumulation of carbon in the atmosphere is less than the fossil fuel contribution because emissions force carbon to redistribute within its global cycle. The oceans, which contain about 60 times more carbon than does the atmosphere, serve as the main sink. Land plants and soils store carbon as well, and rising atmospheric CO$_2$ concentrations may stimulate plant productivity, causing more carbon storage in terrestrial pools; however, human activities such as forest clearing are reducing biomass over large regions and releasing carbon into the atmosphere, in addition to that from fossil fuels.

The carbon cycle's control of changes in CO$_2$ due to fossil fuel emissions is not fully understood. If land use caused substantial carbon releases from terrestrial pools, then estimates of oceanic uptake are inconsistent with observed CO$_2$ increases. In order to resolve this inconsistency, we need to reevaluate and refine our understanding of the carbon cycle and the perturbations of interest from several standpoints:

1. The oceans may revmove carbon from the atmosphere more readily than current models indicate.
2. Other ecosystem responses may compensate for decreases in terrestrial carbon storage due to land use and other activities.
3. Estimates of historical fossil fuel emissions or of releases of carbon from vegetation and soil may be wrong.

All of these points probably contribute to our inability to explain fossil fuel era changes in the carbon cycle, and they are important reasons for skepticism regarding the details of future CO_2 projections derived from the models that are available now. But even recognizing these problems, model exercises are still convincing that atmospheric CO_2 levels can increase substantially by the middle of the next century.

OBSERVED CHANGES IN ATMOSPHERIC CO_2 CONCENTRATION

Keeling and co-workers began accurate, continuous measurements of atmospheric CO_2 concentration at Mauna Loa Observatory, Hawaii in 1958 (21, 22). Figure 2.1 displays monthly average concentrations at Mauna Loa since 1959, with three months of missing values in 1964 estimated by interpolation. Over the 32-year record, the average annual concentration has risen by 12% from 315.83 ppmv (parts per million, by volume) in 1959 to 353.95 ppmv in 1990. The average annual rate of increase was about 1.2 ppmv, or 0.4% per year. These data approximate average atmospheric changes reasonably well.

Shorter records from a number of stations worldwide replicate the general trend in Keeling's Mauna Loa CO_2 measurements (8). Figure 2.2 summarizes monthly average concentrations derived from continuous measurements by the U.S. National Oceanic and Atmospheric Administration at Mauna Loa, Hawaii; Barrow, Alaska; American Samoa; and at the South Pole. These records illustrate

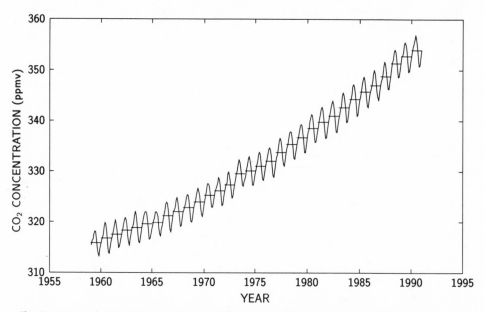

Fig. 2.1. Atmospheric CO_2 concentration at Mauna Loa Observatory, Hawaii (22). Lines connect monthly average values to form the oscillating curve. Horizontal bars indicate annual average concentrations.

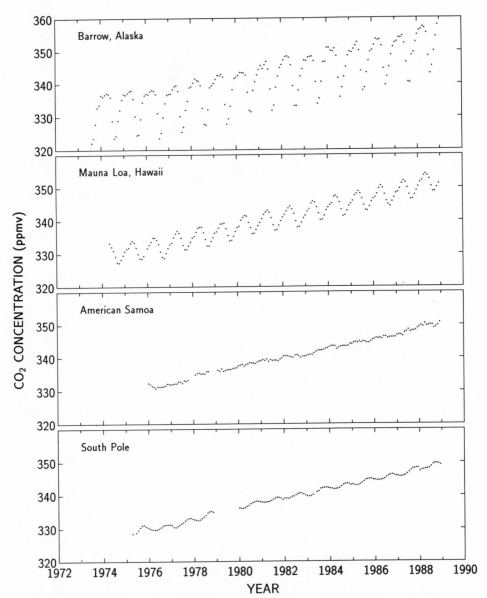

Fig. 2.2. Atmospheric CO$_2$ concentration at four monitoring stations where the National Oceanic and Atmospheric Administration makes continuous measurements (24). The dots represent monthly average values.

geographic patterns in atmospheric CO$_2$ variations. All four contain a steadily increasing trend. A pronounced annual cycle is also present, and there is a weaker, irregular three- to four-year cycle that is associated with El Niño or Southern Oscillation events (2).

Seasonal exchanges between the atmosphere and terrestrial ecosystems cause most of the annual cycle in CO$_2$ concentration, although the oceans are involved as well (22, 43). The amplitude and phase of the annual cycle depend on latitude (Fig. 2.2). Carbon dioxide concentrations increase during the fall and winter in the Northern Hemisphere and decline during spring and summer. This cycle is reversed and of smaller amplitude in the Southern Hemisphere. The amplitude is largest at high latitudes, where exchanges with tundra and boreal ecosystems are strongly seasonal. In the Southern Hemisphere, where there is less land and where exchanges between the atmosphere and terrestrial ecosystems are more constant through the year, the amplitude of the cycle is comparatively small.

The amplitude of the annual cycle at Mauna Loa has increased (22). The increase was pronounced through the 1970s and subsided during the 1980s. The stimulation of terrestrial primary productivity by rising atmospheric CO_2 levels may have contributed to this change in the amplitude of the annual cycle, but Keeling and co-workers point out that the increase should not have ended in the 1980s if this were the primary cause or the only factor involved. They suggest that changes in air temperature may have further altered primary productivity and affected decomposition rates as well.

Keeling et al. (22) analyzed variations in both atmospheric CO_2 concentration and the abundance of ^{13}C in the atmosphere associated with El Niño events. The ^{13}C variations suggest that changes in terrestrial carbon pools, which are depleted in ^{13}C compared to the atmosphere, are responsible for the three- to four-year variations in the atmosphere. Weak summer monsoons in Southeast Asia accompany El Niño events so that widespread drought may be responsible for this influence of terrestrial ecosystems on atmospheric CO_2.

Concentrations in air bubbles trapped in polar ice are the best indicators of atmospheric CO_2 levels prior to the start of accurate measurements. An ice core extracted at Siple Station, Antarctica (15, 29), shows the history of CO_2 from the middle of the eighteenth century into the Mauna Loa record (Fig. 2.3); the agreement with modern measurements is remarkable. The Siple ice data indicate that eighteenth-century CO_2 concentration was about 280 ppmv; the 1990 concentration at Mauna Loa—353.95 ppmv—represents a 26% increase.

The releases of carbon into the atmosphere by fossil fuel use since 1950 can be estimated from fuel production data compiled by the U.N. Statistical Office (44). Marland et al. (27) document the analysis for 1950 through 1986. Coal, crude petroleum, and natural gas are considered. The quantity of each is multiplied by a conversion factor to calculate the mass of carbon released into the atmosphere as CO_2 when the fuel is used. Keeling (19) estimated CO_2 emissions from 1860 through 1949 from earlier U.N. data. Cement manufacturing also releases CO_2; Marland et al. (27) estimate these from U.S. Department of Interior, Bureau of Mines data.

Figure 2.4 displays a composite history of CO_2 emissions from fossil fuels and cement manufacturing, assembled by Marland et al. (27). The total release between 1860 and 1989 was 212 petagrams (Pg; 1 Pg $= 1 \times 10^{15}$ g). Coal use contributed 54.8% of this total, petroleum 32.8%, gas 10.9%, and cement manufacturing 1.5%. The CO_2 release in 1989 was 5.966 Pg carbon (C). The annual release increased monotonically since 1983 at the rate of about 2.7% per year.

Houghton et al. (17) report estimates of carbon releases from vegetation and soils since 1800. Their analysis tracks the area, age, and carbon content of disturbed

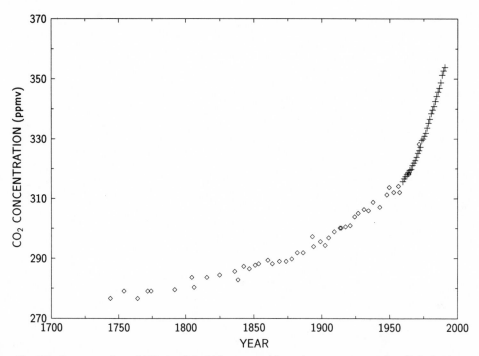

Fig. 2.3. Concentration of CO_2 in air bubbles trapped in an ice core extracted at Siple Station, Antarctica ($\diamondsuit$) (15, 29) and as recorded at Mauna Loa since 1959 ($+$).

regions, using response functions to specify changes in carbon stocks in different ecosystem types. They explicitly account for the oxidation rates of fuel wood and wood products. Recent estimates derived by this approach (18) indicate that land use decreased carbon storage in vegetation and soil by about 170 Pg since 1800. Estimates of the net flux of carbon into the atmosphere in 1980 due to land use range from 0.4 Pg $C \cdot yr^{-1}$ to 2.5 Pg $C \cdot yr^{-1}$ (10, 18).

The Siple ice-core data indicate that the atmospheric concentration of CO_2 in 1860, at the beginning of significant fossil fuel use, was about 289 ppmv. The 1989 annual average concentration at Mauna Loa was 352.75 ppmv. The difference in these concentrations corresponds to a 135.8 Pg increase in the carbon content of the atmosphere, about 67% of the carbon released from fossil fuels. Although the uncertain contribution of carbon to the atmosphere from terrestrial ecosystems is not considered, this is a broad measure of the tendency of the global carbon cycle to redistribute carbon when CO_2 is added to the atmosphere

THE GLOBAL CARBON CYCLE

Arrhenius (1) suggested, as early as the turn of the century, that CO_2 added to the atmosphere by burning fossil fuels might accumulate there and raise the average atmospheric concentration of the gas. But while direct measurements and ice-core data confirm that CO_2 concentration is increasing, the connection to fossil fuel use is complicated and not fully understood.

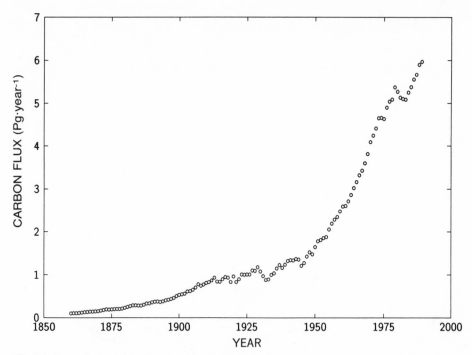

Fig. 2.4. Past releases of carbon into the atmosphere from fossil fuels. These estimates are derived from United Nations energy data (27).

The global carbon cycle, which controls the response of atmospheric CO_2 to fossil fuel emissions and to perturbations such as land-use or climatic change, is composed of numerous reservoirs and fluxes involving processes with characteristic response times ranging from seconds—for example, in the case of gas exchange in photosynthesis and plant respiration—to many centuries for the weathering of rocks and the turnover of sediments (4, 9). But many parts of the carbon cycle respond so slowly that they have not contributed significantly to fossil fuel era changes and will not be important during the next century or two (5).

It is important to note, however, that while some pools and fluxes do not respond to CO_2 increases, they are still involved in the carbon cycle's control of atmospheric responses to fossil fuel emissions and other perturbations. For example, oceanic primary productivity is not likely to increase as a result of rising CO_2 because CO_2 is abundant in surface waters and does not limit photosynthesis. But falling organic matter creates a gradient in the concentration of total carbon from surface through deep waters that affects oceanic carbon uptake.

Figure 2.5 summarizes the pools and fluxes within the carbon cycle that dominate the response of atmospheric CO_2 to fossil fuel emissions (4, 28, 37) for conditions prior to significant fossil fuel emissions. The oceans contain by far the largest pool of actively turning over carbon; however, there are significant pools in land plants and soil as well. The fluxes into the atmosphere from terrestrial ecosystems and from the oceans are comparable, and thus the turnover of carbon in the oceans is much slower than on land.

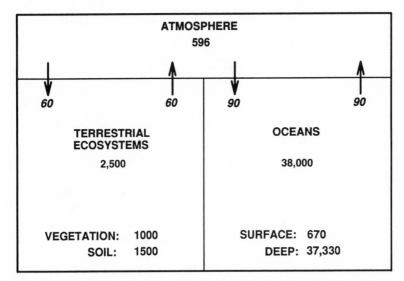

Fig. 2.5. The major reservoirs and fluxes of the global carbon cycle that determine the response of atmospheric CO$_2$ concentration to fossil fuel releases. The oceans are the primary sink for carbon from the atmosphere. Vegetation and soil may contribute carbon to the atmosphere in addition to that from fossil fuels because of disturbance by human activities or, under other circumstances, serve as a carbon sink in addition to the oceans. Estimates of pool sizes and fluxes correspond to approximate equilibrium prior to significant fossil fuel emissions.

The exchange of CO$_2$ between the atmosphere and surface waters is proportional to the difference in their CO$_2$ partial pressures so that an increase in atmospheric CO$_2$ forces carbon into the oceans (6, 20). It is much less clear that rising atmospheric CO$_2$ causes net carbon uptake by terrestrial ecosystems. Laboratory and field studies in controlled environments provide some evidence that carbon assimilation by photosynthesis increases with higher CO$_2$ concentrations, but the effect has not been demonstrated in natural ecosystems over large areas (3, 11). Furthermore, land-use activities, particularly forest clearing, cause net carbon releases in many regions in addition to fossil fuel emissions (45).

The oceans contain about 38,000 Pg of carbon. Broecker et al. (7) summarize estimates of the annual invasion of carbon into the surface waters of the oceans. Their best estimate is 82.3 Pg/year, within values ranging from 69.3 to 104 Pg/year. Three sets of processes determine the rate of CO$_2$ uptake by the oceans: (1) CO$_2$ exchange across the air–sea boundary, (2) the incorporation of surface water carbon into compounds other than CO$_2$, and (3) mixing and circulation, which transport carbon from surface water into deeper layers where it is away from atmospheric exchange (6).

The exchange of CO$_2$ across the atmosphere–ocean interface is controlled by temperature, wind stress, and processes such as turbulent mixing. The partial pressure of CO$_2$ in the atmosphere is proportional to the ratio of the masses of CO$_2$ and dry air, whereas the partial pressure of CO$_2$ in surface waters depends on the chemical equilibria between carbon compounds and transport into deeper water. As CO$_2$ is added to surface waters, chemical equilibration is rapid, compared with transport and mixing processes. The reactions of primary concern are:

$$CO_2(aq) + H_2O \rightleftharpoons H_2CO_3$$

$$H_2CO_3 \rightleftharpoons HCO_3^- + H^+ \qquad (2.1)$$

$$HCO_3^- \rightleftharpoons CO_3^{2-} + H^+$$

so that while the concentration of total carbon is

$$[\textstyle\sum C] = [CO_2(aq)] + [H_2CO_3] + [HCO_3^-] + [CO_3^{2-}] \qquad (2.2)$$

the carbon flux from surface waters back into the atmosphere is proportional only to $[CO_2(aq)]$ (20).

Mixing and circulation remove carbon from surface waters into deeper layers where it may be sequestered for some time. Analyzing ^{14}C measurements collected by the Geochemical Ocean Sections Study, Stuiver et al. (42) find that the average residence time of carbon in the deep ocean out of contact with the atmosphere is about 500 years. While some water is isolated from the surface for 1700 years, deepwater resides in the Atlantic for only about 275 years.

Land plants contain about 560 Pg of carbon (34), and soils contain about 1500 Pg in actively turning-over dead organic matter (36). Prior to disturbance by human activities, carbon in vegetation may have approached 1000 Pg (33). Plants assimilate carbon from the atmosphere by photosynthesis, and decomposition and fire release CO_2 from dead organic matter (12). For reasonably undisturbed ecosystems and over sufficiently long periods, uptake and loss fluxes approximately balance. The balance certainly shifts on shorter time scales with changes in nutrient availability, climate, and sporadic natural disturbances.

Land plants assimilate about 60 Pg C·yr^{-1} of carbon from the atmosphere (34). This *net primary production*—the difference between gross assimilation and the return of CO_2 to the atmosphere by plant respiration—is responsible for longer-term accumulation of carbon in vegetation and ultimately in dead organic matter. The rapid variations in gross photosynthesis and autotrophic respiration that occur within hours and days do not affect the steadily increasing trend in atmospheric CO_2 concentration. But changes in the relative magnitudes of cumulative net production and losses of organic matter from litter and soil pools due to decomposition and fire can be significant.

While the oceans apparently act as a net sink for carbon from the atmosphere in any given year, it seems likely that imbalances between terrestrial income and loss fluxes might cause either net increases or decreases in vegetation and soil pools, depending on environmental conditions, nutrient availability, and sporadic disturbances. A widely distributed 0.5-Pg C change in terrestrial carbon pools within a year is likely to be undetectable.

Disturbance to terrestrial ecosystems by land-use change alters the standing levels of carbon and the dynamics of cycling (12, 17). With land-use change such as forest clearing, carbon is immediately released into the atmosphere by burning or, after a delay, by decomposition. In many instances, after forest clearing, ground vegetation (e.g., annual crops, pasture, or other herbaceous types) is established rather than trees, and by management this altered land cover, with carbon storage lower than in the original ecosystems, may be maintained indefinitely. Following clearing, net productivity generally exceeds losses, and disturbed ecosystems may act as sinks for some time. Abandoned areas recover toward natural conditions and, during recovery, may also serve as carbon sinks. Human management of the landscape has thus created a complex pattern of sources and sinks in different regions that has changed through time.

Plants respond to atmospheric CO$_2$ variations by rapidly changing enzyme activities and stomatal aperture. With increases in atmospheric CO$_2$, average stomatal conductance can probably be decreased, thus reducing transpiration and increasing *water-use efficiency*—the ratio of assimilated CO$_2$ to transpired water. Rapid changes in photosynthesis and water-use efficiency due to changes in stomatal control under enhanced CO$_2$ have been observed in laboratory experiments (39, 41), and some researchers indicate that CO$_2$-enhanced growth can be observed in natural vegetation (26, 35). Nutrient stress could negate growth due to rising CO$_2$, but most experiments show growth enhancements even under nutrient limitation (30, 31).

In addition to the major carbon exchanges between the atmosphere and the oceans and between the atmosphere and vegetation and soil that appear to dominate the response of atmospheric CO$_2$ concentration to fossil fuel releases, numerous other carbon fluxes may also be significant. Furthermore, atmospheric exchanges and the turnover of carbon in the oceanic and terrestrial components of the carbon cycle depend on climate: Climatic change caused by increases in greenhouse gas concentrations may very well feed back on the controls of atmospheric composition in the global carbon cycle (38)

CARBON CYCLE MODELS AND CO$_2$ PROJECTIONS

Observational studies of the phenomena that control CO$_2$ concentration are often impractical because of the time scales involved (decades to centuries), the vast extent and spatial heterogeneity of Earth systems, and the small perturbations to natural levels that fossil fuel releases cause in very large reservoirs such as the oceans. Furthermore, many mechanistically important variables cannot be measured directly. Mathematical models are an important means of contending with these limitations of direct studies. They are unsurpassed as tools for synthesis and integration of diverse concepts and data.

Given reliable estimates of past fossil fuel emissions, an accounting of carbon uptake from the atmosphere by the oceans is the most important requirement in analyzing past changes in CO$_2$ concentration or in estimating future CO$_2$ increases due to different levels of further fossil fuel use. But historic changes in terrestrial carbon storage, whether due to natural phenomena or human activities, cannot be ignored, nor can future changes unless the magnitude of fossil fuel releases increases substantially.

Fossil fuel use transfers carbon into the atmosphere from reservoirs that are otherwise connected to this more active section of the cycle only on geological time scales. This allows us to estimate changes in CO$_2$ concentration by simulating the redistribution of fossil fuel carbon within the atmosphere, oceans, and terrestrial ecosystems as if it were added to a closed system from an external source. Furthermore, the atmosphere is responsible for most of the coupling between carbon pools in terrestrial ecosystems and those in the oceans; rivers and runoff also transfer carbon from land into the oceans, but these links are weak by comparison. Terrestrial and oceanic carbon models can therefore be decoupled for many purposes. This is a practical advantage because the data and formalisms used in each are quite different. Operationally, the solutions to terrestrial models are summarized as a net flux into the atmosphere and treated as an additional atmospheric input in the solution of models of the atmosphere and oceans.

Oeschger et al. (32) proposed a model of carbon turnover in the atmosphere and oceans that is so widely applied that it has become a benchmark against which other models are referenced and compared (13, 25). In the Oeschger model, a diffusion equation with constant diffusivity represents the vertical transport of carbon in a globally averaged water column. The diffusivity parameter and the invasion flux of carbon from the atmosphere are set for agreement between the equilibrium distributions of carbon and radiocarbon implied by the model and idealized profiles derived from observations.

Let c_a represent the mass of carbon in the atmosphere and c_s the total carbon content of surface seawater (Eq. 2.3). In most globally averaged models, the top 75 m of water is assumed to exchange carbon with the atmosphere as if it were well-mixed. The changes in chemical equilibria forced by adding carbon to surface waters can be accounted for by solving the appropriate equilibrium conditions, constrained by sufficient specified values (20). A nonlinear function $p_s(c_s)$ expresses the dependence of the partial pressure of CO_2 dissolved in surface water on their total carbon content. Many models use a linear approximation to this relationship.

In this framework, a coupled system of equations describes the turnover of carbon:

$$\frac{d}{dt} c_a = -k_{as} c_a + k_{sa} p_s(c_s) + f(t) \tag{2.3}$$

$$\frac{d}{dt} c_s = k_{as} c_a - k_{sa} p_s(c_s) + SK \frac{\partial}{\partial z} \hat{c}(z, t) \Big|_{z=0} \tag{2.4}$$

$$\frac{\partial}{\partial t} \hat{c}(z, t) = K \frac{\partial^2}{\partial z^2} \hat{c}(z, t). \tag{2.5}$$

The depth below the bottom of the surface water reservoir is designated z, and $\hat{c}(z, t)$ is the concentration of carbon at depth z. The parameters k_{as} and k_{sa} are rate coefficients for the surface water invasion and evasion fluxes, respectively, K is the diffusivity coefficient, and S the area. The forcing function $f(t)$ includes all inputs into the atmosphere from fossil fuels and other sources. This system of equations is solved from equilibrium initial conditions:

$$c_a(t_0) = c_a^0$$

$$c_s(t_0) = c_s^0$$

$$\hat{c}(z, t_0) = \hat{c}^0(z)$$

$$\hat{c}(0, t_0) = \hat{c}_s(t_0)$$

for specified forcing functions $f(t)$.

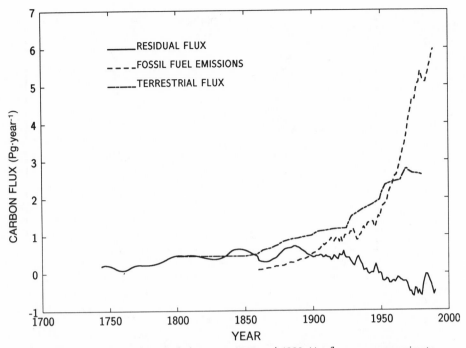

Fig. 2.6. Changes in the carbon cycle between 1745 and 1990. Net fluxes are approximate 1989 values.

Since past changes in terrestrial carbon storage are very uncertain, it is instructive to calculate the net flux into or out of the atmosphere required to match the solution of a model of carbon cycling in the atmosphere and oceans to atmospheric CO$_2$ observations (Fig. 2.3) (14, 23, 40). This procedure takes the model solution from initial equilibrium conditions to the present in a way that is consistent with our best understanding of past changes in atmospheric CO$_2$ concentration; future responses to fossil fuel emissions and other sources can then be simulated. The residual net flux suggests the course of past exchanges with terrestrial ecosystems.

Figure 2.6 summarizes changes in the major carbon reservoirs from 1745 through 1989 simulated by a box-diffusion model with diffusivity $K = 4000 \, \mathrm{m^2 \cdot yr^{-1}}$ as estimated from radiocarbon data. The net residual flux is assumed to be derived from terrestrial ecosystems into the atmosphere. The increase in the carbon content of the atmosphere is 76% of the 212 Pg of carbon released from fossil fuels and 58% of the total release from fossil fuels and from the residual flux assumed to be from vegetation and soils. The 66-Pg residual release required to match simulated oceanic uptake and observed changes in the atmosphere is substantially less than the decreases in terrestrial carbon storage estimated from land-use data.

Houghton et al. (17) developed three reconstructions of changes in terrestrial carbon storage, based on land-use and demographic data, with synoptic functions specifying variations in carbon storage following disturbances in different ecosystems. Figure 2.7 displays the net residual flux implied by the box-diffusion

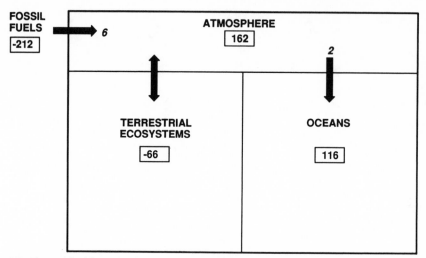

Fig. 2.7. The residual flux into the atmosphere required to match the solution to a box-diffusion model of carbon cycling in the atmosphere and oceans forced by past fossil fuel emissions to atmospheric CO_2 observations. Boxed values are cumulative changes in reservoir contents in a model solution with the atmosphere constrained by the Siple ice-core and Mauna Loa CO_2 measurement records. The fossil fuel emissions and estimates of terrestrial carbon releases by Houghton et al. (17) are also shown for comparison.

model and the nominal net terrestrial release estimate derived by Houghton et al. (17).

Working Group I of the Intergovernmental Panel on Climate Change (IPCC) considered four scenarios of future fossil fuel emissions (16). Figure 2.8 displays these scenarios for CO_2. Scenario A is meant to approximate emissions if recent trends in fossil fuel use continue. Scenarios B, C, and D represent the emissions expected if fossil fuel use is curtailed to varying degrees. These four scenarios only sample the range of potential future fossil fuel emissions. For example, there are sufficient recoverable fossil fuels to support emissions at substantially higher rates than in scenario A. Uncertainty as to future fossil fuel use contributes most to uncertainty in future CO_2 levels.

Figure 2.9 summarizes atmospheric CO_2 projections for the fossil fuel emissions scenarios in Figure 2.8. These projections were derived using a box-diffusion model as described above. In each case, model solutions begin at equilibrium initial conditions in 1740. From 1740 through 1989, the total flux of carbon into the atmosphere was due to fossil fuel emissions and the residual flux (Fig. 2.7). Fossil fuel emissions are the only atmospheric input beyond 1989. This procedure places the atmosphere–ocean model in a 1990 state that is consistent with atmospheric observations and that contends with estimated historical fossil fuel emissions. But some carbon is transferred from the atmosphere into an unspecified sink during the period of the Muana Loa record. This sink is ignored beyond the end of the CO_2 measurement records in 1990; oceanic uptake is the only means of removing carbon from the atmosphere after 1990.

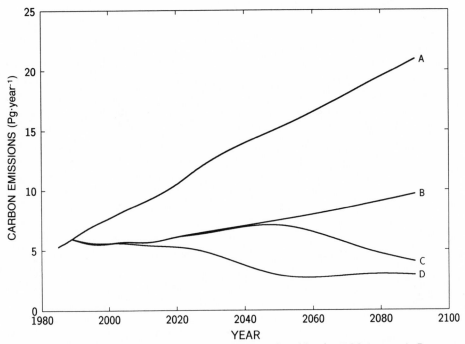

Fig. 2.8. Scenarios of future fossil-fuel CO_2 emissions analyzed by the IPCC. Letters A–D are explained in the text.

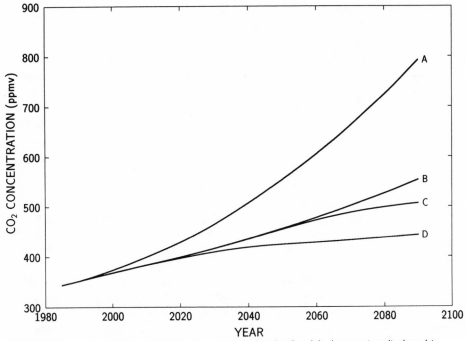

Fig. 2.9. Projected atmospheric CO_2 concentrations for the fossil fuel scenarios displayed in Figure 2.8.

49

CONCLUSIONS

Simulated responses to the four IPCC emissions scenarios (in Figure 2.9) indicate the serious potential of further fossil fuel use to increase CO_2 concentration and the difficulty of curtailing such increases. If fossil fuel emissions continue to increase at about the current rate (scenario A), then atmospheric CO_2 concentration may reach twice preindustrial levels of about 280 ppmv by the middle of the next century. Even scenario B, which assumes substantial action to curtail increases in fossil fuel use, can double CO_2 concentration over preindustrial levels by the end of the next century. Although they certainly slow the rate of CO_2 increase, emissions reductions as in scenarios C and D do not cause decreases in atmospheric CO_2 concentration. However, while these projections suggest the potential for fossil fuel emissions to increase CO_2, our inconsistent understanding of past changes challenges their veracity.

If estimates of past fossil fuel emissions are reasonably accurate, then either the oceans remove carbon from the atmosphere more readily than models suggest, or another sink (probably in terrestrial ecosystems) influences atmospheric CO_2 increases. This is a large sink, if land use has released as much carbon as the estimates based on demographic and land-use data suggest, and future CO_2 increases depend significantly on its characteristics. Thus the CO_2 projections in Figure 2.9 apparently overestimate the CO_2 increases implied by the IPCC scenarios. If substantial terrestrial releases have occurred, the overestimation is significant.

Even though forest clearing, particularly in the tropics, seems to be reducing global biomass, essentially all atmospheric and oceanic data and analyses point to terrestrial ecosystems as a sink in addition to the oceans. The stimulation of plant productivity by rising atmospheric CO_2 concentration seems to be a leading candidate for creating more terrestrial carbon storage, but the mechanisms for maintaining such a sink are unclear, and its response as CO_2 continues to increase or if significant climatic change occurs is uncertain. While numerous experiments are clarifying the effects of increasing CO_2 on individual plants in controlled environments, the expected ecosystem responses in the natural environment are very poorly understood, and a great deal of further research is needed.

It is very important to clarify the influence of other limiting factors on plant productivity. If a limiting factor such as nutrient or moisture availability will eventually restrict ecosystem response to a CO_2 fertilization effect, then atmospheric CO_2 concentration may begin to increase more rapidly as such limiting factors become significant. In other words, the ultimate capacity of terrestrial ecosystems to store carbon may be an important factor in determining atmospheric CO_2 levels a few decades from now.

Carbon storage in roots, and its turnover, are very poorly understood. Only a few primary measurements have been made in forests. There is also substantial carbon in microbial biomass, but it is very difficult to assess the magnitude of that pool separate from that of the dead organic matter pool. Laboratory experiments frequently demonstrate that plants grown at higher CO_2 concentration develop larger root structures so that carbon pools below ground are candidates for a significant response to rising CO$_2$ levels. At this point, such responses are virtually undetectable in field circumstances.

Many disturbances, such as fire, tend to reduce carbon storage and almost certainly are responsible for maintaining lower-stature ecosystems such as grasslands over large regions where climatic conditions would otherwise permit ecosystems with higher-carbon standing stocks such as forests. But disturbance events also initiate recovery sequences during which disturbed ecosystems are net sinks for carbon from the atmosphere. A CO$_2$ fertilization effect can cause increases in average carbon storage in ecosystems undergoing recurring cycles of disturbance and recovery because plants grow more quickly through the recovery period between disturbances.

In addition to gaining a better understanding of the processes responsible for a terrestrial carbon sink, global data on the distribution of ecosystems and their properties needs drastic improvement. Global inventories of carbon in vegetation and soils are very uncertain. These are derived by multiplying estimates of mean carbon density per unit area in each major biome by estimates of the areal extents of those biomes. Mean densities ignore the large variability in carbon storage and fluxes across biomes. That variability is complex, and the normal distribution frequently does not describe it well. Furthermore, most global summaries of terrestrial carbon are based on mean densities derived from a small number of samples or from the judgment of researchers with field experience in particular regions. Where data are used, they are frequently from unrepresentative study sites. Ecologists tend to collect measurements in unusual or particularly pristine ecosystems.

Compartment models that are now used to analyze the carbon cycle lack adequate representations of the processes that are responsible for plant growth and the decomposition of dead organic matter; this forces too much dependence on uncertain carbon inventories for model calibration. Currently available vegetation models that simulate carbon dynamics based on the characteristics of the plants involved and their interactions treat only small land units, are computationally demanding, and have not been fully perfected for several important ecosystem types. Such models, however, can be used to analyze biome-level carbon exchanges by Monte Carlo sampling, and global-scale simulations appear feasible in the foreseeable future. Models need to exploit correlations between carbon storage and environmental and edaphic conditions in order to account for spatial variability, but this requires better global data on temperature, precipitation, ground radiation, topography, and soil characteristics.

Ocean mixing and circulation depend on climate, and the ability of the oceans to sequester carbon from the atmosphere may change drastically in response to climatic change caused by rising greenhouse gas concentrations. This important feedback in the carbon cycle and climate systems can be diagnosed only by using models that explicitly treat the dependence of ocean mixing and circulation on climate; such models are just in the early stages of development. A major expansion of work on coupled atmosphere–ocean general circulation models is needed in order to determine the degree to which climatic changed feed-back on ocean carbon cycling may either exacerbate or moderate the greenhouse problem.

ACKNOWLEDGMENTS

Research sponsored by the U.S. Department of Energy, Carbon Dioxide Research Program, Atmospheric and Climate Research Division, Office of Health and Environmental Research, under contract DE-AC05-84OR21400 with Martin Marietta Energy Systems, Inc.

REFERENCES

1. Arrhenius, S. A. 1903. *Lehrbuch der Kosmischen Physik*, Vol. 2. Hirzel, Leipzig.

2. Bacastow, R. B., Adams, J. A., Keeling, C. D., Moss, D. J., Whorf, T. P., and Wong, C. S. 1980. Atmospheric carbon dioxide, the Southern Oscillation, and the weak 1975 El Niño. *Science* **210**: 66–8.

3. Bazzaz, F. A. 1990. The response of natural ecosystems to the rising global CO_2 levels. *Annu. Rev. Ecol. Systematics* **21**: 167–96.

4. Bolin, B., Degens, E. T., Duvigneaud, P., and Kempe, S. 1979. The global bio-geochemical carbon cycle. In *The Global Carbon Cycle*, ed. B. Bolin, E. T. Degens, S. Kempe, and P. Ketner, pp. 1–56. John Wiley, Chichester.

5. Bolin, B. 1986. How much CO_2 will remain in the atmosphere? In *The Greenhouse Effect, Climatic Change, and Ecosystems*, ed. B. Bolin, B. R. Doos, J. Jaber, and R. A. Warrick, pp. 93–155. John Wiley, New York.

6. Broecker, W. S., Takahashi, T., Simpson, H. J., and Peng, T.-H. 1979. Fate of fossil fuel carbon dioxide and the global carbon budget. *Science* **206**: 409–18.

7. Broecker, W. S., Peng, T.-H., and Engh, R. 1980. Modeling the carbon system. *Radiocarbon* **22**: 565–98.

8. Conway, T. J., Tans, P., Waterman, L. S., Thoning, K. W., Masarie, K. A., and Gammon, R. H. 1988. Atmospheric carbon dioxide measurements in the remote global troposphere, 1981–1984. *Tellus* **40**: 81–115.

9. Degens, E. T., Kempe, S., and Spitzy, A. 1984. Carbon dioxide: A biogeochemical portrait. In *The Environment and the Biogeochemical Cycles*, pp. 127–215. Springer-Verlag, Berlin.

10. Detwiler, R. P., and Hall, C. A. S. 1988. Tropical forests and the global carbon cycle. *Science* **239**: 42–7.

11. Eamus, D., and Jarvis, P. G. 1989. The direct effects of increase in the global atmospheric CO_2 concentration on natural and commercial temperate trees and forests. *Adv. Ecol. Res.* **19**: 1–55.

12. Emanuel, W. R., Killough, G. G., Post, W. M., and Shugart, H. H. 1984. Modeling terrestrial ecosystems in the global carbon cycle with shifts in carbon storage capacity by land-use change. *Ecology* **65**: 970–83.

13. Emanuel, W. R., Fung, Y.-S., Killough, G. G., Moore, B., and Peng, T.-H. 1985. Modeling the global carbon cycle and changes in the atmospheric carbon dioxide levels. In *Atmospheric Carbon Dioxide and the Global Carbon Cycle*, ed. J. R. Trabalka, pp. 141–73. DOE/ER-0239, Carbon Dioxide Research Division, U.S. Department of Energy, Washington, D.C.

14. Enting, I. G. and Mansbridge, J. V. 1987. The incompatibility of ice-core CO_2 data with reconstructions of biotic CO_2 sources. *Tellus* **39**: 318–25.

15. Friedli, H., Lotscher, H., Oeschger, H., Siegenthaler, U. and Stauffer, B. 1986. Ice core record of $^{13}C/^{12}C$ ratio of atmospheric carbon dioxide in the past two centuries. *Nature* **324**: 237–38.

16. Houghton, J. T., Jenkins, G. J., and Ephramus, J. J. (eds.). 1990. *Climate Change.* Cambridge University Press, Cambridge.

17. Houghton, R. A., Hobbie, J. E., Melillo, J. M., Moore, B., Peterson, B. J., Shaver, G. R., and Woodwell, G. M. 1983. Changes in the carbon content of terrestrial biota and soils between 1860 and 1980: A net release of CO$_2$ to the atmosphere. *Ecol. Monogr.* **53**: 235–62.

18. Houghton, R. A., and Skole, D. L. 1991. Carbon. In *The Earth as Transformed by Human Action*, ed. R. L. Gurner, S. C. Clark, R. W. Kages, J. F. Richards, J. T. Mathews, and W. B. Meyer, pp. 393–408. Cambridge University Press, Cambridge.

19. Keeling, C. D. 1973. Industrial production of carbon dioxide from fossil fuels and limestone. *Tellus* **25**: 174–98.

20. Keeling, C. D. 1973. The carbon dioxide cycle: Reservoir models to depict the exchange of atmospheric carbon dioxide with the oceans and land plants. In *Chemistry of the Lower Atmosphere*, ed. S. I. Rasool, pp. 251–328. Plenum, New York.

21. Keeling, C. D., Bacastow, R. B., Bainbridge, A. E., Ekdahl, C. A., Guenther, P. R., Waterman, L. S., and Chin, J. F. 1976. Atmospheric carbon dioxide variations at Mauna Loa Observatory, Hawaii. *Tellus* **28**: 538–51.

22. Keeling, C. D., Bacastow, R. B., Carter, A. F., Piper, S. C., Whorf, T. P., Heimann, M., Mook, W. G., and Roeloffzen, H. 1989. A three-dimensional model of atmospheric CO$_2$ transport based on observed winds: 1. Analysis of observational data. In *Aspects of Climate Variability in the Pacific and the Western Americas*, ed. D. H. Peterson, pp. 165–236. American Geophysical Union, Washington, D.C.

23. Killough, G. G., and Emanuel, W. R. 1981. A comparison of several models of carbon turnover in the ocean with respect to their distributions of transit time and age and response to atmospheric CO$_2$ and ^{14}C. *Tellus* **33**: 274–90.

24. Komhyr, W. D., Gammon, R. H., Harris, T. B., Waterman, L. S., Conway, T. J., Taylor, W. R., and Thoning, K. W. 1985. Global atmospheric CO$_2$ distribution and variations from 1968–1982 NOAA/GMCC CO$_2$ flask sample data. *J. Geophys. Res.* **90**: 5567–96.

25. Kratz, G. 1985. Modelling the global carbon cycle. In *The Handbook of Environmental Chemistry*, Vol. 1: *The Natural Environment and the Biogeochemical Cycles*, ed. O. Hutzinger, Part D, pp. 29–81. Springer-Verlag, Berlin.

26. La Marche, V. C., Jr., Graybill, D. A., Fritts, H. C., and Rose, M. R. 1984. Increasing atmospheric carbon dioxide: Tree-ring evidence for growth enhancement in natural vegetation. *Science* **225**: 1019–21.

27. Marland, G., Boden, T. A., Griffin, R. C., Huang, S. F., Kanciruk, P., and Nelson, T. R. 1989. Estimates of CO$_2$ emissions from fossil fuel burning and cement manufacturing, based on the United Nations Energy Statistics and the U.S. Bureau of Mines cement manufacturing data. ORNL/CDIAC-25. Oak Ridge National Laboratory, Oak Ridge, Tenn.

28. Moore, B. 1985. The oceanic sink for excess atmospheric carbon dioxide. In *Wastes in the Ocean*, ed. I. W. Duedall, D. R. Kester, and P. K. Park, pp. 95–125. John Wiley, New York.

29. Neftel, A., Moor, E., Oeschger, H., and Stauffer, B. 1985. Evidence from polar ice cores for the increase in atmospheric CO$_2$ in the past two centuries. *Nature* **315**: 45–7.

30. Norby, R. J., O'Neill, E. G., and Luxmoore, R. J. 1986. Effects of atmospheric CO$_2$ on the growth and mineral nutrition of *Quercus alba* seedlings in nutrient poor soil. *Plant Physiol.* **82**: 83–89.

31. O'Neill, E. G., Luxmoore, R. J., and Norby, R. J. 1987. Elevated atmospheric CO$_2$ effects on seedling growth, nutrient uptake, and rhizosphere bacterial populations of *Liriodendron tulipifera* L. *Plant and Soil* **104**: 3–11.

32. Oeschger, H., Siegenthaler, U., and Gugelman. A. 1975. A box diffusion model to study the carbon dioxide exchange in nature. *Tellus* **27**: 168–92.

33. Olson, J. S. 1974. Terrestrial ecosystem. In *Encyclopae dia Britannica*, 15th edition, pp. 144–49. Helen Hemingway Benton, Chicago.

34. Olson, J. S., Watts, J. A., and Allison, L. J. 1983. *Carbon in Live Vegetation of Major World Ecosystems*. ORNL-5862. Oak Ridge National Laboratory, Oak Ridge, Tenn.

35. Parker, M. L. 1987. Recent abnormal increase in tree-ring widths: A possible effect of elevated atmospheric carbon dioxide. In *Proceedings of the International Symposium on Ecological Aspects of Tree-Ring Analysis*, ed. G. C. Jacoby and J. W. Hornbeck, CONF-8608144, Carbon Dioxide Research Division, U.S. Department of Energy, Washington, D.C.

36. Post, W. M., Emanuel, W. R., Zinke, P. J., and Stangenberger, A. G. 1982. Soil carbon pools and world life zones. *Nature* **298**: 156–59.

37. Post, W. M., Peng, T.-H., Emanuel, W. R., King, A. W., Dale, V. H., and DeAngelis, D. L. 1990. The global carbon cycle. *Am. Sci.* **78**: 310–26.

38. Post, W. M., Chavez, F., Mulholland, P. J., Pastor, J., Peng, T.-H., Prentice, K. and Webb, T., III. 1992. Climatic feedbacks in the global carbon cycle. In *The Science of Global Change*, ed. D. A. Dunnette and R. J. O'Brien, pp. 392–412. American Chemical Society, Washington, D.C.

39. Rogers, H. H., Bingham, G. E., Cure, J. D., Smith, J. M., and Surano, K. A. 1983. Responses of selected plant species to elevated carbon dioxide in the field. *J. Environ. Qual.* **12**: 42–44.

40. Siegenthaler, U., and Oeschger, H. 1987. Biospheric CO_2 emissions during the past 200 years reconstructed by deconvolution of ice core data. *Tellus* **39**: 140–54.

41. Strain, B. R., and Cure, J. D. (eds.). 1985. *Direct Effects of Increasing Carbon Dioxice on Vegetation*. DOE/ER-0238. Carbon Dioxide Research Division, U.S. Department of Energy, Washington, D.C.

42. Stuiver, M., Quay, P. D., and Ostlund, H. G. 1983. Abyssal water carbon-14 distribution and the age of the world oceans. *Science* **219**: 849–51.

43. Tucker, C. J., Fung, I. Y., Keeling, C. D., and Gammon, R. H. 1986. Relationship between atmospheric CO_2 variations and a satellite-derived vegetation index. *Nature* **319**: 195–99.

44. United Nations. 1986. *Energy Statistics Yearbook 1986*. Statistical Office, U.N. Department of International Economic and Social Affairs, New York.

45. Woodwell, G. M., Hobbie, J. E., Houghton, R. A., Melillo, J. M., Moore, B., Peterson, B. G., and Shaver, G. R. 1983. Global deforestation: Contribution to atmospheric carbon dioxide. *Science* **222**: 1081–86.

3

Solar Ultraviolet Radiation: Effect on Rates of CO_2 Fixation in Marine Phytoplankton

OSMUND HOLM-HANSEN and DAN LUBIN

OVERVIEW

Experiments with temperature controlled incubators with sharp cut-off filters were used to determine the magnitude of inhibition of photosynthetic rates in Antarctic phytoplankton by natural solar ultraviolet radiation (UVR). There was little or no inhibition by UVR when mean irradiance during the incubation period was below 5–10 Watts m^{-2} of solar UVR. At UVR irradiances greater than this threshold value the magnitude of photoinhibition of phytosynthesis increased proportionally with increased UVR. At the highest UV irradiances (30 Watts m^{-2}), photosynthetic rates in all-day incubations were enhanced by approximately 15%, 80%, and 250% when wavelengths less than 305 nm, 323 nm, and 378 nm, respectively, were absorbed by filters. When data from all samples obtained from within the upper mixed layer were compared, the mean assimilation number (0.96) for samples in which both UV-A and UV-B were excluded was significantly different than the assimilation number for samples in which just UV-B was excluded (0.57); both of these treatments were significantly different than the assimilation number for the controls in quartz vessels (0.38). In situ incubation experiments suggest that UVR in the upper few meters of the water column may reduce photosynthetic rates by 50%, but the magnitude of the inhibition decreases rapidly with depth, with no effect being detectable at 20 m. In contrast to these data with Antarctic phytoplankton, similar experiments with tropical phytoplankton did not show any significant inhibitory effect of solar UVR on photosynthetic rates.

Results from radioactive transfer model calculations are also presented so that our experimental data on photoinhibition of photosynthesis by UVR can be considered in the context of the variation in the photoregime which phytoplankton normally experience as a function of latitude and season. These model calculations also show the magnitude of the increase in UV-B radiation as a result of the well-developed ozone hole in 1987.

As the amount of primary production occurring in the oceans by phytoplankton is roughly comparable to that by terrestrial plants, the assimilation of CO_2 by marine phytoplankton represents one of the major 'sinks' for removing carbon dioxide from the atmosphere. It is thus often assumed that marine phytoplankton will have a strong mitigating impact on the greenhouse effect caused by increased concentrations of CO_2 in the atmosphere (23). It has recently been pointed out by Broecker (2), however, that under steady-state conditions marine phytoplankton will not have any significant impact on lowering the concentrations of anthropogenic CO_2 in the atmosphere. This conclusion is based on the assumption that inorganic carbon is not limiting for growth of marine phytoplankton.

Any environmental factor that significantly affects the rate of global primary production in the oceans, however, will represent a perturbation from the steady state and will have some impact on atmospheric CO_2 concentrations and potentially on the magnitude of the greenhouse effect. The formation of the ozone hole over Antarctica in recent years (6) has caused much concern that the resulting increase of ultraviolet radiation UVR (20) might have a calamitous effect on phytoplankton and subsequently on the entire food web in the southern ocean (5, 8, 31). Recent data from NASA (11) indicating a decade-long decline in upper atmospheric ozone concentrations in the northern hemisphere has emphasized that the depletion of ozone in the atmosphere is of global concern and not necessarily restricted to the south polar region. If the resulting increased levels of UVR causes significantly lower rates of primary production on a global scale, it could potentially exacerbate the build-up of CO_2 in the atmosphere.

In this paper we are concerned with the effects of UVR on rates of primary production by marine phytoplankton, including both the impact of present-day ambient levels of UVR and the consequences of increasing levels of UVR resulting from decreasing concentrations of ozone in the stratosphere. Attention will be focused on the Antarctic ecosystem where the seasonal formation of the ozone hole represents the area of most rapid change in spectral UVR incident upon the earth, but some additional data are presented for tropical latitudes for comparison.

BACKGROUND REGARDING ENVIRONMENTAL ULTRAVIOLET RADIATION (UVR)

All plants and animals, when exposed to sufficiently high fluences of natural solar radiation, are damaged or inhibited by energy in the UV region of the spectrum. Although primary concern has focused on the effects of UV-B radiation (280–320 nm), there is considerable information that UV-A radiation (320–400 nm) can also be deleterious to organisms. Most organisms show photoadaptive mechanisms whereby they minimize UVR-induced damage to sensitive cell components by synthesis of UV-blocking compounds (4). No organisms, however, seem to completely block all UVR through the synthesis of such compounds. Hence most photoautotrophic organisms appear to be living under continual UVR stress. These organisms include marine phytoplankton, as studies have shown that solar UVR can decrease photosynthetic rates at depths to at least 20 m (19, 28). It therefore seems likely that any higher levels of UV-B irradiance, due to ozone depletion, will have some impact on growth processes of phytoplankton. The questions that must be addressed in this context include the following. First, what is the magnitude and degree of "reversibility" of cellular damage due to increased levels of UV-B; and second, can phytoplankton photoadapt sufficiently so as to show no significant effect of increased levels of UV-B radiation?

Under normal conditions, the flux of UV-B incident upon the Earth is significantly less in the polar regions as compared to tropical areas. This is due primarily to the fact that (a) there is generally 40% more atmospheric ozone in polar regions as compared to equatorial regions (1); and that (b) the angle of solar elevation is low in the polar regions, and hence solar radiation has a greater path length through the atmosphere. From a historical perspective, the Antarctic ecosystem would therefore receive relatively low levels of UV-B radiation; hence, it is commonly assumed that microbial cells in the Antarctic will be genetically adapted to low ambient levels of UVR. The other main consideration supporting the view that the Antarctic ecosystem is "fragile" in regard to UVR is that with development of the ozone hole, the rate of change of the photoregime to which cells are exposed (i.e. the UV-B irradiation levels as well as the ratio of UV-B to UV-A radiation) will be very great as compared to other geographic areas.

Some of the above relationships comparing ambient levels of UV-B, UV-A, and *photosynthetically available radiation* (PAR; 400–700 nm) are best comprehended by examining radiative transfer model calculations of surface solar radiation as a function of latitude, season, and presence of the ozone hole. Clear-sky surface irradiance, computed using the radiative transfer model of Frederick and Lubin (7), are shown in Figures 3.1 and 3.2. These calculations were performed using zonally averaged atmospheric ozone abundances measured by the Nimbus-7 Total Ozone Mapping Spectrometer (TOMS) (30). As the magnitude of radiation-induced damage is usually dependent on both the incident irradiance level and the duration of the exposure, Figures 3.1 and 3.2 show the local noon irradiance levels and integrated daily dose, respectively, for UV-B, UV-A, and PAR. The effect of a well-developed ozone hole on surface UV-B levels is also shown in Figures 3.1A and 3.2A; for the calculations here we considered the zonally averaged ozone amounts which were measured by Nimbus-7 TOMS on September 21, 1987, as reported Krueger et al. (18). On this day, the ozone column abundance decreased from nearly 300 Dobson units (DU) at 60°S to 175 DU at the South Pole (90°S). Changes in stratospheric ozone abundance do not have any significant effect on levels of UV-A or PAR radiation. It is seen from Figures 3.1A and 3.2A that UV-B irradiance levels under a well-developed ozone hole at the vernal equinox are still considerably lower than the levels normally occurring (with no ozone depletion) three months later at the summer solstice.

EFFECT OF UVR ON PHOTOSYNTHETIC RATES OF PHYTOPLANKTON

Our experimental approach to obtain ecologically relevant data on effects of UVR on phytoplankton photosynthetic rates utilized natural phytoplankton assemblages exposed to sunlight with and without screening off various spectral regions of the UVR. We used both in situ incubation techniques and temperature-controlled experimental incubators exposed to incident solar radiation. In situ incubations are preferred to the use of incubators because the irradiance conditions are

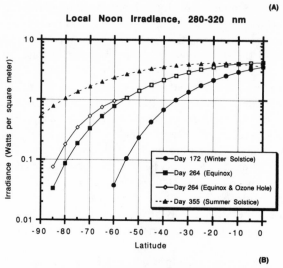

(A)

Local Noon Irradiance, 280-320 nm

Irradiance (Watts per square meter)

- ●— Day 172 (Winter Solstice)
- ■— Day 264 (Equinox)
- ○— Day 264 (Equinox & Ozone Hole)
- ▲— Day 355 (Summer Solstice)

Latitude

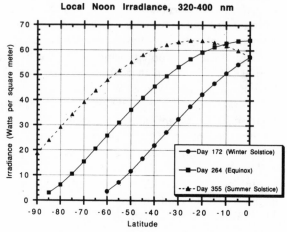

(B)

Local Noon Irradiance, 320-400 nm

Irradiance (Watts per square meter)

- ●— Day 172 (Winter Solstice)
- ■— Day 264 (Equinox)
- ▲— Day 355 (Summer Solstice)

Latitude

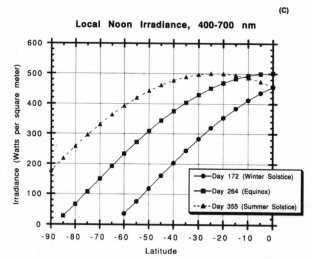

(C)

Local Noon Irradiance, 400-700 nm

Irradiance (Watts per square meter)

- ●— Day 172 (Winter Solstice)
- ■— Day 264 (Equinox)
- ▲— Day 355 (Summer Solstice)

Latitude

completely natural in the former except for any experimental filters covering the samples. When incubators are exposed to sunlight, it is not possible to simulate both the spectral irradiance and total flux of UVR and PAR found at various depths in the water column. Much of our work, however, relied on incubator experiments because (a) ship time, weather conditions, or sea state did not permit day-long in situ incubations; and (b) incubators allow greater flexibility in the use of filters and length of exposure period. Rates of CO_2 fixation were determined by standard radiocarbon techniques during incubation periods of 6–10 h as described in Helbling et al. (10). Studies in Antarctic waters were done at various time periods between September and February during 1988–91; studies in tropical waters were done during March–April of 1991.

ANTARCTIC PHYTOPLANKTON

The results of our incubator experiments using filters to absorb radiation below 305, 323, and 360 nm (Pyrex, Mylar, and Plexiglas, respectively) are summarized in Figure 3.3A. It is apparent that the use of Mylar increased the mean photosynthetic assimilation number from 0.38 (samples in quartz glass vessels) to 0.56 (a mean increase of 47%); the use of the Plexiglas filter increased the assimilation number to 0.96 (a mean increase of 146% as compared to the controls in quartz). The results from incubations with Mylar and Plexiglas filters were significantly different from those with Pyrex and quartz vessels and also different among themselves (Tukey test). There was no significant difference between the samples incubated in Pyrex and quartz vessels.

It is seen in Figure 3.3 that the assimilation numbers of Antarctic phytoplankton are low (close to 1.0 for the noninhibited sample screened by Plexiglas) as compared to the assimilation numbers of tropical phytoplankton, which are close to 6.0 (Fig. 3.3B). The relatively low assimilation numbers of Antarctic phytoplankton are due to two major factors: (a) The cellular carbon/Chl-a ratios in Antarctic phytoplankton are low (generally 30–100) (12), as compared to tropical phytoplankton where the corresponding ratio is often over 100 (17); and (b) the low temperatures in Antarctic waters (-1.8 to $+2.0°C$ for the above experiments) result in low specific growth rates, as expected from thermodynanic considerations (22). The range of assimilation numbers shown in Figure 3.3 is typical of most data from the Antarctic (13, 29) or tropical regions (17) and represents normal, healthy phytoplankton assemblages.

Weather conditions in Antarctica are characterized by a predominance of overcast days, during which the incident solar radiation may be only about 15% of that on a sunny day. The effect on photosynthetic rates when the incident solar UVR varies from low to high is shown in Figure 3.4. Photosynthetic rates in the

Fig. 3.1. Results of atmospheric radiative transfer calculations, showing local noon solar irradiance incident upon the Earth as a function of latitude and season in the Southern Hemisphere at vernal equinox under the severe Antarctic ozone depletion of 1987. (A) UV-B radiation; (B) UV-A radiation; (C) photosynthetically available radiation, 400–700 nm.

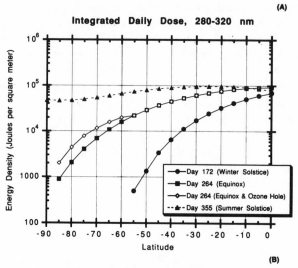

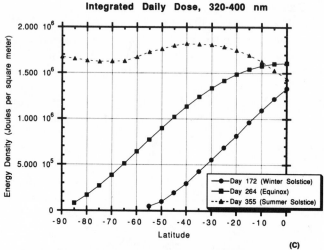

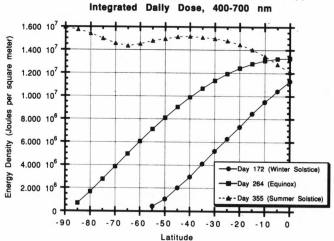

three treatments (Pyrex, Mylar, plastic film) do not show any significant differences as compared to the control samples in quartz when the ambient UVR (295–385 nm) was less than 5–10 Watts·m^{-2}. Above this threshold value, the magnitude of enhancement of photosynthetic rates is directly related to the elevated levels of UV irradiance. The maximum enhancement values by screening off wavelengths less than 305, 323, and 378 nm are approximately 17%, 80%, and 250%, respectively.

In addition to the filters mentioned in the experiments described above, some incubator experiments used various Schott long-pass filters to get more definition in the relationship between photosynthetic rate and spectral irradiance impinging upon the cells. The results from these experiments (Fig. 3.5) indicate that energy in the UV-B portion of the spectrum accounts for approximately 50% of the total inhibition caused by all wavelengths less than 378 nm. The shorter UV-B wavelengths (around 300 nm) have relatively little impact on photosynthetic rates.

Data from the above experiments have been used to construct an action spectrum for inhibition of photosynthesis in Antarctic phytoplankton by solar UVR (Fig. 3.6). It is seen from Figure 3.6B that the shorter the wavelength of UVR, the greater its effectiveness in inhibiting CO_2 fixation rates when expressed as response per unit energy. The relative effectiveness of energy at a wavelength of 296 nm is almost two orders of magnitude greater than the effectiveness of energy at the longer UV-A wavelengths. Although the shorter UV-B wavelengths are potentially the most effective in reducing photosynthetic rates, the shorter UV-B wavelengths apparently are not responsible for much of the total inhibition of photosynthesis by solar UVR (Fig. 3.5). The reason for this is that the flux of solar UVR incident upon the Earth decreases very rapidly below 320 nm; irradiances at 320 nm are almost four orders of magnitude greater than that at 300 nm (Fig. 3.7).

Data in Figure 3.8 show photosynthetic rates of natural phytoplankton assemblages when incubated in situ under three different spectral regimes. Rates determined in the quartz vessels (with no supplemental filter) are referred to as "100%" for all samples, and the rates determined in those samples where wavelengths shorter than 305 nm (Pyrex filger) or 360 nm (Plexiglas filter) were removed are shown as a percentage of those in the quartz vessels. Elimination of the shorter UV-B wavelengths (samples with Pyrex filter) resulted in approximately 30% higher rates of photosynthesis in samples collected and incubated close to the surface, with the effect diminishing rapidly with depth; by 10 m depth there was no difference in the samples contained within quartz or Pyrex vessels. The effects of removing all energy below 360 nm (Plexiglas filter) resulted in much higher rates of incorporation of radiocarbon. As compared to the corresponding data from quartz vessels, the rates were approximately doubled in samples close to the surface; they were approximately 10% higher at 10 m depth, and showed

Fig. 3.2. Results of atmospheric radiative transfer calculations showing integrated daily doses of solar radiation incident upon the Earth as a function of latitude and season in the Southern Hemisphere, in addition to increased UV-B radiation at vernal equinox under the severe Antarctic ozone depletion of 1987. (A) UV-B radiation; (B) UV-A radiation; (C) photosynthetically available radiation, 400–700 nm.

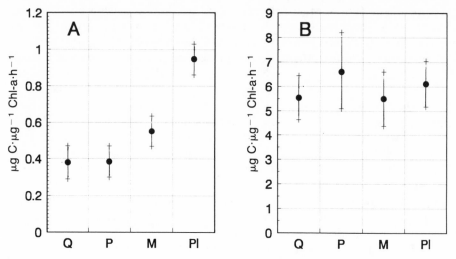

Fig. 3.3. Comparison of photosynthetic assimilation numbers of Antarctic and tropical phytoplankton when incident solar radiation is filtered through quartz (Q), Pyrex (P), Mylar (M), or Plexiglas (Pl). Solid circles represent the mean value of all samples, and the vertical bars the 95% confidence intervals. (A) Antarctic phytoplankton, total of 132 samples; (B) tropical phytoplankton, total of 61 samples. All samples were collected within the upper mixed layer. [From Helbling et al. (10).]

no detectable differences at a depth of 20 m. These data suggest that the shorter wavelengths of UV-B radiation (i.e., 280–306 nm) depress photosynthetic rates much less than do longer-UVR wavelengths—results consistent with all our incubator experiments discussed above.

TROPICAL PHYTOPLANKTON

Our data on the UVR effects of phytoplankton in tropical waters are limited to experiments performed in on-deck incubators cooled with surface seawater during a cruise across the Equator in March–April of 1991. Equipment and techniques were identical to those employed in the Antarctic. Screening out UV-B and UV-A radiation increased photosynthetic rates in only a few experiments, but the percent enhancement by removal of all UVR was less than 20%, despite the fact that UVR values during the incubations were between 31 and 49 Watts·m^{-2}. When all samples are combined from the upper mixed layer (UML) of the water column and analyzed for differences between the various photoregimes (Fig. 3.3B), there were no significant differences among treatments (ANOVA test).

When water samples were obtained from depths 25–30 m below the UML and exposed to incident solar radiation, the assimilation numbers were 5–10% of those obtained from surface waters. This is in contrast to our observations in Antarctic waters, where assimilation numbers of samples taken from within the upper 75 m of the water column and incubated under incident solar radiation often show near-comparable values as compared to surface samples.

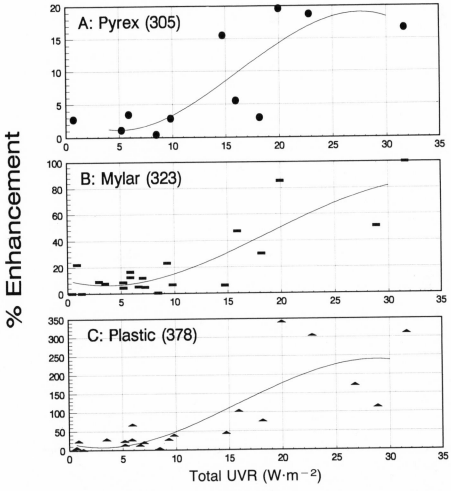

Fig. 3.4. Percentage of enhancement of photosynthetic rates as a function of mean incubation values of UVR incident upon the samples in quartz vessels and when various spectral regions of UVR are removed by Pyrex, Mylar, or a plastic film with 50% transmission at 378 nm. Data are from years 1988, 1989, and 1991. Note the different ordinate values in A, B, and C. The lines through the data points were computer-drawn, using a third-degree polynomial function. [From Helbling et al (10).]

DISCUSSION

Our data strongly suggest that phytoplankton in Antarctic waters are more sensitive to UVR than are phytoplankton at lower latitudes, and hence ozone-related increases in UV-B radiation may be expected to have some impact on primary productivity. We do not know why Antarctic phytoplankton are apparently more sensitive to UVR than are tropical phytoplankton, but sensitivity could be related to (a) genomic differences, or (b) the degree to which cells are photoadapted to the high-light environment found in the UML of the water column.

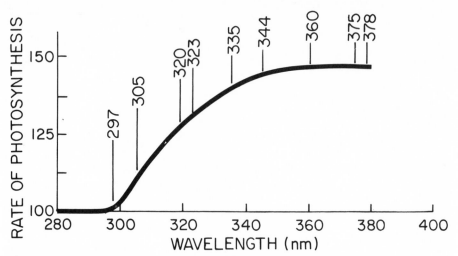

Fig. 3.5. Magnitude of inhibition of photosynthesis of Antarctic phytoplankton by solar ultraviolet radiation (UVR) that has been selectively "cut off" at various wavelengths by the use of plastic or glass filters. The rate of photosynthesis in quartz control vessels has been set at 100%. The bold line, which has been generalized from all our data, represents the increase in photosynthetic rate relative to that in the quartz vessels. The numbers above the line indicate the spectral cutoff of the various filters used in our experiments. [From Holm-Hansen (15).]

In contrast to tropical waters, which generally have a relatively shallow (20–30 m) and stable UML, Antarctic waters are characterized by relative instability of the upper column and frequent deep-mixing by storms. As both particulate and dissolved materials are generally quite uniformly distributed throughout the UML, it appears that wind-induced mixing keeps phytoplankton cells circulating vertically within the UML at some undertermined rate. The extent of the UML in most Antarctic waters is close to 50 m (21), which means that the mean irradiance to which cells will be exposed is quite low, resulting in low-light adapted cells (25). Even when rich blooms develop in shallow UMLs (20 m or less) in protected shelf waters close to the Antarctic Continent, it appears that the phytoplankton are low-light-adapted. The reason for this is that nutrients are so high in Antarctic waters that blooms commonly reach levels of >20 μg Chl-a·L^{-1}, resulting in rapid attenuation of solar radiation so that the 1% light level is commonly at 10 m or less. The overall effect of these conditions in Antarctic waters may prevent cells from remaining under high solar irradiance long enough to become physiologically fully adapted to high light conditions. The situation in tropical waters is very different, where inorganic nutrients generally limit growth rates and the potential phytoplankton biomass in the stable and relatively shallow UMLs. Under such conditions, cells receive high solar irradiance levels throughout the UML and, hence, have sufficient time to become physiologically adapted to higher solar radiation, including UVR.

To what extent do our results provide an answer to the question raised by Roberts (24) in his article title: "Does the ozone hole threaten Antarctic life?"

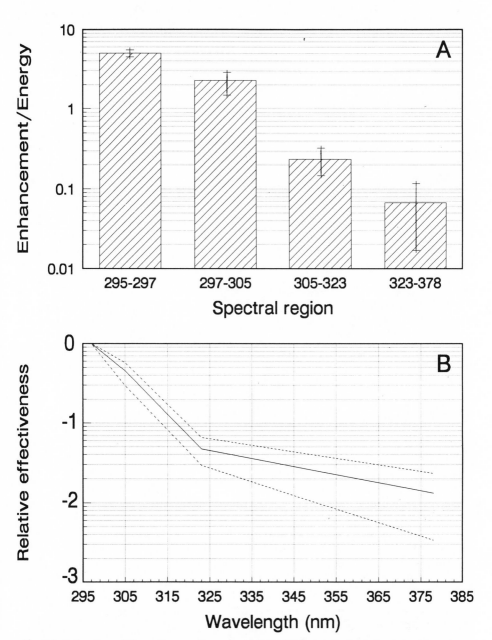

Fig. 3.6. Estimation of the action spectrum for UVR inhibition of photosynthesis in Antarctic marine phytoplankton by comparison of photosynthetic enhancement results with data on incident spectral irradiance. (A) Each vertical bar shows the ratio of the average percent photosynthetic enhancement resulting from four cut-off filters divided by the integrated radiation flux for the spectral regions 295–297, 297–305, 305–323, and 323–378 nm. Vertical lines indicate the minimum and maximum values. (B) Line showing the relative effectiveness of UVR in photoinhibition (expressed as response per unit energy), where the effectiveness is set at 1.0 at 296 nm. Note that the ordinate is a log scale. Dashed lines represent minimum and maximum values. [From Helbling et al. (10).].

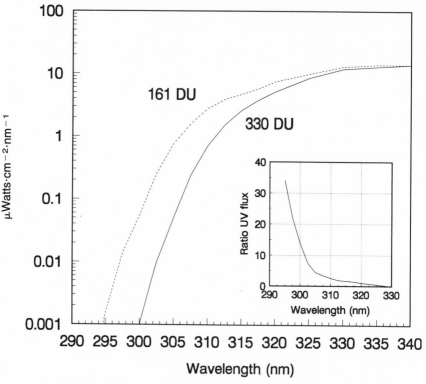

Fig. 3.7. Spectral UVR irradiance at McMurdo Station in the Antarctic as influenced by ozone concentrations in the atmosphere. *Solid line:* Normal UVR measured on October 20, 1989, with the ozone concentration measured at 330 Dobson units (DU). *Dashed line:* UVR measured on October 20, 1989, with an ozone concentration of 161 DU, showing enhanced flux of shorter wavelengths in the UV-B portion. Note that ordinate values are on a log scale. *Inset:* Data from 1988 at McMurdo Station showing the ratio of the UV flux on October 8 (ozone = 263 DU) to the UV flux on October 13 (ozone = 386 DU). (Data from C. R. Booth.)

This question could be extended to the rest of the world's oceans should there be global depletion of ozone. The primary consideration in this context is the extent to which the total rate of primary production is decreased by increasing levels of UV-B. Although the inhibition of photosynthetic rates by UVR appears pronounced in surface water, the percent inhibition decreases rapidly with depth as the result of rapid attenuation of UV-B in seawater, with no inhibition detectable by 20 m. However, other studies in the same area with comparable Chl-a concentrations have shown that net photosynthesis occurs down to at least 50 m. The UVR-induced "loss" of potential photosynthate throughout the entire euphotic zone would appear to be less than 15% of the total primary production. At least half of this loss is the result of UV-A radiation, which is not affected by ozone depletion. The remainder is due to UV-B, but a significant portion of this is due to the flux of UV-B under normal ozone conditions. The most dramatic aspect of the increased UV-B radiation resulting from ozone depletion is the large increase in irradiance at the shorter wavelenghts (see inset, Fig. 3.7). Although the relative

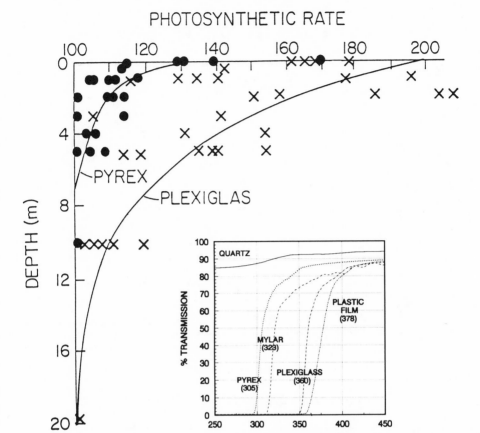

Fig. 3.8. Relative rates of in situ photosynthesis in the upper 20 m of the water column, when natural phytoplankton samples are exposed to varying proportions of UV radiation. Samples were incubated in (a) quartz glass, which transmits nearly all UV radiation (photosynthetic rates for these samples are shown as "100%" for all depths); (b) Pyrex glass (●), with 50% transmission at about 305 nm; or (c) quartz glass screened with a Plexiglas filter (×), with 50% transmission at about 360 nm. The curves were drawn by hand. Samples were collected close to Anvers Island during November–December, 1988. *Inset:* Spectral transmission characteristics of the sample containers. [From Holm-Hansen et al. (14).]

effectiveness of these wavelengths is high (see Fig. 3.6), the fluences at these wavelengths are so low (Fig. 3.7) that it is unlikely that these wavelengths would have a dramatic effect on photosynthetic rates during one-day incubations. Some of our experiments in the Antarctic were done close to Palmer Station (Anvers Island; 64°46′S, 64°04′W) in October–November when the ozone hole was intermittent overhead. There are no indications in our data (e.g., Figs 3.4, 3.5) that short-wavelength UV-B radiation played a dominant role in the magnitude of the overall rate of inhibition of photosynthesis by UVR.

Our data on percent inhibition of photosynthesis caused by solar UVR represent a "worst case" scenario for the following reasons: First, some incubator

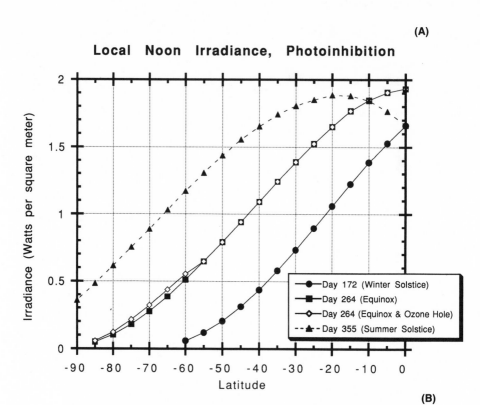

(A)

Local Noon Irradiance, Photoinhibition

Legend:
- Day 172 (Winter Solstice)
- Day 264 (Equinox)
- Day 264 (Equinox & Ozone Hole)
- Day 355 (Summer Solstice)

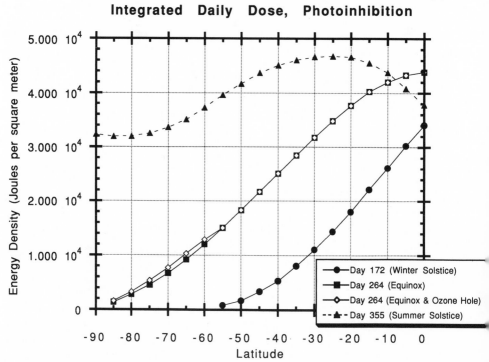

(B)

Integrated Daily Dose, Photoinhibition

Legend:
- Day 172 (Winter Solstice)
- Day 264 (Equinox)
- Day 264 (Equinox & Ozone Hole)
- Day 355 (Summer Solstice)

experiments utilized neutral-density filters to reduce solar irradiance to all treatments as it is not feasible to simulate the spectral irradiance fields characteristic of increasing depths in the water column. This means that the ratio of UV-B to UV-A (normalized to PAR) incident upon the phytoplankton cells in the incubator will be greater than the corresponding ratio in the water column. The relative ratio of UV-A to UV-B absorbed by cells is important in photorepair mechanisms, as the longer wavelengths have been shown to be effective in photorepair of DNA damage in Antarctic phytoplankton (16). As the ratio of UV-B to UV-A is artifically high in some of the incubator experiments, it would thus tend to maximize the net inhibition by UVR. Second, as mentioned earlier, phytoplankton cells in the UML are continually "circulating" vertically, owing to physical mixing processes. As a result the cells experience alternating periods of high to low irradiance, coupled with increasing ratios of UV-A to UV-B when the cells are transported downward. In our in situ incubation techniques, however, samples are held at one depth throughout the incubation period of 6–10 h. This also tends to give us higher estimates of UVR-induced inhibition than would occur if the cells were not enclosed in a fixed container.

Another aspect of the question concerning the possible impact of ozone-related increased UVR is how great this change of UV-B irradiance is as compared to normal UV-B levels the cells might encounter as a function of seasonal or latitudinal variations in solar irradiation. Figures 3.1A and 3.2A show that both the local noon irradiance and the integrated daily dose of UV-B under the ozone hole at the equinox are much lower than the levels prevailing three months later at the summer solstice. Although the ozone hole is usually a seasonal phenomenon, with maximal ozone depletion occurring between September–October, in 1991 the ozone hole persisted to a lesser degree into the month of December. In such a case the curves showing UV-B values (Figs 3.1A, 3.2A) might reach the solstice values early in December. Whether or not this "early" exposure to elevated UV-B levels is particularly damaging to phytoplankton as compared to the effect produced at the time of the solstice depends upon the autecology of individual species. This would include (a) possible differential sensitivity to UVR at various cellular stages and (b) the length of time required for acquiring protective mechnaisms against UVR, including screening pigments and both dark and light repair capabilities. It should also be noted that while some Antarctic phytoplankton species (as determind by traditional microscopic techniques) are restricted to polar waters, many species are found as far north as 30°S, while other species are cosmopolitan in distribution (3, 9, 26). Figures 3.1A and 3.2A show that such northerly-living phytoplankton would receive much higher doses of UV-B than would be possible anywhere in Antarctica.

Figure 3.9 shows radiative-transfer-model calculations of seasonal and

Fig. 3.9. Radiative transfer model calculations showing photoinhibition-weighted radiation incident upon the Earth as a function of latitude and season in the Southern Hemisphere, in addition to the increased photoinhibition-effective radiation at vernal equinox resulting from the Antarctic ozone depletion of 1987. These curves have been obtained by convolving incident clear-sky spectral irradiance with the photoinhibition action spectrum shown in Figure 3.6B. (A) Photoinhibition-effective irradiance at local noon; (B) integrated daily dose.

latitudinal variation in UVR weighted by the action spectrum for photoinhibition shown in Figure 3.6B. It is seen that the magnitude of the photoinhibitory-effective radiation due to the formation of the ozone hole is less than about 10% of the normal value, expressed either for local noon irradiance (Fig. 3.9A) or for integrated daily dose (Fig. 9B). When ozone-related increases in UV-B radiation in the Antarctic are viewed in such a seasonal and latitudinal perspective, it seems unlikely that the enhanced short UV-B wavelengths would reduce primary production by more than a few percent. As this would be far less than the estimated interannual variability due to extent of sea ice, cloud cover, and so on, it does not appear that the ozone hole will result in any collapse of the southern ocean ecosystem. With the apparent progressive loss of ozone in the Northern Hemisphere, the question that Roberts (24) posited for the Antarctic is relevant for lower latitudes as well. Based on our data from tropical waters (Fig. 3.3B), it seems that phytoplankton exposed to the highest ambient fluences of natural solar UVR (Fig. 3.1A,B) possess sufficient adaptive mechanisms to minimize any significant effect on photosynthetic rates. As it is unlikely that atmospheric ozone concentrations at mid to low latitudes will be depleted to the same extent as over the Antarctic Continent, the possibility for severe impact by enhanced UV-B on photosynthetic rates of phytoplankton in temperate or tropical waters seems minimal.

The above discussion has been restricted to the effects of UVR on photosynthetic rates during 6 to 10-h incubation periods. In considering the possible impact of enhanced short UV-B wavelengths on phytoplankton, one must also examine longer-term effects that may ultimately feed back to affect overall rates of primary production. The most obvious possibility in this regard is the ability of UVR to cause structural damage in macromolecules such as DNA, with the result that cells lose all ability for cellular division. Laboratory studies with the unicellular green alga *Chlorella pyprenoidosa* have shown that gamma radiation can completely inhibit cellular division without causing any detectable change in the rate of photosynthesis or metabolic incorporation patterns of newly assimilated CO_2 for several days (Ref. 32; Holm-Hansen, unpublished). If UVR causes similar metabolic effects in Antarctic phytoplankton, the impact would not be detected by our short-term CO_2 incorporation studies. In reference to Robert's question (24) regarding the possible impact of UVR on the Antarctic ecosystem, it is therefore necessary also to consider the impact of enhanced UV-B on genetic material and the ability of cells to undergo mitotic divisions. The DNA-weighted irradiance under the ozone hole in the Antarctic is greatly enhanced, as shown by the data in Figure 3.10. The biological effectiveness of solar UVR in the Antarctic in causing DNA lesions and loss of cell viability in isolated cultures of phytoplankton has been reported by Karentz et al. (16).

The seasonal and latitudinal variation in spectral DNA-weighted UVR radiation in the Antarctic is shown in Figure 3.11, with and without the presence of an ozone hole. It is seen that the relative magnitude of the change in DNA-effective radiation is considerably greater than the change in photoinhibition-weighted radiation (Fig. 3.9), in regard to both local noon irradiance (Fig. 3.11A) and integrated daily dose (Fig. 3.11B). We do not know the threshold value of UV-B radiation for inhibition of cell division in natural Antarctic

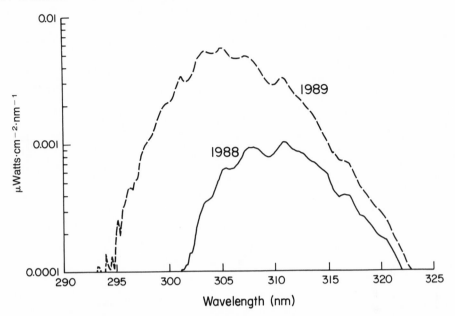

Fig. 3.10. DNA-weighted spectral irradiance recorded at McMurdo Station, Antarctica, under normal ozone conditions (330 DU; October 20, 1988), and under the ozone hole (161 DU; October 20, 1989). (Data from C. R. Booth.)

phytoplankton assemblages, nor do we have any data regarding the loss of viability as a function of the time–irradiance relationships. If a brief exposure to high solar UVR causes a loss of viability in a significant percentage of the phytoplankton cells, the effect could be cumulative throughout the depth of the UML, as cells are circulated throughout its depth. In such a case, a deep-mixed water column would not mitigate this effect, and the impact on overall rates of primary production might be considerable.

Most of the biologically oriented studies on the impact of UVR in the Antarctic have been concerned with the effects of photosynthetic rates, and on total primary production which supports the entire food web in the southern ocean. UVR, however, can potentially damage cells by many different mechanisms, and we do not know the biochemical basis for the effect of photosynthetic rates as discussed in this paper. The major explanations of UV-induced photoinhibition include DNA lesions, effects on protein synthesis, interference with membrane function with subsequent impairment in active uptake processes, and oxidation of photosynthetic pigments, especially those associated with photosystem II. We do not know the dose–response relationships of these possible mechanisms that may alter photosynthetic rates, nor the extent to which the damage would be reversible. In order to support or modify the tentative conclusions suggested by our data on the possible impact of enhanced levels of UV-B on primary production in the Antarctic, it is necessary to understand better the mechanisms involved, the extent of differential sensitivity of UVR by phylogenetic groups and species of phytoplankton, and the temporal aspects of physiological adaptation to elevated fluence of UVR.

Local Noon Irradiance, DNA Action Spectrum

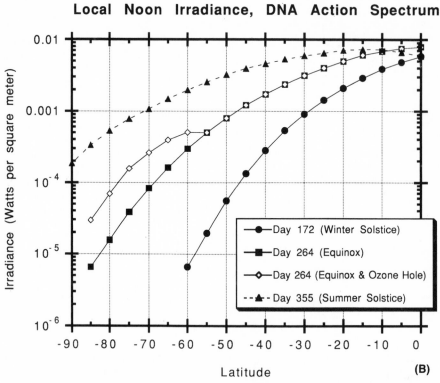

Integrated Daily Dose, DNA Action Spectrum

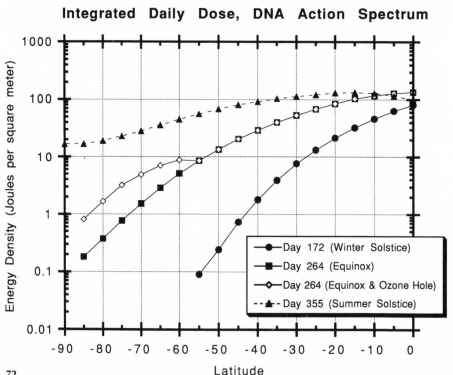

REFERENCES

1. Brasseur, G., and Solomon, S. 1984. *Aeronomy of the Middle Atmosphere.* ed. D. Reidel, Dordrecht, 441 pp.

2. Broecker, W. S. 1991. Keeping global change honest. *Global Biogeochem. Cycles* **5**: 191–92.

3. Cassie, V. 1963. Distribution of surface phytoplankton between New Zealand and Antarctic, December 1957. *Sci. Rep. Commonwealth Transantarctic Exped.* **1955–1958**: 1–11.

4. Dunlap, W. C., Chalker, B. E., and Oliver, J. K. 1986. Bathymetric adaptations of reef-building corals at Davies Reef, Great Barrier Reef, Australia. III. UV-B absorbing compounds. *J. Exp. Mar. Biol. Ecol.* **104**: 239–48.

5. El-Sayed, S. Z. 1988. Fragile life under the ozone hole. *Nat. Hist.* **97**: 72–80.

6. Farman, J. C., Gardiner, B. G., and Shanklin, J. D. 1985. Large losses of total ozone in Antarctica reveal seasonal CO_x/NO_x interaction. *Nature* **315**: 207–10.

7. Frederick, J. E., and Lubin, D. 1988. The budget of biologically active ultraviolet radiation in the earth–atmosphere system. *J. Geophys. Res.* **93**: 3825–32.

8. Gribbin, J. 1988. *The Hole in the Sky.* Bantam Books, New York.

9. Hasle, G. R. 1976. The biogeography of some marine planktonic diatoms. *Deep-Sea Res.* **23**: 319–38.

10. Helbling, E. W., Villafane, V., Ferrario, M., and Holm-Hansen, O. 1992. Impact of natural ultraviolet radiation on rates of photosynthesis and on specific marine phytoplankton species. *Mar. Ecol. Prog. Ser.* **80**: 89–100.

11. Herman, J. R., McPeters, R., Stolarski, R., Larko, D., and Hudson, R. 1991. Global average ozone change from November 1978 to May 1990. *J. Geophys. Res.* **96**: 17, 297–305.

12. Hewes, C. D., Sakshaug, E., Reid, F. M. H., and Holm-Hansen, O. 1990. Microbial autotrophic and heterotrophic eucaryotes in Antarctic waters: Relationships between biomass and chlorophyll, adenosine triphosphate and particular organic carbon. *Mar. Ecol. Prog. Ser.* **63**: 27–35.

13. Holm-Hansen, O., and Mitchell, B. G. 1990. Spatial and temporal distribution of phytoplankton and primary production in the western Bransfield Strait region. *Deep-Sea Res.* **38**: 961–80.

14. Holm-Hansen, O., Mitchell, B. G., and Vernet, M. 1989. UV radiation in Antarctic waters: Effect on rates of primary production. *Antarctic J. U.S.* **24**: 177–8.

15. Holm-Hansen, O. 1990. Effects of UV-B and UV-A on phytosynthetic rates of Antarctic phytoplankton. *Antarctic J. U.S.* **25**: 176–77.

16. Karentz, D., Cleaver, J. E., and Mitchell, D. L. 1991. Cell survival characteristics and molecular responses of Antarctic phytoplankton to ultraviolet-B radiation. *J. Phycol.* **27**: 326–41.

17. Kiefer, D. A., Olson, R. J., and Holm-Hansen, O. 1976. Another look at the nitrite and chlorophyll maxima in the central North Pacific. *Deep-Sea Res.* **23**: 1199–1208.

18. Krueger, A. J., Ardenuy, P. E., Sechrist, F. S., Penn, L. M., Larko, D. E., Doiron, S. D., and Galimore, R. N. 1987. *The 1987 Airborne Antarctic Ozone Experiment: The Nimbus-7 TOMS Data Atlas.* NASA Reference Publication.

Fig. 3.11. Radiative transfer model calculations showing DNA-effective radiation incident upon the Earth as a function of latitude and season, in the Southern Hemisphere, in addition to the increased DNA-effective radiation at vernal equinox resulting from the Antarctic ozone depletion of 1987. These curves have been obtained by convolving incident clear-sky spectral irradiance with the spectral DNA-absorption values as described by Setlow (27). (A) DNA-effective irradiance at local noon; (B) integrated daily dose.

19. Lorenzen, C. J. 1979. Ultraviolet radiation and phytoplankton photosynthesis. *Liminol. Oceanogr.* **24**: 1117–20.

20. Lubin, D., and Frederick, J. E. 1991. The ultraviolet radiation environment of the Antarctic Peninsula: The roles of ozone and cloud cover. *J. Appl. Meteor.* **30**: 478–93.

21. Mitchell, B. G., Brody, E. A., Holm-Hansen, O., McClain, C., and Bishop, J. 1991. Light limitation of phytoplankton biomass and macronutrient utilization in the Southern Ocean. *Limnol. Oceanogr.* 1662–77.

22. Neori, A., and Holm-Hansen, O. 1982. Effect of temperature on rate of photosynthesis in Antarctic phytoplankton. *Polar Biol.* **1**: 33–38.

23. Revelle, R. 1990. Letter in Forum section. *Issues Sci. Technol.* **7**: 21–22.

24. Roberts, L. 1989. Does the ozone hole threaten Antarctic life? *Science* **244**: 288–89.

25. Sakshaug, E., and Holm-Hansen, O. 1986. Photoadaption in Antarctic phytoplankton: Variations in growth rate, chemical composition and P versus I curves. *J. Plank. Res.* **8**: 459–73.

26. Semina, H. J. 1979. The geography of plankton diatoms of the Southern Ocean. *Nova Hedwig.* **646**: 341–56.

27. Setlow, R. B. 1974. The wavelengths in sunlight effective in producing skin cancer: A theoretical analysis. *Proc. Natl. Acad. Sci. USA* **71**: 3363–66.

28. Smith, R. C., Baker, K. S., Holm-Hansen, O., and Olson, R. 1980. Photoinhibition of photsynthesis in natural waters. *Photochem. Photobiol.* **31**: 585–92.

29. Tilzer, M. M., Bodungen, B. V., and Smetacek, V. 1985. Light-dependence of phytoplankton photosynthesis in the Antarctic Ocean: Implications for regulating productivity. In *Antarctic Nutrient Cycles and Food Webs*, ed. W. R. Siegfried, P. R., Condy, and R. M. Laws, 60–9. Springer-Verlag, New York.

30. TOMS. 1989. *World Meterorological Organization, Scientific Assessment of Stratospheric Ozone: 1989.* Global Ozone Research and Monitoring Project Report No. 20. Geneva.

31. Voytek, M. A. 1989. *Ominous Future Under the Ozone Hole.* Environmental Defense Fund, Washington, D.C.

32. Zill, P. L., and Tolbert, N. E. 1958. Effect of ionizing radiation and ultraviolet radiations on photosynthesis. *Arch. Biochem.* **76**: 196–203.

4

Global Geochemical Cycles of Carbon

JAMES C. G. WALKER

The processes that might control the amounts of oxygen and carbon dioxide in the atmosphere are analyzed in terms of reservoir amounts and the rates of transfer of material between reservoirs. Exchange of material between the atmosphere and the biotic reservoir by photosynthesis and by respiration is rapid, but the biotic reservoir is too small, compared with the amount of oxygen in the atmosphere and the amount of carbon in oceans and sedimentary rocks, for these rapid processes to have a lasting impact on the atmospheric concentrations of oxygen and carbon dioxide. The residence time of oxygen in the atmosphere is about 5 million years. The concentration appears to be controlled by a balance between the rate of consumption of oxygen in the oxidative weathering of reduced constituents of sedimentary rocks and the rate at which net photosynthetic oxygen is left behind in the atmosphere as photosynthetically produced organic matter sequestered in sedimentary rocks. In the very long term, the amount of reduced material in sedimentary rocks is controlled by exchange with the interior of the Earth.

On a time scale longer than decades, the carbon dioxide content of the atmosphere is determined by equilibrium with the ocean, and this equilibrium depends on the rates of supply of carbon and neutralizing cations by volcanism and weathering and the rates of removal in the precipitation of carbonate minerals. On time scales of millions of years or longer, the carbon dioxide concentration of the atmosphere is determined by the balance between weathering rates and precipitation rates. On time scales of billions of years there is an important change of carbon between the surface of the Earth and the interior. Exchange with biota affects atmospheric carbon dioxide only on a time scale of decades to centuries.

In this chapter, I shall explore the processes that influence and control the amounts of oxygen and carbon in the atmosphere. To give the reader a preview of my conclusions, I shall be arguing that the rates of photosynthesis exercise very little control on the amounts of oxygen and carbon in the atmosphere. With that said, I should add a few precautionary remarks: I am interested in the longer term, not just the next year or the next few decades, but in the next few generations as well. Nevertheless, there are tremendous uncertainties in the field of global geochemical cycles. We cannot do controlled experiments. The experiment, of course, that we would like to do is to increase the global rate of photosynthesis or to kill all the green plants and see how the system responds; and then do another experiment. Such experiments are not possible. We can only conduct thought experiments and simulations.

RESERVOIRS AND RATES

My arguments are very simple, when dealing with the conservation of matter, the production of oxygen, the related production of organic matter, and the destruction of oxygen and organic matter at equal molar rates, one uses principles based on reservoir sizes and the rates of transfer of material between reservoirs. Figure 4.1 shows a reservoir with a source of material and a sink of material (24). This reservoir could be the amount of carbon in the atmosphere or the ocean. Here it is a leaky bucket, as you can see. Now if the system is in balance, the rate of the source is equal to the rate of the sink, and the residence time of the material in the reservoir is given by the amount of material in the reservoir divided by the rate of either the source or the sink. The residence time is very important because it gives an indication of how fast the amount of material in the reservoir can change. As I mentioned, the material might be atmospheric oxygen. The same argument applies to the relationship of population or standing crop to primary

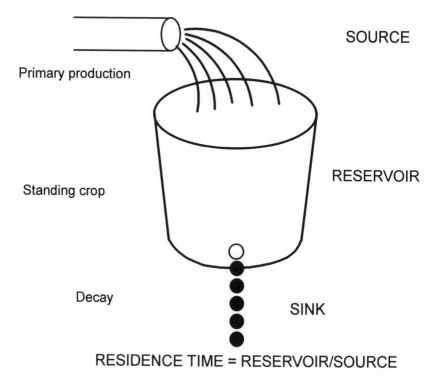

Fig. 4.1. If the amount of material in a reservoir is constant, the source equals the sink. The residence time is defined as the amount of material in the reservoir divided by either the source rate or the sink rate.

production and decay. In analyzing the geochemical cycles of oxygen and carbon, we need to think in terms of the sizes of reservoirs and the rates of transfer of material between reservoirs.

OXYGEN

Now take that idea and think of atmospheric oxygen. The analysis presented in this section is based on the work of several authors (11–13, 21–23, 26, 28). Figure 4.2 shows a simple budget of the oxygen in the atmosphere. The reservoirs are indicated here by boxes, and the processes that transfer material between reservoirs are indicated by ovals. The atmosphere contains 2.8×10^{19} moles oxygen; it is coupled to a biota reservoir, which includes humus and dissolved organic carbon in the ocean, the terrestrial biota, and the marine biota. The biota is a much smaller reservoir, only 2×10^{17} moles, less than 1% of the size of the atmospheric reservoir (21, 31). Shown is the rate of global photosynthesis adding material both to the atmosphere and to the biotic reservoir and a balancing rate of respiration and decay removing material from those two reservoirs. For both reservoirs the source is photosynthesis and the sink is respiration and decay.

Now let us calculate some residence times. The residence time of atmospheric oxygen is something like 4000 years. That is how fast atmospheric oxygen can respond, in principle, to changes in the rate of photosynthesis. It cannot respond any faster. The value is derived by dividing the photosynthetic flux into the oxygen reservoir amount. On the other hand, the residence time of material in the organic carbon reservoir at the surface of the Earth, the biota, is just 20 years, a value

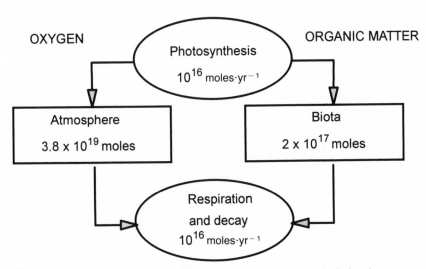

Fig. 4.2. Oxygen in the atmosphere and organic matter in the biota, including humus, are connected by the processes of photosynthesis and respiration and decay. In this and the following figures, reservoirs appear as rectangles, and processes that transfer material between reservoirs appear as ovals.

midway between the life of a tree and the life of bacteria. The short residence time implies that the amount of organic matter in the biota, the reservoir, can change on a time scale of decades, as we well know.

The diagram of Figure 4.2 is deceptively quantitative. There are varying degrees of knowledge and precision in a diagram like this. We know the concentration of oxygen in the atmosphere very precisely; it does not vary measurably with position or with time. We know the mass of the atmosphere, so we know the mass of oxygen in the atmosphere. The amount of material in the biota is known with much less precision, because the components of the biotic reservoir are heterogeneous. There are pronounced differences between the Everglades and the Sahara. But in principle, the amount of organic matter at the surface of the Earth could be determined with arbitrary precision. If we made enough measurements, if we conducted a thorough bookkeeping of the entire globe, we could find out precisely how much reduced organic matter there is in the biotic reservoir. And in principle also, the rates of photosynthesis and decay could be determined with arbitrary precision if enough measurements were made, although at present they are also poorly known. So the reader must understand that a global budget mixes things that are very well known and things that are approximately known.

However, what we really want to know does not appear at all on a diagram like this—that is, what controls the amount of oxygen in the atmosphere or the amount of carbon dioxide. What controls the amount is not revealed by reservoir sizes and rates. It depends also on how the rates vary with conditions. How does the rate of photosynthesis vary with the amount of oxygen in the atmosphere; how does the rate of respiration and decay vary with the amount of oxygen in the atmosphere? Now those numbers cannot be determined by measurements to arbitrary degrees of precision because they are global rates. That is where we would need to do that experiment I told you about. Let us double the amount of oxygen in the atmosphere and see how the global rate of photosynthesis changes. That kind of experiment cannot be performed because we cannot utilize direct measurement or observation to determine how global rates vary with conditions. That means that any conclusions I reach for global geochemical cycles will remain uncertain and, to a considerable extent, speculative or controversial. Thus there are different degrees of knowledge in the field; least known are how the rates and feedbacks vary with conditions. But it is the variation in rates and feedbacks with conditions that controls the amount of oxygen in the atmosphere.

Nonetheless, the *relative* sizes of these two reservoirs are informative and not uncertain. There is no question that there is very much less organic matter in the biota than there is oxygen in the atmosphere. Photosynthesis adds material to the two reservoirs at equal molar rates. Respiration removes material from the two reservoirs at equal molar rates. What is controlling what? Think of a bucket and a bathtub (Fig. 4.3). The speed of the pump hardly affects the level of the water in the bathtub, but it has a large effect on the level of the water in the leaking bucket (26). In other words, when two reservoirs of unequal size are connected, the rate of material between the reservoirs controls the size of the small reservoir, not the size of the large reservoir. Referring to Figure 4.2, then, I conclude that the rate of photosynthesis controls the amount of material in the biota reservoir, and hardly affects the amount of material in the oxygen reservoir.

The speed of the pump sets the water
level in the bucket, not the bathtub

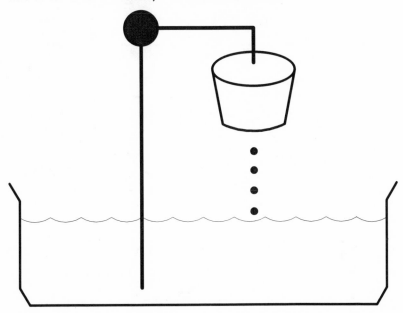

Fig. 4.3. When reservoirs of markedly different size are connected, the rate of the process that transfers material between reservoirs determines the amount of material in the small reservoir, not the large reservoir.

Let us suppose that one could engineer a much more efficient photosynthetic organism, so much so that when the system came back into balance, there was twice as much organic matter in the world. One would have increased the amount of oxygen in the atmosphere by less than 0.5%. Not a very important change. So the rate of photosynthesis determines, insofar as we know, the amount of organic matter in the world, not the amount of oxygen. Something else is controlling the amount of oxygen in the atmosphere, because the oxygen reservoir is so much larger than the biotic reservoir. Suppose one conducted that doomsday experiment and killed off all the plants in the world, but allowed respiration and decay to continue (29). In 20 years or so, all the organic matter would have been consumed, but there would have been very little reduction in the amount of oxygen in the atmosphere. On these grounds I argue that photosynthesis does not control the amount of oxygen in the atmosphere (3).

What controls the amount of oxygen in the atmosphere is the connection to another reservoir, a much larger reservoir. The much larger reservoir is the reduced material in sedimentary rocks. Figure 4.4 shows the original budget with atmosphere, biota, photosynthesis, and respiration, but I have added another component of the overall system, a reservoir with a large amount of reduced material in it. The reduced matter in sedimentary rocks is mainly what is called *kerogen*—old organic matter, as well as sulfide minerals and reduced iron minerals.

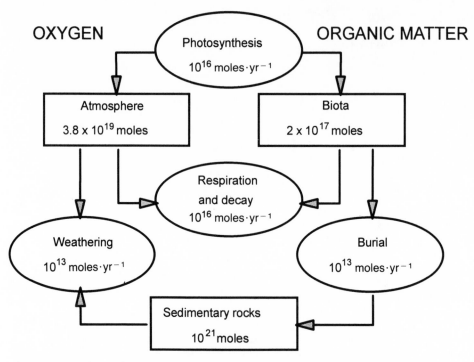

Fig. 4.4. Biogeochemical cycles affecting atmospheric oxygen. The reservoirs include atmospheric oxygen, the reduced material in the biota, and the reduced constituents of sedimentary rocks. Transfer processes in addition to photosynthesis and respiration and decay include oxidative weathering of the reduced constituents of sedimentary rocks and the burial of reduced organic matter in new sediments.

It has a size of 10^{21} moles—a lot larger than the atmospheric reservoir. The atmospheric reservoir is coupled to the rock reservoir by the processes of weathering and oxidation. Weathering of sedimentary rocks, and oxidation of the reduced minerals those rocks contain, occur at an approximate rate of 10^{13} moles per year (1). This process, which draws oxygen out of the atmosphere, is balanced by a process in which reduced material is added to the reservoir when new sedimentary rocks form. Some of the organic matter is extracted from the short-term cycle into the long-term cycle, so that the organic matter or the reducing power that flows down into sedimentary rocks is not available to consume oxygen in the short term.

Although these budgets are shown as being in balance, the data are by no means good enough to establish that they are in fact in balance. The convention of balance is an arbitrary convention. We do not really know whether atmospheric oxygen is increasing or decreasing at present. Any change is too slow to detect because the residence time is some 3.8 million years. If the process of photosynthesis and burial were to stop, it would take 4 million years for this process of weathering to remove the oxygen from the atmosphere. The amount of oxygen in the atmosphere cannot change more rapidly than over millions of years because the

OXYGEN BUDGET

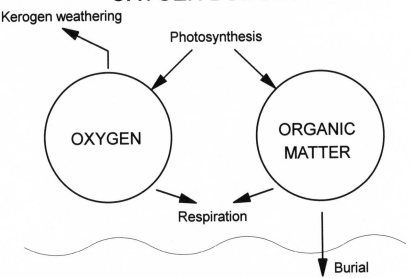

Fig. 4.5. A simple view of the processes controlling the oxygen content of the atmosphere. Oxygen is consumed by the weathering of reduced constituents of sedimentary rocks, principally kerogen. This sink must be balanced, in the long term, by burial of photosynthetically produced organic matter in new sedimentary rocks.

reservoir is so large and the rates of the processes that affect atmospheric oxygen are so small. Figure 4.4 shows the most important elements of the global geochemical cycle affecting atmospheric oxygen. The answer to the question of what controls the amount of oxygen in the atmosphere is probably related to this cycle of burial of reduced material and weathering of sedimentary rocks. How the system might work in its simplest version appears in Figure 4.5.

This figure shows atmospheric oxygen and surface organic matter. Photosynthesis adds material to these two reservoirs at equal molar rates; respiration consumes material in both reservoirs. The weathering of ancient organic carbon consumes oxygen; that sink for oxygen is balanced by the burial of fresh organic matter. In general terms, if atmospheric oxygen increases, the rate of burial will decrease, the rate of weathering will increase, and there will be a negative feedback that will reduce the amount of oxygen in the atmosphere. The processes that control the amount of oxygen in the atmosphere include the rate of weathering and how it depends on the amount of oxygen in the atmosphere and the rate of burial of organic matter and how it depends on the amount of oxygen in the atmosphere. We do not know how these rates depend on the amount of oxygen in the atmosphere, but we can assume that they vary with oxygen in the right sense to buffer the amount of oxygen in the atmosphere. (For alternative suggestions, see Refs 16, 30.)

Actually it is a little more complicated still, because there is another reservoir, the interior of the Earth (Fig. 4.6). The interior of the Earth is predominantly reduced, as the reader knows. The core of the Earth is metallic iron; there is no

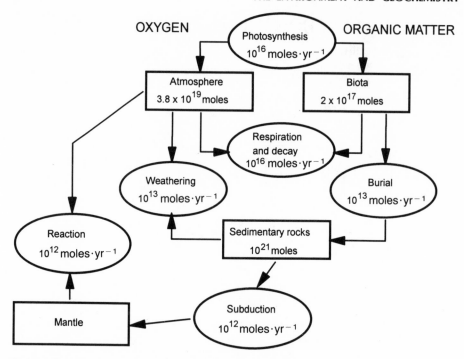

Fig. 4.6. The oxygen content of the atmosphere is controlled by a hierarchy of biogeochemical cycles that couple reservoirs of increasing size by exchange processes of decreasing rate.

free oxygen down there. The reduced interior does indeed react with oxidants at the surface of the Earth, consuming oxidants at the rate of 10^{12} moles per year (7, 8). Consumption of oxidants is balanced by transport into the interior of the Earth of reduced material, principally organic carbon in sedimentary rocks carried down in the subduction zones at an equal rate of 10^{12} moles per year, insofar as we know. The sedimentary rock reservoir has a residence time of 10^9 years; its size is 10^{21} moles, and the rates of mantle exchange if 10^{12} moles·yr^{-1}. In 10^9 years the reservoir size can be changed by exchange of oxidant and reductant with the Earth's interior. This is a subject not very well explored or understood, but the picture I want to convey is one of a nested hierarchy of processes that affect the oxidation state of the surface layers of the Earth and ultimately the amount of oxygen in the atmosphere. If we want to know why there is 21% oxygen in the atmosphere, we need to understand this hierarchy of processes with different rates and different residence times, and we need to know how all of these rates depend on conditions—in particular, the amount of oxygen in the atmosphere. When we understand the whole system, we can answer the question of why there is 21% oxygen in the atmosphere. The answer, in my opinion, does not have much to do with the peculiarities of photosynthesis. I can answer the question asked, at least implicitly by Tolbert in Chapter 1: What would be the effect on atmospheric oxygen of changing the rate of photosynthesis? Although the precise answer is that we do not really know, I believe the effect would be small. In any event, any change would take millions of years.

CARBON

Carbon is in many ways more complicated than oxygen, and therefore, it is convenient to lay out the principles of this kind of study by considering oxygen first. The analysis of this section is based on the work of several authors (2, 4, 12, 13, 16, 22, 26, 28, 29). The case for photosynthetic control is more promising for carbon because there is less carbon dioxide in the atmosphere than there is in the biota. Maybe the size of the biota controls the amount of carbon in the atmosphere. Figure 4.7 shows that there are 0.5×10^{17} moles of carbon dioxide in the atmosphere, and 20 of the same units in the biota. A change in the rates of photosynthesis and respiration can affect the amount in the atmosphere, on a short time scale, but of course, atmosphere and biota are connected to a much larger reservoir in the ocean. On a time scale of more than a few decades, what really controls the amount of carbon bioxide in the atmosphere is the partitioning of carbon between the large oceanic reservoir and the atmosphere plus biota reservoir, not just the exchange between atmosphere and biota.

Let us now calculate residence times (Fig. 4.8). There are 0.5×10^{17} moles of carbon dioxide in the atmosphere and 2×10^{17} moles in the biota. Carbon dioxide has a residence time of five years against removal by photosynthesis. Carbon in the biota has a residence time of 20 years, as in Figure 4.2, but atmospheric carbon dioxide is also exchanging rapidly with the surface ocean. The residence time is only seven years against transfer into the surface ocean. And if we are interested in problems on a time scale longer than a few decades, we have to think also

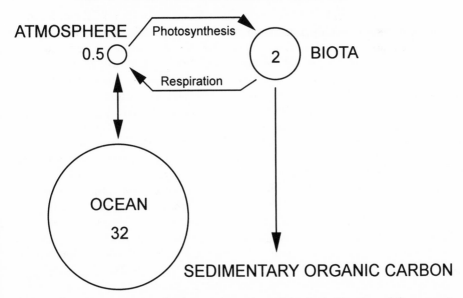

CARBON RESERVOIRS

ATMOSPHERE / Photosynthesis

0.5 ⃝   ⃝ 2 BIOTA

Respiration

OCEAN

32

SEDIMENTARY ORGANIC CARBON

Fig. 4.7. The major reservoirs of carbon at the surface of the Earth. The sizes of the reservoirs are expressed in units of 10^{16} moles.

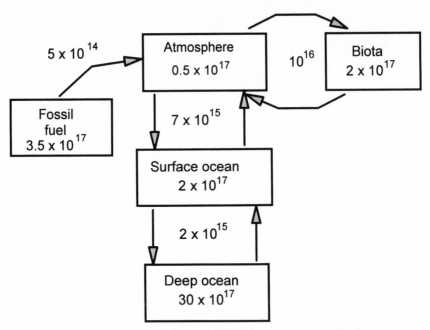

Fig. 4.8. Carbon dioxide in the atmosphere is affected by exchange with the biota, surface ocean, and fossil fuel reservoirs. The surface ocean reservoir, in turn, is connected to the deep ocean reservoir.

about coupling to the deep ocean, where the carbon has a residence time of a thousand years or so.

Consider, now, some perturbation of the system—for example, the doomsday perturbation that suddenly stops photosynthesis (29). In 20 years or so, all the carbon in the biota reservoir will be released to the atmosphere, leading initially to a large increase in the amount of carbon dioxide in the atmosphere. But in no time at all, in terms of human generations, that extra carbon dioxide will work its way down into the very large deep sea reservoir where the addition of 2×10^{17} moles to the 30×10^{17} moles already there will have little effect. The system will not end up with a lot of extra carbon dioxide in the atmosphere, even if photosynthesis stops completely. The figure shows also the fossil fuel rate, which is smaller than the rate of photosynthesis. The fossil fuel reservoir, which reflects how much carbon we might ultimately release, is larger than the biota reservoir.

The budget of Figure 4.8 can be misleading because there is in fact a limit to how much carbon dioxide the ocean can absorb (5). This limit arises from considerations of charge balance. Seawater has to remain electrically neutral, which means that the sum of the charges on all the positive species in seawater minus the sum of the charges on the abundant negative species has to balance the charge on the dissolved carbon species (equations 4.1 and 4.2).

4.1 The partitioning of carbon between ocean and atmosphere depends on the charge balance:

$$2\,[Ca^{2+}] + 2\,[Mg^{2+}] + [Na^+] + [K^+] - [Cl^-] - 2\,[SO_4^{2-}]$$

$$= [HCO_3^-] + 2\,[CO_3^{2-}]$$

4.2 To dissolve more carbon dioxide in the ocean, positive charges—for example, calcium carbonate—must be added:

$$CaCO_3 + CO_2 + H_2O = 2\,Ca^{2+} + 2\,HCO_3^-$$

The system cannot add carbon dioxide to seawater without also adding positively charged material, because the carbon species in seawater are negatively charged carbonate and bicarbonate. This is a complicating factor. Therefore, the analysis for carbon is not quite as simple as the analysis for oxygen because one must keep track of positively charged species as well as carbon dioxide. In particlar, to dissolve more carbon dioxide in the ocean, it is necessary to *dissolve* calcium carbonate, not precipitate it. If you are thinking of genetic engineering to affect the rate at which phytoplankton precipitate calcium carbonate as a solution to the greenhouse problem, do not precipitate more calcium carbonate, but less. This is very important: The precipitation of calcium carbonate from seawater releases carbon dioxide to the atmosphere, not the reverse. The partitioning of carbon between ocean and atmosphere depends on charge balance and the cation concentration, and the amount of carbon dioxide in the atmosphere depends on that partitioning.

The geochemical cycles of carbon also include sedimentary rocks; therefore, in Figure 4.9 I have lumped together atmosphere, biota, and ocean in a reservoir of

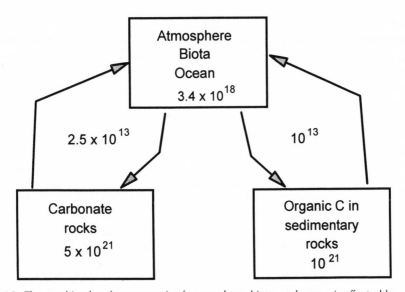

Fig. 4.9. The combined carbon reservoir of atmosphere, biota, and ocean is affected by exchange of carbon with much larger reservoirs consisting of sedimentary carbonate rocks and organic carbon in sedimentary rocks.

total size of 34.5×10^{17} moles. This is coupled to a reservoir of sedimentary carbonate rocks equal to 5×10^{21} moles and the reservoir of organic carbon in sedimentary rocks of 10^{21} moles. The figure shows the rates of transfer of material. Carbon is extracted from atmosphere, biota, and ocean into carbonate rocks at a rate of 2.5×10^{13} moles $\cdot$ yr^{-1} and as organic matter into sedimentary rocks at a rate of 10^{13} moles $\cdot$ yr^{-1}. The carbon is released from those rock reservoirs at more or less equal rates by the processes of weathering and oxidation of the sedimentary constituents. The residence time of carbon in the sedimentary reservoirs is about 2×10^{8} years, which is the average length of time that sedimentary rocks last in the crust of the Earth before being destroyed by erosion. The residence time of surface carbon against exchange with the rocks is about 10^{5} years.

The controls on the amount of oxygen in atmosphere, biota, and ocean lie in the processes coupling the surface reservoir to the much larger rock reservoir. The question as to what determines how much carbon is in atmosphere, biota, and ocean relates to how the rates of weathering and precipitation vary with the partial pressure of carbon dioxide and the cation content of seawater. The controls lie not in the rates of photosynthesis or respiration but in the rates of weathering and precipitation of carbon-bearing minerals. The rates of weathering and precipitation control the amount of carbon in the surface reservoir. Partitioning of the carbon between atmosphere and ocean depends on the charge balance, or the cation content, of the ocean. Photosynthesis plays a role, of course, in determining how carbon partitions between atmosphere and biota, but the reader must remember that the atmosphere is coupled to an enormous oceanic reservoir on a time scale of centuries.

There is a larger and slower cycle also (Fig. 4.10), in which carbon comes out of the Earth's mantle at a rate of 2×10^{12} moles $\cdot$ yr^{-1} and is returned to the Earth's mantle in the form of carbonate minerals in the sea floor (6, 9, 10, 14, 15, 18, 25). It is this cycle, coupling the surface to the mantle, that determines how much carbon there is in sedimentary rocks. The residence time of carbon in the surface reservoir against return to the mantle is about 10^{9} years. For carbon, as for oxygen, there is a hierarchy of cycles. The figure shows the processes that control carbon all the way up the hierarchy. The amount of carbon in the various reservoirs determines ultimately how much carbon is in the atmosphere. When we understand the whole system, we shall be able to answer the question of why there are 300 ppm carbon dioxide in the atmosphere. In my opinion, the answer does not have much to do with the peculiarities of photsynthesis.

I can answer the question asked at least implicitly by Tolbert in Chapter 1 as to the effect of changing the rate of photosynthesis on atmospheric carbon dioxide. The first answer is that we really do not know. The expectation is that there would be short-term response as carbon is exchanged between atmosphere and biota, but no long-term response because in the long term, atmospheric carbon dioxide equilibrates with the very large oceanic reservoir, and the carbon content of this reservoir is determined by interactions with the even larger rock reservoir. It is possible, however, that life, and therefore at least indirectly the rate of photosynthesis, may affect the rock cycle by altering, for example, the rate of weathering (17, 19). In this sense, then, the rate of photosynthesis might affect the amount of carbon dioxide in the atmosphere in the long term.

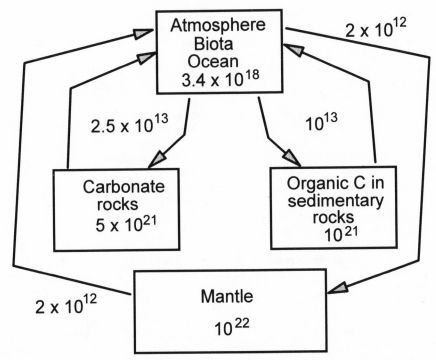

Fig. 4.10. The carbon dioxide content of the atmosphere is controlled by a hierarchy of biogeochemical cycles that exchange carbon with the biota, the ocean, sedimentary rocks, and the Earth's mantle.

CONCLUSIONS

Simple arguments based on conservation of matter suggest that the global rate of photosynthesis does not have a lasting impact on the atmospheric concentrations of oxygen and carbon dioxide. The rate of exchange of material between atmosphere and biota by photosynthesis and respiration and decay is rapid, but the total amount of material in the biotic reservoir is too small compared with the amount of oxygen in the atmosphere or the amount of carbon in oceans and sedimentary rocks for the transfer of material in or out of the biotic reservoir to have a lasting impact on the atmosphere. It is, of course, possible that life affects the rates of the controlling geochemical processes, weathering, and precipitation of sedimentary rocks, but there is little quantitative information about this possibility.

ACKNOWLEDGMENT

This research was supported in part by the National Aeronautics and Space Adminisgration under Grant NAGW-179.

REFERENCES

1. Berner, R.A. 1982. Burial of organic carbon and pyrite in the modern ocean: Its geochemical and environmental significance. *Am. J. Sci.* **282**: 451–73.

2. Berner, R. A., Lasaga, A. C., and Garrels, R. M. 1983. The carbonate–silicate geochemical cycle and its effect on atmospheric carbon dioxide over the past 100 million years. *Am. J. Sic.* **283**: 641–83.

3. Broecker, W. S. 1970. Man's oxygen reserves. *Science* **168**: 1537–38.

4. Broecker, W. S. 1971. A kinetic model for the chemical composition of sea water. *Quatern. Res.* **1**: 188–207.

5. Broecker, W. S., and Peng, T. H. 1982. *Tracers in the Sea.* Lamont-Doherty Geological Observatory, Columbia University, Pallisades, N.Y.

6. Des Marais, D. J. 1985. Carbon exchange between the mantle and the crust and its effect upon the atmosphere: Today compared to the Archean time. In *The Carbon Cycle and Atmospheric CO_2: Natural Variations Archean to Present*, ed. E. T. Sundquist and W. S. Broecker, pp. 602–11. American Geophysical Union, Washington, D.C.

7. Edmond, J. M., Measures, C., Mangum, B., Grant, B., and Sclater, F. R. 1979. On the formation of metal-rich deposits at ridge crests. *Earth Planet. Sci. Lett.* **46**: 19–30.

8. Edmond, J. M., Measures, C., McDuff, R. E., Chan, L. H., and Collier, R. 1979. Ridge crest hydrothermal activity and the balances of the major and minor elements in the ocean: The Galalagos data. *Earth Planet. Sci. Lett.* **46**: 1–18.

9. Gerlach, T. M. 1989. CO_2 from magma-chamber degassing. *Nature* **337**: 124.

10. Harris, N. 1989. Carbon dioxide in the deep crust. *Nature* **340**: 347–48.

11. Holland, H. D. 1973. Ocean water, nutrients, and atmospheric oxygen. In *Proceedings of Symposium on Hydrogeochemistry and Biogeochemistry*, ed. Vol. 1, pp. 68–81. The Clarke Co., Washington, D.C.

12. Holland, H. D. 1978. *The Chemistry of the Atmosphere and Oceans.* Wiley-Interscience, New York.

13. Holland, H. D. 1984. *The Chemical Evolution of the Atmosphere and Oceans.* Princeton University Press, Princeton, N.J.

14. Javoy, M., Pineau, F., and Allegre, C. J. 1982. Carbon geodynamic cycle. *Nature* **300**: 171–73.

15. Javoy, M., Pineau, F., and Delorme, H. 1986. Carbon and nitrogen isotopes in the mantle. *Chem. Geol.* **57**: 41–62.

16. Lasaga, A. C., Berner, R. A., and Garrels, R. M. 1985. An improved model of atmospheric CO_2 fluctuations over the past 100 million years. In *The Carbon Cycle and Atmospheric CO_2: Natural Variations Archean to Present*, ed. E. T. Sundquist and W. S. Broecker, pp. 397–411. American Geophysical Union, Washington, D.C.

17. Lovelock, J. E., and Whitfield, M. 1982. Life span of the biosphere. *Nature* **296**: 561–63.

18. Marty, B., and Jambon, A. 1987. $C/^3He$ in volatile fluxes from the solid Earth: Implications for carbon geodynamics. *Earth Planet. Sci. Lett.* **83**: 16–26.

19. Schlesinger, W. H. 1986. Changes in soil carbon storage and associated properties with disturbance and recovery. In *The Changing Carbon Cycle: A Global Analysis*, ed. J. R. Trabalka and D. E. Reichle, pp. 194–220. Springer-Verlag, New York.

20. Volk, T. 1987. Feedbacks between weathering and atmospheric CO_2 over the last 100 million years. *Am. J. Sci.* **287**: 763–79.

21. Walker, J. C. G. 1974. Stability of atmospheric oxygen. *Am. J. Sci.* **274**: 193–214.

22. Walker, J. C. G. 1977. *Evolution of the Atmosphere.* Macmillan, New York.

23. Walker, J. C. G. 1980. The oxygen cycle. In *The Natural Environment and the Biogeochemical Cycles*, ed. O. Hutzinger, pp. 84–104. Springer-Verlag, Berlin.

24. Walker, J. C. G. 1982. Evolution of the atmosphere's chemical composition. *Impact* **32**: 261–69.

25. Walker, J. C. G. 1983. Carbon geodynamic cycle. *Nature* **303**: 730–1.

26. Walker, J. C. G. 1986. *Earth History: The Several Ages of the Earth*. Jones and Bartlett Publishers, Boston, Mass.

27. Walker, J. C. G. 1985. Global geochemical cycles of carbon, sulfur, and oxygen. *Marine Geol.* **70**: 159–74.

28. Walker, J. C. G., and Drever, J. I. 1988. Geochemical cycles of atmospheric gases. In *Chemical Cycles in the Evolution of the Earth*, ed. B. Gregor and J. B. Maynard, pp. 55–76. John Wiley and Sons, New York.

29. Walker, J. C. G. in press. Feedback processed in the biogeochemical cycles of carbon. In *Science of Gaia*, ed. S. H. Schneider, pp. 000–000. MIT Press, Cambridge, Mass.

30. Watson, A., Lovelock, J. E., and Margulis, L. 1978. Methanogenesis, fires and the regulation of atmospheric oxygen. *Biosystems* **10**: 293–8.

31. Woodwell, G. M., Whittaker, R. H., Reiners, W. A., Likens, G. E., Delwiche, C. C., and Botkin, D. B. 1978. The biota and the world carbon budget. *Science* **199**: 141–6.

II

THE C₃ PHOTOSYNTHETIC CARBON CYCLE

5

Regulation of the C_3 Reductive Cycle and Carbohydrate Synthesis

JACK PREISS

This short report will summarize the reactions of carbon metabolism involved in the C_3 cycle and the phosphate/triose phosphate (P_i/triose-P) shuttle, and in the synthesis of sucrose and starch. The aim is to make the reader aware of these pathways and of the regulatory phenomena prevalent in their integration. The possible effect of increased atmospheric CO_2 will also be dealt with, as well as the possibilities or advantages that could be derived from increased CO_2 fixation by plants

THE C_3 CARBON REDUCTION PATHWAY

Table 5.1 illustrates the reactions of the C_3 carbon reduction cycle (2, 14, 30). All of these reactions are localized in a subcellular organelle called the chloroplast. As shown in the reactions of Table 5.1, the plant can fix six molecules of CO_2 to form one sugar-P unit. If the sugar-P remains in the chloroplast, it can form starch (reactions 9 and 10). The extra carbon can also leave the chloroplast via the P_i/triose-P translocator (8) located in the plastid inner envelope, mainly in the form of dihydroxyacetone-P and in exchange for P_i. In the cytosol, this dihydroxyacetone-P molecule can first form, via gluconeogenic reactions, hexose-P, and then sucrose (3, 34). These reactions will be discussed later.

Nature has devised mechanisms to ensure that the enzyme-catalyzed reactions in the carbon reduction cycle are optimal during photosynthesis and not operative in the dark. Table 5.2 illustrates some of the mechanisms that modulate the activity of the C_3 pathway elucidated so far. Regulation of Rubisco activity has been discussed in many reviews (2, 31, 36). In addition to the activation of Rubisco, a number of enzymes in the pathway are activated in a mechanism that involves ferredoxin. Ferredoxin is reduced during photosynthesis, and in turn it reduces thioredoxin, which then reduces and activates the enzymes (except Rubisco) listed in Table 5.2 (1, 4, 17).

In the light there is also an increase in pH, and some of these enzymes including Rubisco have higher activity at the higher pH (8.3–8.5) prevalent in the stroma

Table 5.1. Reactions of the C_3 Carbon Reductive Cycle.

1. 6 CO_2 + 6 Ribulose-1,5-bis-P + 6 H_2O → 12 3-P-Glycerate
2. 12 3-P-Glycerate + 12 ATP → 12 3-P-Glycerol-P + 12 ADP
3. 12 3-P-Glyceroyl-P + 12 NADPH + 12 H^+ → 12 Glyceraldehyde-3-P + 12 $NADP^+$ + 12 P_i
4. 5 Glyceraldehyde-3-P → 5 Dihydroxyacetone-P
5. 3 Glyceraldehyde-3-P + 3 Dihydroxyacetone-P ↔ 3-Fructose-1,6-bis-P
6. 3 Fructose-1,6-bis-P + 3 H_2O → 3 Fructose-6-P + 3 P_i
7. Fructose-6-P ↔ Glucose-6-P
8. Glucose-6-P ↔ Glucose-1-P
9. Glucose-1-P + ATP → ADPglucose + PP_i
10. ADPglucose → Starch
11. 2 Fructose-6-P + 2 Glyceraldehyde-3-P ↔ 2 Xylulose-5-P + 2 Erythrose-4-P
12. 2 Erythrose-4-P + 2 Dihydroxyacetone-P ↔ 2 Sedoheptulose-1,7-bis-P
13. 2 Sedoheptulose-1,7-bis-P + 2 H_2O → 2 Sedoheptulose-7-P + P_i
14. 2 Sedoheptulose-7-P + 2 Glyceraldehyde-3-P ↔ 2 Ribose-5-P + 2 Xylulose-5-P
15. 2 Ribose-5-P ↔ 2 Ribulose-5-P
16. 4 Xylulose-5-P ↔ 4 Ribulose-5-P
17. 6 Ribulose-5-P + 6 ATP → 6 Ribulose-1,5-bis-P + 6 ADP

Sum^a: 6 CO_2 + 19 ATP + 12 H_2O + 12 NADPH + 12 H^+ → $C_6H_{12}O_6$ + 17 P_i + 19 ADP + 12 $NADP^+$ + PP_i

^{a} *Note*: Reactions 1 through 10 generate one glucosyl unit of starch, and reactions 11 through 17 regenerate ribulose-1,5-bis-P for further CO_2 fixation.

during photosynthesis (2, 7, 14, 30). Furthermore, some enzymes have a requirement for Mg^{2+} and become more active because of the increase in stromal Mg^{2+} concentration that occurs during photosynthesis (2, 7, 14, 30).

The reactions involved in the synthesis of sucrose (3, 34) from fructose-1.6-bis-P are illustrated in Table 5.3. Also seen are the phenomena that are potentially regulatory for sucrose synthesis (3, 34). Reactions 1 and 5 are the rate-limiting reactions and can be regulated by allosteric effectors. Two other processes are also associated with regulation. P_i is needed for the transport of carbon from the chloroplast into the cytosol (where fructose-1,6 bis-P is synthesized), and this exchange of P_i for triose-P is catalyzed by the chloroplast P_i/triose-P translocator. Covalent modification via phosphorylation of the sucrose-P synthase, to render it inactive, has been shown to occur during the night in spinach and in maize

Table 5.2. Potential Light Regulatory Phenomena Associated with the Photosynthetic Carbon Reduction Cycle.

	Ref.
1. Regulation of Rubisco	2, 31, 36
2. Regulation by Ferredoxin/Thioredoxin in the activation of NADP-glyceraldehyde-3-P Dehydrogenase, Fructose-1,6-bis-P 1-Phosphatase, Sedoheptulose-bis-P, 1-Phosphatase and Ribulose-5-P Kinase	1, 4, 17
3. Alkalinization of the stroma	7
4. Increase in stromal [Mg^{2+}]	7

Table 5.3. Sucrose Biosynthesis from Fructose 1,6-bis-P and Its Regulation.

1. Fructose-1,6-bis-P + H$_2$O → Fructose-6-P + P$_i$
2. Fructose-6-P ↔ Glucose-6-P
3. Glucose-6-P ↔ Glucose-1-P
4. Glucose-1-P + UTP → UDPglucose + PP$_i$
5. UDPglucose + Fructose-6-P → Sucrose-6-P + UDP
6. Sucrose-6-P + H$_2$O → Sucrose + P$_i$

Note: AMP and fructose-2,6-bis-P inhibit reaction 1 (34). Glucose-6-P activates and P$_i$ inhibits reaction 5 (34). Diurnal fluctuations of activity are probably caused by the covalent modification of sucrose-P synthase.

leaves (9, 10, 12, 13). In this way, sucrose synthesis decreases at night, an adjustment consistent with the absence of carbon fixation in the dark. Carbon obtained from starch degradation at night can also be used for sucrose synthesis. Activation of the sucrose-P synthase occurs by action of a protein phosphatase (11), whose activity is induced in the light. Currently, the activation of the phosphatase by light is thought to be due to the decrease in P$_i$ concentration during photosynthesis, as P$_i$ is a inhibitor of the protein phosphatase (11). This regulatory system is currently under active investigation.

The reactions involved in starch synthesis (25–28) are illustrated in Table 5.4. Regulation of starch synthesis is mediated by the activation of ADPglucose synthesis by 3-phosphoglycerate (3-PGA) and its inhibition by P$_i$. Moreover, increasing concentrations of 3-PGA can overcome P$_i$ inhibition, and, conversely, increasing concentrations of pP$_i$ can reverse activation by 3-PGA.

Mutants of maize endosperm, *shrunken* 2 and *brittle* 2, of pea embryo (r_b) and of *Arabidopsis thaliana*, indicate that the ADPglucose pyrophosphorylase (ADPGlc PPase) pathway is the dominant pathway (if not the sole one) for starch biosynthesis both in reverse (storage) as well as in leaf tissue. Moreover, transformation of potato tubers with anti-sense mRNA to the structural genes of the potato ADPGlc PPase subunits resulted in minitubers containing only 3% of the starch seen in normal minitubers (20). Thus in at least four different plant systems there is very solid evidence that ADPGlc PPase is the dominant enzyme in the production of ADPglucose and that the mechanism proposed by Akazawa and associates (24) that ADPglucose is synthesized via sucrose synthase action is virtually nonexistent in in vivo conditions.

There is much in vitro evidence which suggests that the synthesis of ADPglucose is regulated via the activation of ADPGlc PPase by 3-PGA and its inhibition by

Table 5.4. Reactions Involved in Starch Biosynthesis and Its Regulation.

1. ATP + Glucose-1-P → ADPglucose + PP$_i$
2. ADPGlucose + (G)$_n$ → α-1,4-Glucosyl-(G)$_n$ + ADP
3. (G)$_{n+x}$ → α-1,6 α-1,4-Glucan (Amylopectin)

Note: Reaction 1 is catalyzed by ADPglucose pyrophosphorylase and is activated by 3-P-glycerate and inhibited by P$_i$. Activation of reaction 1 is also caused, but to a smaller extent, by fructose-6-P, fructose-1,6-P$_2$ and phosphoenol pyruvate.

P_i. Evidence, obtained by in vivo and in situ experiments, showing a direct correlation between the concentrations of 3-PGA and starch, and an inverse correlation between P_i and starch levels, has been reviewed elsewhere (26–28). Recently, Pettersson and Ryde-Pettersson (23) applied modern control theory (5, 15) to develop a kinetic model. The objective of this model was to determine the extent to which stromal metabolites, known to affect the activity of leaf ADPglucose pyrophosphorylase activity in vitro, control the rate of photosynthetic starch production in vivo, under conditions of light and CO_2 saturation. The model consists of the 13 enzyme-catalyzed steps of the photosynthetic reductive pentose phosphate pathway, together with the reactions of starch synthesis and photo-synthate export (P_i/triose-P translocator) from the chloroplast, as output processes from the pentose phosphate pathway. The model takes into account various enzme properties and equilibria. Using this model, the steady-state concentrations of various stromal metabolites, and the corresponding rates of CO_2 fixation and starch production, were calculated as a function of the P_i concentration outside the chloroplast. Essentially, the model agrees with reported experimental data, regarding both metabolite concentrations and the rates of CO_2 fixation and starch synthesis. In the model, ATP and glucose-1-P (substrates of the ADPGlc PPase), and 3-PGA, fructose-6-P, and P_i, make significant contributions to the changes in the rate of starch synthesis caused by increased P_i concentrations outside the chloroplast. At low P_i concentrations, ATP and, to a smaller extent, 3-PGA made the most significant contribution to increased starch synthetic rate. At P_i concentrations higher than 0.12 mM, 3-PGA became the predominant reglator of starch synthesis, with glucose-1-P and fructose-6-P contributing to regulation to a significant, but smaller, extent. Thus, a modern mathematical control analysis has been used to create a kinetic model that is consistent with the data obtained in vivo. The authors of the model conclude that 3-PGA and P_i play an important role in regulating starch synthesis and that ATP, glucose-1-P, and fructose-6-P make significant contributions. Since these metabolites are either substrates or effectors of the ADPGlc PPase, the analysis is consistent with the view that 3-PGA is a positive effector and P_i is a negative effector of ADPglucose synthesis and that the 3-PGA/P_i ratio therefore regulates starch synthesis not only in vitro but also in vivo via its regulatory effect on the ADPglucose pyrophosphorylase.

The availability of chloroplast mutants of phosphoglucose isomerase of *Clarkia xantiana* (18, 21), of phosphoglucomutase (6), and of ADPGlc PPase of *Arabidopsis thaliana* (19, 22) has allowed the analysis of the extent of control that these enzymes exert on chloroplast starch synthesis. Mutant plants with reduced activity of both cytosolic (64%, 36%, 18% of wild-type) and chloroplastic (75%, 50% of wild-type) phosphoglucoisomerase were used to determine the effect of the enzyme level on fluxes toward starch and sucrose synthesis as well as on photosynthetic rate and control coefficients (15) of these enzymes on these pathways (18). Saturating or limiting light intensities were used in the various experiments. The plastid P-glucoiosomerase exerted very little control over starch or sucrose synthesis in low light, but did exert control of starch synthesis in saturating light. Lowering the cytosolic enzyme activity had little effect on either starch or sucrose synthesis in saturating light but increased starch synthetic rate in low light and decreased sucrose synthetic rate. Thus partitioning of carbon between sucrose

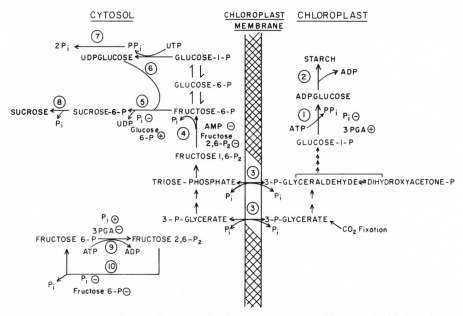

Fig. 5.1. Reactions involved in the synthesis of starch, sucrose, and fructose-2,6-bisphosphate. The reactions are catalyzed by enzymes numbered as follows: 1. ADPglucose pyrophosphorylase; 2. starch synthase; 3. P_i triose translocator; 4. fructose-1,6-bisphosphatase; 5. sucrose phosphate synthase; 6. UDPglucose pyrophosphorylase; 7. inorganic pyrophosphatase; 8. sucrose phosphate phosphatase; 9. phosphofructo-2 kinase; 10. fructose-2,6-bisphosphatase. The known inhibitors $(-)$ and activators $(+)$ of some of the enzymes are indicated by the encircled respective symbols at the reactions they catalyze.

and starch, as seen in Figure 5.1, was affected by variation of the cytosolic phosphoglucoisomerase levels. Further studies (21) confirmed that reduction of plastid phosphoglucoisomerase had little effect in low light but reduced starch synthesis by 50% in saturating light with no corresponding increase in sucrose synthesis. The reduced levels of cytosolic enzyme (18% of wild type) lowered the sucrose synthetic rates and increased the rate of starch synthesis.

Metabolite levels were also affected in these mutants. In the mutant containing only 18% of the wild-type cytosolic phosphoglucoisomerase activity, both fructose-2,6-bisphosphate and 3-PGA levels increased about 100%. Neuhaus et al. (21) have said that their results provide strong evidence that the reduced sucrose synthesis rate is due to the increased fructose-2,6-bis-P level, which causes increased inhibition of cytosolic fructose-1,6-bisphosphatase (for reviews on sucrose synthesis, see Ref. 3 and 34) which is on the pathway toward sucrose synthesis (Fig. 5.1). They also indicate that the data strongly support the view that increased starch synthesis in the mutants with reduced levels of phosphoglucoisomerase is due to activation of the ADPglucose pyrophosphorylase by the increased 3-PGA concentration and 3-PGA/P_i ratio. Thus, these experiments provide further support for the in vivo regulation of starch synthesis by 3-PGA.

These experiments have been extended to the null chloroplast phosphoglucomutase (6) and the low-activity (7% of wild type) of the ADPglucose

Table 5.5. Rate of Photosynthesis, Starch Synthesis, and Sucrose Synthesis in Low
$(75 \ \text{mmol} \cdot \text{m}^{-2} \cdot \text{s}^{-1})$ and High $(600 \ \text{mmol} \cdot \text{m}^{-2} \cdot \text{s}^{-1})$ Irradiance in *Arabidopsis thaliana*.

Plant Mutants	Photosynthesis	Starch Synthesis	Sucrose Synthesis	Sucrose synthesis / Starch synthesis
	Flux (μatom C·mg Chl^{-1}·20 min^{-1})			Ratio
Low irradiance				
Wild type	27.6	7,0	11.9	1.7
ADPGlcPPase 7%	27.2	5.4	14.0	2.8
ADPGlcPPase 50%	28.0	1.8	17.1	10.2
Plastid 50%	27.2	6.9	13.6	2.0
Phosphoglucomutase 0%	23.6	0.3	16.9	—
High irradiance				
Wild type	61.0	14.0	29.0	2.1
ADPGlc PPase 50%	48.3	8.6	26.2	3.1
ADPGlc PPase 7%	31.6	1.3	19.3	16.2
Plastid 50%	53.1	11.5	28.5	2.6
Phosphoglucomutase 0%	28.0	0.4	16.8	—

pyrophosphorylase mutants (19) of *Arabidopsis thaliana*. Neuhas and Stitt (22) utilized the alleles to construct hydrid plants containing, respectively, 50% of wild-type phosphoglucomutase activity and 50% of wild-type ADPglucose pyrophosphorylase activity. The effects of these reduced activities on starch and sucrose fluxes and on CO_2 fixation in low light and high light intensity were measured (see Table 5.5). In low light, a 50% decrease in phosphoglucomutase activity had no significant effect on the above fluxes. However, a 50% and 93% decrease of ADPGlc pyrophosphorylase activity resulted in a 23% and 74% decrease in flux of starch synthesis with a concomitant increase of 17% and 42% increase in sucrose synthetic rate. Thus diminution of ADPglucose synthesis activity not only significantly affected starch synthesis but also affected the partitioning of photosynthetic carbon, causing more to be directed toward sucrose biosynthesis. In high light, a 50% decrease in phosphoglucomutase activity resulted in a 20% decrease in starch synthesis with little effect on sucrose synthesis rate. However, reduction of the ADPglucose-synthesizing activity to 50% and 93% resulted in a 39% and 90% decrease in starch synthesis flux. The flux of photosynthetic carbon under these conditions was not redirected toward sucrose synthesis; rather, the photosynthetic rate was inhibited about 46%. The flux control coefficients (22) for the enzymes for starch synthesis were calculated to determine the distribution of control and compared with previous results obtained with the *Clarkia xantiana* phosphoglucoisomerase; these are seen in Table 5.6. Of the enzmes under study, the flux to starch synthesis is only regulated by ADPglucose pyrophosphorylase in low light. In high light and CO_2, although ADPglucose pyrophosphorylase activity still exerts major control, other enzmes such as plastid phosophoglucomutase and phosphoglucoisomerase do exert control to a small but significant extent.

Table 5.6. Estimated Flux Control Coefficients for Starch Synthesis.

	Flux control coefficient, C^{Starch}	
	Low Light	High Light
Plastid phosphoglucoisomerase	—	0.35
Plastid phosphoglucomutase	0.01	0.21
ADPGlucose pyrophosphatase	0.28	0.64

In summary, various analyses of the starch biosynthetic system in a number of plants, or utilization of in vivo data obtained from different plants with application of the Kacser and Burns control analysis method (15, see also Ref. 16), point out that the major site of regulation of starch synthesis is at ADPglucose pyrophosphorylase and that 3-PGA and P$_i$ are important regulatory metabolites of that enzyme. The most decisive proof that indeed the above regulatory metabolites are functional in vivo would be the isolation of a mutant plant that was altered in starch content and in which the alteration was correlated with the mutant containing an ADPglucose pyrophosphorylase with allosteric properties altered in respect to activation by 3-PGA and/or inhibiton by P$_i$. At present, such an interesting mutation in plants has not been reported; however, such mutants have been reported for the bacteria *Escherichia coli* and *Salmonella typhimurium*. A class of mutants has been found with altered glycogen content which can be correlated with altered allosteric properties of their ADPglucose pyrophos-phorylases (reviewed in Ref. 29). In addition, there are also mutants of *E. coli* in which the enzymes ADPglucose pyrophosphorylase and glycogen synthase are overexpressed (29). Concomitant with this overexpression is a dramatic increase in glycogen synthetic rate. Thus genetic manipulation of either the structural or regulatory genes of the starch biosynthetic enzymes may provide the means for alteration of the starch levels in a plant.

Figure 5.1 also shows all the reactions together to describe, in molecular terms, what is considered to occur in various leaves and what happens at the onset of photosynthesis (25, 33). There is a sudden increase in triose-P due to photosynthetic reduction and fixation of CO$_2$. There is also a need for P$_i$ for photophosphorylation to make ATP in the chloroplast. Thus, there is an exchange of cytosolic P$_i$ for triose-P, which leaves the chloroplast. The increased DHAP, or resultant 3-PGA from the DHAP, inhibits formation of fructose-2,6-P$_2$ by the PFK II reaction. The decrease in P$_i$ in the cytosol also decreases the activity of the PFK II reaction, as P$_i$ is an activator of that reaction. FBPase II activity takes over and hydrolyzes fructose-2,6-P$_2$. As a result of the decrease in fructose-2,6-P$_2$, the cytosolic fructose-1,6-P$_2$ase is no longer inhibited, and therefore, hexose-P synthesis, and thus sucrose biosynthesis, can commence. The formation of sucrose causes the release of 4 P$_i$ which can go to the chloroplast membrane and be exchanged via the translocator for more DHAP or triose-P. When sink requirements are fulfilled and when signals (unknown) indicate no need for further sucrose synthesis, there is no P$_i$ available for transport of carbon to the cytosol and thus the excess triose-P remaining in the chloroplast is utilized to make starch.

EFFECT OF ATMOSPHERIC CO_2 AND O_2

No aspects of the reactions of the C_3 cycle, the triose-P shuttle, and sucrose or starch synthesis or degradation appear to be directly regulated by the CO_2 concentration. The reactions are also not affected by normal O_2 levels, as they are anaerobic and proceed at 5% or 21% O_2 about equally well. Two percent to 5% O_2 must be available for normal dark respiration. The carbohydrate reactions are fueled by photosynthetic light energy and the initial reactions of the C_3 cycle. Differential accumulation of starch and sucrose is an active research area based on starch storage or sucrose transport of sinks. Thus we believe that the regulation of plant growth by atmospheric CO_2 and O_2 concentrations is restricted to the competitive reactions of Rubisco which deviate carbon flow from the carbohydrates of the C_3 cycle to carbohydrate oxidation by the C_2 cycle.

It has already been shown in a number of cases that plants do thrive in higher CO_2, but other factors such as nutrient requirements become more critical or limiting. There may be a need for more P_i for recycling of carbon, and certainly an increase in leaf starch has been documented. This does not seem to be a bonus, as it has been shown that with an increase in chloroplast starch there is also a concomitant decrease in photosynthetic rate. It is believed that the excess starch may be harming, obstructing, or disrupting, possibly in a physical manner, the thylakoid membrane function in photosynthesis. Would it therefore be advantageous to devise a leaf cell that would have minimal starch and allow all carbon fixed to be transformed to sucrose for transport? The answer to this question is, not likely. It has been shown that *Arabidopsis* mutants unable to make starch, because of a defect in ADPglucose pyrophosphorylase or chloroplast phosphoglucomutase, are not able to grow well in a normal 12-h photoperiod. They grow as well as the normal *Arabidopsis* in continuous light, most probably because of the continuous production of sucrose during photosynthesis. Also the photosynthetic rate in these mutants is reduced compared to that in the wild-type plant. If one wanted to take advantage of the possible increase in carbon available to the plant, one should possibly concentrate on increasing the sink demands of the reserve tissues such as endosperm or tuber or fruit. This may be done by recombinant DNA techniques that possibly increase the biosynthetic genes present in the reserve tissues: for example, those of ADPglucose pyrophosphorylase. Indeed, experiments done in collaboration with scientists at Monsanto Chemical Company, St. Louis, Mo., and reported at the Third International Plant Molecular Biology meeting in Tucson, Ariz., in October 1991, showed that transfection of potato plants with a bacterial mutant ADPglucose pyrophosphorylase gene results in the production of tubers containing 25–50% more starch per plant (32). Also reported at the same meeting was the increase of about 10-fold in the amount of starch in tobacco callus when transfected by the same bacterial mutant ADPglucose pyrophosphorylase gene (32).

In addition to increasing starch content, one could also consider altering the starch quality by varying the amount of branching enzyme in the reserve tissue or possibly starch/glycogen synthase by transformation with the respective genes. A number of people associated with the starch industry have indicated a desire for increased starch content in such crops as maize, barley, potato, or even tomato.

It is estimated that even a small 5% increase in starch would increase farm income considerably. It would then be of interest, assuming that other problems could be surmounted, to determine whether advantage could be taken of the increased atmosopheric CO_2.

REFERENCES

1. Anderson, L. E. 1979. Interaction between photochemistry and activity of enzymes. In *Encyclopedia of Plant Physiology*, New Series, Vol. 6: *Photosynthesis II*, ed. M. Gibbs and E. Latzko, pp. 271–81. Springer-Verlag, Berlin.

2. Andrews, T. J., and Lorimer, G. H. 1987. Control of photosynthetic sucrose formation. In *The Biochemistry of Plants*, Vol. 10, ed. M. D. Hatch and N. K. Boardman, pp. 132–218. Academic Press, New York.

3. ap Rees, T. 1987. Compartmentation of plant megabolism. In *The Biochemistry of Plants*, Vol. 12, ed. D. D. Davies, pp. 87–115. Academic Press, New York.

4. Buchanan, B. B. 1979. Ferredoxin-linked carbon dioxide fixation in photosynthetic bacteria. In *Encyclopedia of Plant Physiologicy*, New Series, Vol. 6: *Photosynthesis II*, ed. M. Gibbs and E. Latzko, pp. 416–24. Sprinber-Verlag, Berlin.

5. Burns, J. A., Cornish-Bowden, A., Groen, A. K., Heinrich, R., Kacser, H., Porteous, J. W., Rapoport, S. M., Rapoport, T. A., Stucki, J. W., Tager, J. M., Wanders, R. J. A., and Westerhoff, H. V. 1985. Control analysis of metabolic systems. *TIBS* **10**: 16.

6. Caspar, T., Huber, S. C., and Somerville, C. 1986. Alterations in growth, photosynthesis and respiration in a starch mutant of *Arabidopsis thaliana* (L.) Heynh. deficient in chloroplast phosphoglucomutase activity. *Plant Physiol.* **79**: 1–7.

7. Heldt, H. W. 1979. Light-dependent changes of stromal H^+ and Mg^{2+} concentrations controlling CO_2 fixation. In *Encyclopedia of Plant Physiology*, New Series, Vol. 6: *Photosynthesis II*, ed. M. Gibbs and E. Latzko, pp. 202–207. Springer-Verlag, Berlin.

8. Heldt, H. W., and Flügge, U. I. 1987. Subcellular transport of metabolites in plant cells. In *The Biochemistry of Plants*, Vol. 10, ed. D. D. Davies, pp. 50–85. Academic Press, New York.

9. Huber, J. L. A., Huber, S. C., and Nielsen, T. H. 1989. Protein phosphorylation as a mechanism for regulation of spinach leaf sucrose-phosphate synthase activity. *Arch. Biochem. Biophys.* **270**: 681–90.

10. Huber, J. L., Hite, D. R. C., Outlaw, W. H., Jr., and Huber, S. C. 1991. Inactivation of highly activated spinach leaf sucrose-phosphate synthase by dephosphorylation. *Plant Physiol.* **95**: 291–97.

11. Huber, S. C., and Huber, J. L. 1990. Activation of sucrose-phosphate synthase from darkened spinach leaves by an endogenous protein phosphatase. *Arch. Biochem. Biophys.* **282**: 421–26.

12. Huber, S. C., and Huber, J. L. 1990. In vitro phosphorylation and inactivation of spinach leaf sucrose-phosphate synthase by an endogenous protein kinease. *Biochim. Biophys. Acta* **1091**: 393–400.

13. Huber, S. C., and Huber, J. L. 1991. Regulation of maize leaf sucrose-phosphate synthase by protein phosphorylation. *Plant Cell Physiol.* **32**: 319–26.

14. Jensen, R. G. 1980. Biochemistry of the chloroplast. In *The Biochemistry of Plants*, Vol. 1, ed. N. E. Tolbert, pp. 274–313. Academic Press, New York.

15. Kacser, H. 1987. Control of metabolism. In *The Biochemistry of Plants*, Vol. 11, ed. D. D. Davies, pp. 39–67. Academic Press, New York.

16. Kacser, H., and Burns, J. A. 1973. Control of flux. *Symp. Soc. Exp. Biol.* **27**: 65–107.

17. Knapf, D. B. 1989. The regulatory role of thioredoxin in chloroplasts. *TIBS* **14**: 433–34.

18. Kruckeberg, A. L., Neuhas, H. E., Feil, R., Gottlieb, L. D., and Stitt, M. 1989. Decreased-activity mutants of phosphoglucose isomerase in the cytosol and chloroplast of *Clarkia xantinana*. Impact on mass-action ratios and fluxes to sucrose and starch and estimation of flux control coefficients and elasticity coefficient. *Biochem. J.* **261**: 457–67.

19. Lin, T. P., Caspar, T., Somerville, C., and Preiss, J. 1988. A starch deficient mutant of *Arabidopsis thaliana* with low ADPglucose pyrophosphorylase activity lacks one of the two subunits of the enzyme. *Plant Physiol.* **88**: 1175–81.

20. Müller-Röber, B., Sonnewald, U., and Willmitzer, L. 1992. Inhibition of the ADP-glucose pyrophosphorylase in transgenic potato leads to sugar-storing tubers and influences tuber formation and expression of tuber storage genes. *EMBO J.* **11**: 1229–38.

21. Neuhaus, H. E., Kruckeberg, A. L., Feil, R., and Stitt, M. 1989. Reduced-activity mutants of phosphoglucose isomerase in the cytosol and chloroplast of *Clarkai xantiana*. II. Study of the mechanisms which regulate photosynthate partitioning. *Planta* **178**: 110–22.

22. Neuhaus, H. E., and Stitt, M. 1990. Control analysis of photosynthate partitioning: Impact of reduced activity of ADPglucose pyrophosphorylase or plastid phosphoglucomutase on the fluxes to starch and sucrose in *Arabidopsis*. *Planta* **182**: 445–54.

23. Pettersson, G., and Ryde-Pettersson, U. 1989. Metabolites controlling the rate of starch synthesis in the chloroplast of C₃ plants. *Eur. J. Biochem.* **179**: 169–72.

24. Pozueta-Romero, J., Frehner, M., Viale, A. M., and Akazawa, T. 1991. Direct transport of ADPglucose by an adenylate translocator is linked to starch biosynthesis in amyloplasts. *Proc. Natl. Acad. Sci. USA* **88**: 5769–73.

25. Preiss, J. 1984. Starch, sucrose biosynthesis and partition of carbon in plants are regulated by orthopnosphate and triose-phosphates. *TIBS* **9**: 24–27.

26. Preiss, J. 1988. Biosynthesis of starch and its regulation. In *The Biochemistry of Plants*, Vol. 14, ed. J. Preiss, pp. 181–254. Academic Press, New York.

27. Preiss, J. 1990. Biology and molecular biology of starch synthesis and its regulation. In *Oxford Survey of Plant Molecular and Cellular Biology*, Vol. 7, ed. B. Miflin, pp. 59–114. Oxford University Press, Oxford.

28. Preiss, J., and Levi, C. 1980. Starch biosynthesis and degradation. In *The Biochemistry of Plants*, Vol. 3, ed. J. Preiss, pp. 371–423. Academic Press, New York.

29. Preiss, J., and Romeo, T. 1989. Physiology, biochemistry and genetics of bacterial glycogen synthesis. In *Advances in Microbial Physiology*, Vol. 30, ed. A. H. Rose and D. W. Tempest, pp. 183–238.

30. Robinson, S. P., and Walker, D. A. 1981. Photosynthetic carbon reduction cycle. In *The Biochemistry of Plants*, Vol. 8, ed. M. D. Hatch and N. K. Boardman, pp. 194–236. Academic Press, New York.

31. Salvucci, M. E., Portis, Jr. A. R., and Ogren, W. E. 1985. A soluble class protein catalyzes ribulosebisphosphate carboxylase/oxygenase activation in vivo. *Photosynth. Res.* **7**: 193–201.

32. Stark, D. M., Timmerman, K. P., Barry, G. F., Preiss, J., and Kishore, G. M. 1992. Regulation of the amount of starch in plant tissues by ADPglucose pyrophosphorylase. *Science* **258**: 287–92.

33. Stitt, M. 1990. Fructose-2,6-bisphosphate as a regulatory molecule in plants. *Annu. Rev. Plant Physiol. Plant Mol. Biol.* **41**: 153–85.

34. Stitt, M., Huber, S., and Kerr, P. 1987. Control of photosynthetic sucrose formation. In *The Biochemistry of Plants*, Vol. 10, ed. M. D. Hatch and N. K. Boardman, pp. 328–409. Academic Press, New York.

35. Werneke, J. M., Chatfield, J. M., and Ogren, W. L. 1988. Catalysis of ribulosebisphosphate carboxylase/oxygenase activation by the product of a Rubisco activase cDNA clone expressed in *Escherichia coli*. *Plant Physiol.* **87**: 917–20.

6

The Regulation of Photosynthetic Electron Transfer and Its Role in Determining the Rate and Efficiency of Photosynthesis

DONALD R. ORT

The purpose of this brief discussion is to consider how the regulation of photosynthetic electron transfer may impact on the rate and efficiency of net photosynthesis under agricultural or natural habitats. More specifically, the assignment includes the identification of "factors" that contribute to the control of electron transfer that could be targets for genetic manipulation toward improving biomass accumulation. As such, my focus will be on outlining the concepts and general mechanisms that govern photosynthetic electron transfer. Unfortunately, this will be at the expense of tracing the origin of ideas and development of the crucial supporting data.

Traditionally, consideration of the control of photosynthetic electron transfer in higher plant chloroplasts has been dominated by discussions concerning the oxidation of plastoquinol by the cytochrome b_6f complex. This quinol-metalloprotein oxidoreductase is structurally and functionally analogous to complexes found in the photosynthetic and respiratory membranes from both prokaryotes and eukaryotes that oxidize a low-potential quinol and reduce a high-potential metalloprotein (1). As depicted in the scheme of a higher plant thyladoid membrane shown in Figure 6.1, this complex oxidizes plastoquinol generated by photosystem II and reduces plastocyanin oxidized by photosystem I. The mechanism of both plastoquinone reduction and plastoquinol oxidation, in which intermediate reductive states are maintained bound to specific catalytic sites, ensures absolute asymmetry in proton uptake in keeping with the role of transmembrane proton asymmetries in energy coupling (8). This role that plastoquinol oxidation plays in energy coupling is especially important in considering the special role that it also plays in controlling the rate and efficiency of electron transfer.

The underlying mechanism for this control can be qualitatively understood through recognizing that plastoquinol oxidation by the cytochrome b_6f complex is the only appreciably reversible step in the sequence of photosynthetic electron transfer reactions.

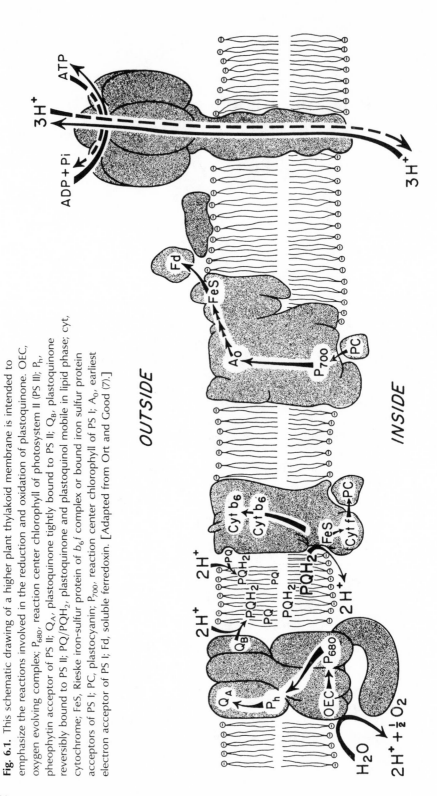

Fig. 6.1. This schematic drawing of a higher plant thylakoid membrane is intended to emphasize the reactions involved in the reduction and oxidation of plastoquinone. OEC, oxygen evolving complex; P_{680}, reaction center chlorophyll of photosystem II (PS II); P_h, pheophytin acceptor of PS II; Q_A, plastoquinone tightly bound to PS II; Q_B, plastoquinone reversibly bound to PS II; PQ/PQH₂, plastoquinone and plastoquinol mobile in lipid phase; cyt, cytochrome; FeS, Rieske iron-sulfur protein of b_6f complex or bound iron sulfur protein acceptors of PS I; PC, plastocyanin; P_{700}, reaction center chlorophyll of PS I; A_0, earliest electron acceptor of PS I; Fd, soluble ferredoxin. [Adapted from Ort and Good (7).]

104

$$PQH_2 \leftrightarrow PQ + 2H^+ + 2e^-$$

As hydrogen ions are accumulated within the thylakoid vesicle as a result of light-driven electron transfer, the equilibrium increasingly favors the reverse reaction [i.e., the formation of plastoquinol (PQH_2) by the reduction of plastoquinone (PQ) on the cytochrome b_6f complex]. Thus, as the electrochemical potential of accumulated protons becomes large, the rate of net oxidation of plastoquinol by the cyctochrome b_6f complex and therefore the rate of electron transfer becomes slower and, on an absorbed photon basis, less efficient. On thermodynamic grounds, the rate of the net reaction would be zero should the electrochemical potential of accumulated protons become equivalent to the redox free energy available in the electron transport couple involved with the proton uptake—in this case, PQ/PQH_2 and the one-electron redox components of the cytochrome b_6f complex.

Up to this point, only the role of the chemical potential of the accumulated protons has been considered in the thermodynamic control of photosynthetic electron transfer. However, for the sake of completeness, mention should be made of the control that can be exerted by the transmembrane electric field associated with proton accumulation on the rate of net plastoquinol oxidation. Above, the reaction resulting in the oxidation of plastoquinol was viewed as an electrically neutral process in which no separation of charge across the thylakoid membrane takes place. If this were actually the case, a transmembrane electrical field would not be expected to influence the rate of plastoquinol oxidation. However, the existence of an electrogenic reaction associated with the oxidation of plastoquinol by the cytochrome b_6f complex is now experimentally well established (1). A variety of models have been proposed to account for this electrogenic event, and, although the alternative models differ in fundamental aspects of their mechanism, they share both the general name Q-cycle and the feature that one of the electrons released from plastoquinol during its oxidation is shuttled back across the membrane establishing a separation of charges (Fig. 6.1). The existence of an electrogenic process in the rate-limiting reaction of the chloroplast electron-transfer sequence predicts that a sufficiently large membrane potential (positive inside) would slow the rate of plastoquinol oxidation and, consequently, the rate of electron flux through the chain. Indeed, control of the rate of electron transfer by the electrical component of the accumulated protons has been clearly demonstrated (6).

My purpose in reviewing what might be termed the "thermodynamic control" of photosynthetic electron transfer is to illustrate that this occurs as an inevitable energetic consequence of plastoquinol oxidation by the cyctochrome b_6f complex. As such it does not qualify as a target for genetic manipulation that would hope to enhance photosynthetic rate and efficiency and thereby increase biomass production. Only a few years ago, this discussion would now conclude with the idea that the regulation of electron transfer and the efficiency of light utilization are not candidates to be considered in discussions focused on improved photosynthetic performance and productivity. However, as a result of the increasingly more sophisticated investigations that have been done in recent years on intact plants and on plants in natural habitats, evidence for a previously unrecognized

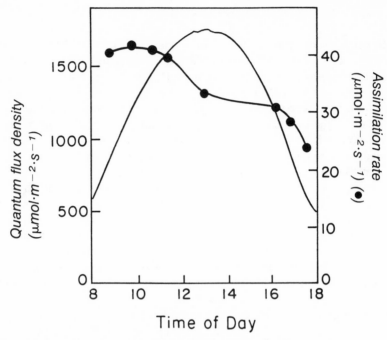

Fig. 6.2. Profile of irradiance and net photosynthesis of attached sunflower leaves as a function of the time of day. These data are typical of those observed for irrigated sunflower plants on sunny days during the seed-filling stage. [Modified from Wise et al. (11).]

regulation of light energy utilization has emerged. Recent data from several laboratories provide strong experimental evidence for the existence of intricate mechanisms able to down-regulate photosynthetic efficiency as illumination levels exceed photosynthetic capacity (e.g., Refs. 4, 7, 10). Although my laboratory was by no means the first to begin thinking along these lines, I would like to illustrate the phenomenon of *down-regulation* of photosynthesis efficiency with results that we obtained while studying the response of photosynthesis to the daily cycling of leaf water potential in field-grown sunflower.

On clear days, mid-afternoon decreases in leaf water potential of about 1.5 mega-Pascals (MPa) and in net photosynthesis of up to 50% were typical for irrigated sunflower plants during seed filling. For instance, Figure 6.2 shows the typical diurnal pattern of photosynthesis that we observed. The highest rate of photosynthesis for the day was reached in the mid-morning, and thereafter the rate of net photosynthesis declined even though irradiance increased during mid-morning and into the early afternoon. At 9 h, the irradiance was near 900 μmol·m²·s⁻¹ (Fig. 6.2), and the leaf water potential was about -0.6 MPa. By late afternoon (17 h), irradiance was once again near 900 μmol·m⁻²·s⁻¹ but the leaf water potential had declined to -1.5 MPa and the rate of photosynthesis was less than 70% of the 9-h value. During the decline of net photosynthesis over the course of a cloudless day (Fig. 6.2), a strong and nearly linear relationship was maintained between stomatal conductance and photosynthesis (Fig. 6.3). Since

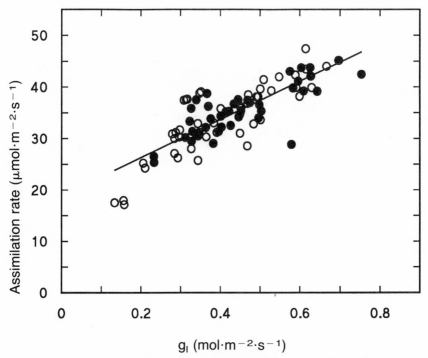

Fig. 6.3. Relationship between leaf conductance (g_l) and light-saturated assimilation rate measured at atmospheric CO_2 levels in irrigated (●) and droughted (○) field-grown sunflower plants. There is a positive and significant correlation between g_l and the rate of light-saturated assimilation ($r = .817$, $P < .0001$).

direct control by stomatal aperture over the rate of photosynthesis is exerted through the control that stomata have on intercellular CO_2 concentration (C_i), the lower C_i values observed in the afternoon when the stomatal conductance was lowest were anticipated (Fig. 6.4). However, further analysis showed that there was also a strong positive correlation between C_i and leaf conductance (g_l), indicating that the actual response of assimilation to C_i is obscured in Figure 6.4 by the large changes in g_l. We extended the analyses using multiple linear regression in order to try to separate the effects of conductance and intercellular CO_2. Simple correlation analysis showed that g_l and C_i were the only variables correlated with significant variation in light-saturated assimilation; that is, there were no otherwise hidden correlations between assimilation and vapor pressure deficit, time of day, or leaf temperature. Thus, multiple regression procedures could be used to factor out the effect of conductance and reveal the underlying response of assimilation to C_i. The results can be visualized using a partial residual plot (3), as shown in Figure 6.5. In this plot, with the contaminating effects of conductance removed, the response of photosynthesis to C_i is unambiguously negative ($r = -.82$, $P < .0001$).

To begin to understand the basis for the inverse response of assimilation to C_i shown in Figure 6.5, we looked at what was happening to the intrinsic efficiency of photosynthesis by measuring the light-limited quantum yield of CO_2 assimilation

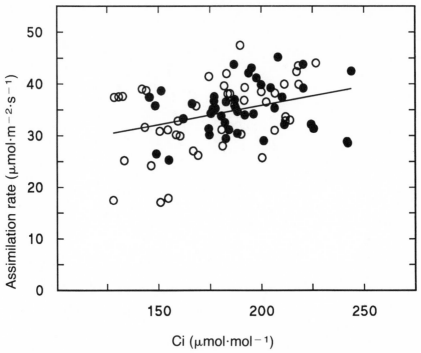

Fig. 6.4. Light-saturated assimilation versus intercellular CO_2 concentration (C_i) measured at atmospheric CO_2 levels in field-grown sunflower plants. The relationship between the rate of light-saturated assimilation and C_i shows a positive and significant correlation ($r = .336$, $P = .0012$). No statistically significant difference was found between the data from irrigated (●) and droughted (○) plants.

over the course of the day in field-grown sunflower. Figure 6.6 shows a declining quantum efficiency that qualitatively mimics the decline in the rate of net photosynthesis over the course of the day. In this case we cannot say that these are typical data, as several replicas of this experiment gave somewhat different results. It is likely these differences reflect procedural aspects of the measurement because, as others have reported (5), this depression of the quantum yield induced by high light reverses quite rapidly at the low light intensities used to make quantum yield determinations. Furthermore, the half-time of this reversal, which may be as short as several minutes, is influenced to an unknown degree by factors such as the intensity and duration of light (including, therefore, time of day) and possibly by leaf water status and conductance. In fact, the rapidly reversing character of this effect of high light intensity on the quantum efficiency of CO_2 assimilation is very pertinent to this discussion. Seemingly comparable experiments in which growth-chamber-grown plants are exposed to higher than growth-chamber light intensities are often observed to respond with marked depression in quantum yield that can be traced to light-dependent damage to photosystem II. Indeed, light-dependent damage to the photosynthetic apparatus has become to be known as *photoinhibition*. While the light-dependent decline in photosynthetic efficiency seen in Figure 6.6 might also be termed photoinhibition, its rapid reversal

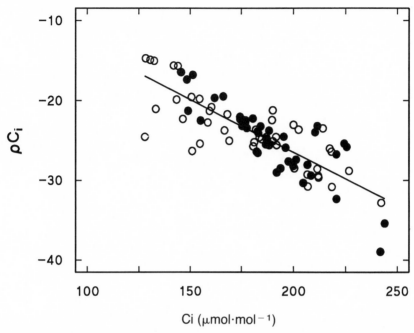

Fig. 6.5. A partial residual plot showing the response of light-saturated assimilation to C_i after the removal of the conductance response by multiple linear analysis. The partial residual ρ_{C_i} = residual + βC_i, where β is the regression coefficient for C_i, and the residual is the difference between the predicted assimilation rate and the actual assimilation rate.

suggests, and a broad range of experiments done in other laboratories demonstrates (5), that this inhibition does not involve damage but is instead regulatory in nature.

Currently there is a good deal of activity and a similar level of controversy concerning the mechanism of this down-regulatory effect on photosynthetic efficiency. Most of the ideas presently under discussion involve radiationless deexcitation of antennae chlorophyll as a central feature of the mechanism; that is, fewer of the photons that are absorbed into the antennae chlorophyll arrays are transferred to the reaction centers where the photochemistry that leads to electron transfer occurs. It is postulated that large transmembrane pH differences, that are associated with high irradiance levels (which would be particularly large if the rate of ATP utilization in carbon dioxide assimilation were less than the capacity of the thylakoid membranes to generate ATP), induce a special, but wholly undefined state in the antennae complexes, with an enhanced probability of radiationless deactivation competing with excitation energy transfer to the reaction center. The enhanced probability of radiationless deactivation would also compete with radiative deactivation—that is, with fluorescence. Although this competition is very important from the standpoint of utilizing chlorophyll fluorescence to monitor down-regulation, it is negligible in energy terms since such a small proportion of the absorbed photons are reemitted as light.

Although at this time vanishingly little is known about the mechanistic basis, it seems quite clear that chloroplasts have the capability to down-regulate the rate and efficiency of net photosynthesis. It seems equally clear that significantly lower

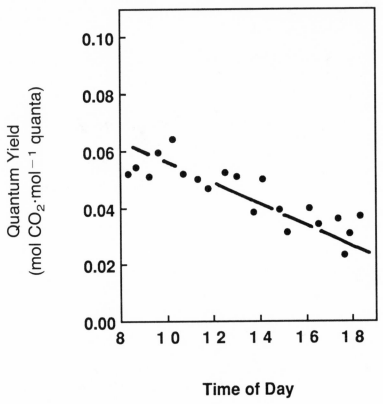

Time of Day

Fig. 6.6. The light-limited quantum yield of photosynthesis measured on leaf discs punched from field-grown sunflower at different times of day. (From unpublished data of Dr. Robert Wise and Denise Sparrow.)

rates and efficiencies associated with down-regulation are commonplace under agricultural conditions (Refs. 2, 9, and references therein). More difficult is the question of whether this down-regulatory mechanism represents an opportunity to improve the photosynthetic performance of plants. With our present state of knowledge we are convinced only by our intuition that down-regulation is an important protective mechanism when the photon flux exceeds photosynthetic capacity—as it may have in our field sunflowers in the afternoon when irradiances were high and stomata were partially closed to avoid further water loss. If this general line of thinking is correct, then it may be that the mechanism of diverting more photons away from photochemistry than is required to avoid damage to the photosynthetic apparatus is too conservative for some situations. This is possibly the case when C_i levels were observed to actually increase as photosynthesis declined over the course of the day (e.g., Refs. 2, 12) perhaps because down-regulation of light utilization overcompensated for the decrease in stomatal conductance. My predilection is that, although there may be some opportunities in the future for high-input agriculture, it seems that season-long biomass production in forests and other natural stands of vegetation will ultimately be limited by mineral and water resources, and will thus be comparatively insensitive to any reduction in the daily efficiency of photosynthesis that may occur at high irradiance due to down-regulation.

REFERENCES

1. Cramer, W. A., Black, M. T., Widger, W. R., and Girvin, M. E. 1987. Comparative structure and function of the b cytochromes of the $b-c_1$ and b_6-f complexes. In *The Light Reactions*, ed. J. Barber, pp. 446–93. Elsevier Science Publishing, Amsterdam.

2. Cheeseman, J. M., Clough, B. F., Carter, D. R., Lovelock, C. E., Eong, O. J., and Sim, R. G. 1991. The analysis of photosynthetic performance in leaves under field conditions: A case study using *Bruguiera* mangroves. *Photosyn. Res.* **29**: 11–22.

3. Cheeseman, J. M., and Wickens, L. K. 1986. Control of Na^+ and K^+ transport in *Spergularia marina* ii. Effects of plant size, tissue ion content and root–shoot ratio at moderate salinity. *Physiol. Plant.* **67**: 7–14.

4. Demmig, B., Winter, K., Kruger, A., and Czyban, F.-C. 1987. Photoinhibition and zeaxanthin formation in intact leaves. *Plant Physiol.* **84**: 218–24.

5. Foyer, C., Furbank, R., Harbinson, J., and Horton, P. 1990. The mechanisms contributing to photosynthetic control of electron transport by carbon assimilation in leaves. *Photosyn. Res.* **25**: 83–100.

6. Graan, T., and Ort, D. R. 1983. Initial events in the regulation of electron transfer in chloroplasts: The role of the membrane potential. *J. Biol. Chem.* **258**: 2831–6.

7. Harbinson, J., Genty, B., and Baker, N. R. 1989. Relationship between the quantum efficiencies of photosystems I and II in pea leaves. *Plant Physiol.* **90**: 1029–34.

8. Ort, D. R. 1986. Energy transduction in oxygenic photosynthesis: An overview of structure and mechanism. In *Photosynthetic Membranes and Light Harvesting Systems*, ed. L. A. Staehelin and C. J. Arntzen, pp. 143–96. Springer-Verlag, New York.

9. Ort, D. R., and Good, N. E. 1989. Textbooks ignore photosystem II-dependent ATP formation: Is the Z scheme to blame? *Trends Biochem. Sci.* **13**: 467–9.

10. Weis, E., and Berry, J. 1987. Quantum efficiency of photosystem II in relation to 'energy'-dependent quenching of chlorophyll fluorescence. *Biochim. Biophys. Acta* **894**: 198–208.

11. Wise, R. R., Frederick, J. R., Alm, D. M., Kramer, D. M., Hesketh, J. D., Crofts, A. R., and Ort, D. R. 1990. Investigation of the limitations to photosynthesis induced by leaf water deficit in field-grown sunflower (*Helianthus annuus* L.). *Plant Cell Environ.* **13**: 923–31.

12. Wise, R. R., Sparrow, D. H., Ortiz-Lopez, A., and Ort, D. R. 1991. Biochemical regulation during the mid-day decline of photosynthesis in field-grown sunflower. *Plant Sci.* **74**: 45–52.

III

PHOTORESPIRATION AND RESPIRATION

7

Energy Utilization by Photorespiration

WILLIAM L. OGREN

Photosynthetic efficiency is considerably reduced by the competing process of photorespiration. The interaction of photosynthesis and photorespiration is mediated by the bifunctional enzyme ribulose bisphosphate carboxylase/oxygenase (Rubisco), with CO_2 and O_2 being mutually competitive substrates (9). The stiochiometry of the two reactions is determined by the kinetic properties of Rubisco and varies significantly among the broad taxonomic groups: C_3 higher plants, green algae, cyanobacteria, and photosynthetic bacteria (4, 5). The stoichiometries of the photosynthetic carbon reduction (PCR) and photorespiratory cycles have also been determined (2, 16); thus the energy required to reduce CO_2 to sugar can be easily determined for various atmospheric conditions.

Since the rate-limiting step in photorespiration is the oxygenase activity of Rubisco (9), reduced photorespiration and thus reduced energy waste require that the kinetic properties of Rubisco be altered so as to favor the carboxylase and/or reduce the oxygenase activity (11, 12). Again, knowledge of the energy stoichiometry of photosynthesis and photorespiration permits the calculation of how much photorespiratory energy might be conserved if the kinetic properties of Rubisco could be appropriately altered.

The rising atmospheric CO_2 concentration is reducing photorespiratory activity in plants to a significant extent. The impact of reduced photorespiration on the rate and extent of atmospheric change, due to increased plant growth and carbon accumulation, is not as readily determined as the effect of rising CO_2 on energy loss due to photorespiration. Exposure of crop plants to supraphysiological CO_2 concentrations often leads to short-term increases in photosynthesis followed by an accumulation of starch and reduced positive response to increased CO_2 (7, 8). The lack of a sustained photosynthetic response raises the question of the relevance of high CO_2 experiments on plants adapted to low CO_2, and the accuracy of global CO_2 models that might incorporate such plant responses.

THE ENERGY COST OF PHOTORESPIRATION

From the classical PCR cycle (2, Fig. 7.1), the carboxylation of RuBP and reduction of 1 mole CO_2 to carbohydrate requires 3 moles ATP and 2 moles

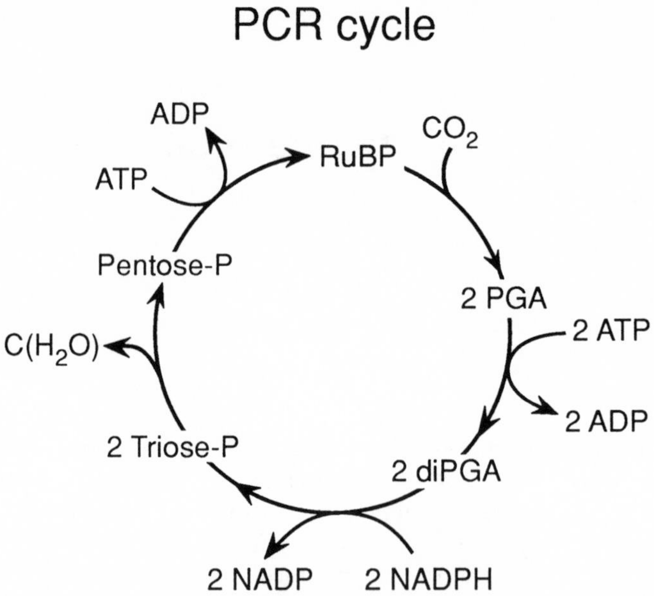

PCR cycle

$$RuBP + CO_2 + 3 ATP + 2 NADPH \longrightarrow C(H_2O) + RuBP$$

Fig. 7.1. Energetics of the photosynthetic carbon reduction (PCR) cycle.

NADPH. Since the energy in 1 reduced pyridine nucleotide is equivalent to that in 3 ATP, photosynthetic reduction of 1 mole CO_2 requires the equivalent of 9 moles ATP. In photorespiration, RuBP is both carboxylated and oxygenated, with the ratio of the two activities determined by the kinetic properties of Rubisco, by the concentrations of CO_2 and O_2, and by the temperature (6, 9). The relative rates of RuBP carboxylation and oxygenation, and thus the relative rates of photosynthesis and photorespiration, are given by Equation 7.1 (9):

$$\frac{v_c}{v_o} = \frac{V_c K_o [C]}{V_o K_c [O]} \tag{7.1}$$

where V and K represent the Rubisco V_{max} and K_m for the substrates CO_2 (C) and O_2 (O). Thus the carboxylation/oxygenation ratio is proportional to $V_c K_o / V_o K_c$, which is the *Rubisco specificity factor* (5), and the $[CO_2]/[O_2]$ molar ratio. In the current atmosphere (340 ppm CO_2, 21% O_2) at 25°C, the ratio of carboxylation to oxygenation is about 4:1 (9, 11).

The products of RuBP oxygenase activity are PGA, which directly enters the PCR cycle, and P-glycolate (abbreviated P-GLL in figures). P-glycolate is hydrolyzed in the chloroplast, then excreted and metabolized in a series of reactions in the C_2 cycle or glycolate pathway in the peroxisomes and mitochondria (Ref. 16, Fig. 7.2). Photorespiratory CO_2 is evolved in this series of reactions along with an equimolar amount of NH_3. The final product of this metabolism is glycerate

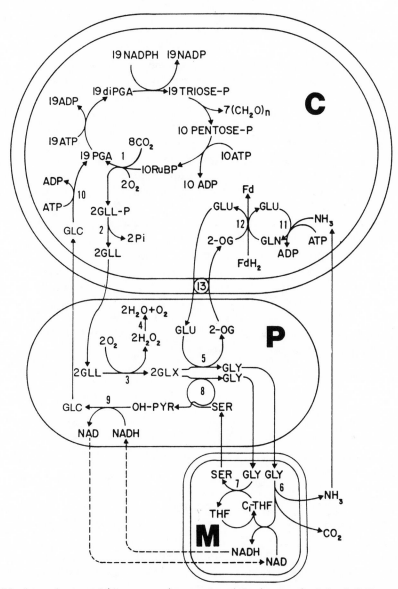

Fig. 7.2. Approximate stoichiometry and energetics of C_3 photosynthesis in air (340 ppm CO_2, 21% O_2) at 25°C. Reactions occur in the chloroplast (C), peroxisome (P), and mitochondrion (M).

(GLC), which enters the chloroplast and is phosphorylated to PGA, which then reenters the PCR cycle. The significant extra energy cost associated with this pathway arises from the release of NH_3. Reassimilation of 1 mole NH_3 into glutamate, the amino donor for the conversion of glyoxylate to glycine in the pathway, requires 1 mole each of reduced ferredoxin (equivalent to 1 NADPH, or 3 ATP) and ATP. Glycine oxidation, the source of photorespiratory CO_2 and NH_3, provides 1 mole NADH from NAD, but an equimolar amount of NADH is required at a later step in the pathway, the reduction of hydroxypyruvate to glcycerate (Fig. 2).

The approximate stoichiometry of photosynthesis/photorespiration in air (350 ppm CO_2, 21% O_2) at 25°C as well as the energy required under these conditions, if shown in Figure 7.2. To carboxylate/oxygenate and regenerate 10 moles RuBP requires 31 ATP, 19 NADPH, and 1 reduced ferredoxin. From Figure 7.1, carboxylation and regeneration of 10 moles RuBP in the absence of photorespiration requires 30 ATP and 20 NADPH. Since NADPH and reduced ferredoxin are energetically equivalent, it is evident that the energy required to turn over the PCR cycle in the presence or absence of photorespiration is virtually the same. The only difference in energy consumption is that for every 10 turns of the cycle, one additional ATP is required when photorespiration is occurring. Thus photorespiration does not dissipate any more energy than does photosynthesis, and does not function to protect the leaf from excess energy as has been suggested elsewhere (16).

While photorespiration does not require a significant amount of additional energy to regenerate RuBP, the net assimilation of CO_2 per unit energy is greatly reduced by the substitution of O_2 for CO_2 and the release of photorespiratory CO_2. The energy lost can be determined by calculating the number of ATP equivalents per unit of reduced carbohydrate. From Figures 7.1 and 7.2, the stoichiometry of C_3 photosynthesis in ATP equivalents in the absence (Eq. 7.2) and presence (Eq. 7.3) of photorespiration is:

$$10 \text{ RuBP} + 10 \text{ } CO_2 + 90 \text{ ATP} \rightarrow 10 \text{ } C(H_2O) + 10 \text{ RuBP} \qquad (7.2)$$

$$10 \text{ RuBP} + 8 \text{ } CO_2 + 2 \text{ } O_2 + 91 \text{ ATP} \rightarrow 7 \text{ } C(H_2O) + 1 \text{ } CO_2 + 10 \text{ RuBP} \qquad (7.3)$$

Factoring Equation 7.3 gives 10 $C(H_2O)$ fixed:

$$14.3 \text{ RuBP} + 11.4 \text{ } CO_2 + 2.8 \text{ } O_2 + 130 \text{ ATP}$$

$$\rightarrow 10 \text{ } C(H_2O) + 1.4 \text{ } CO_2 + 14.3 \text{ RuBP} \qquad (7.4)$$

Thus, reducing 10 moles CO_2 to carbohydrate requires 90 moles ATP without photorespiration and 130 moles ATP with photorespiration. The energy wasted by photorespiration is thus 31%. To determine the effect of altered CO_2 concentration on the amount of energy wasted by photorespiration, one can perform similar analyses. For example, when the atmospheric CO_2 concentration was a low as 250 ppm, the stoichiometry of carboxylation to oxygenation was about 3:1. The number of ATP equivalents needed to reduce 1 mole CO_2 is 14.6, compared to 9.0 in the absence of photorespiration (Fig. 7.3). Thus, at 250 ppm CO_2, 21% O_2, and 25°C, the energy wasted by photorespiration is 38%. If the CO_2 concentration were to reach 1000 ppm CO_2, the carboxylation/oxygenation ratio would increase to about 12:1, and the amount of energy wasted would fall to 12% (Fig. 7.4). The dependence of energy lost by photorespiration on atmospheric CO_2 concentration is summarized in Figure 7.5. These values are altered by environmental variables such as temperature, which alters the carboxylation/oxygenation ratio (6, 9). Drought stress reduces stomatal conductance, and this increases photorespiratory energy loss by reducing the leaf internal CO_2 concentration.

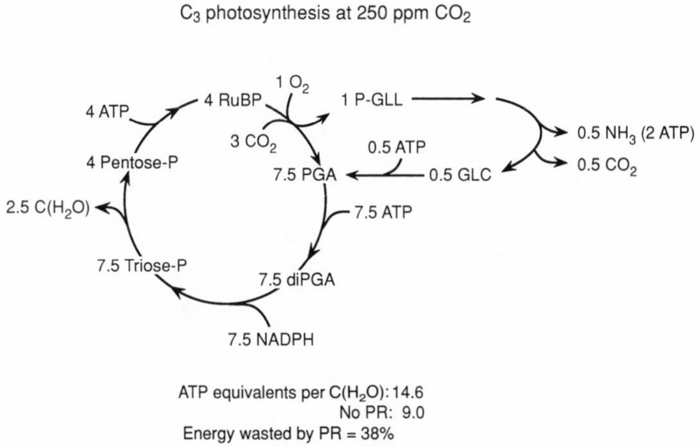

C$_3$ photosynthesis at 250 ppm CO$_2$

ATP equivalents per C(H$_2$O): 14.6
No PR: 9.0
Energy wasted by PR = 38%

Fig. 7.3. Stoichiometry of C$_3$ photosynthesis and energy use in 250 ppm CO$_2$, 21% O$_2$, at 25°C.

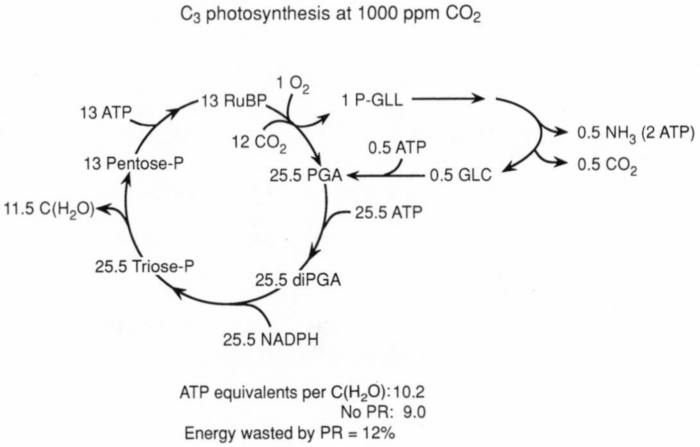

C$_3$ photosynthesis at 1000 ppm CO$_2$

ATP equivalents per C(H$_2$O): 10.2
No PR: 9.0
Energy wasted by PR = 12%

Fig. 7.4. Stoichiometry of C$_3$ photosynthesis and energy use in 1000 ppm CO$_2$, 21% O$_2$, at 25°C.

RUBISCO SPECIFICITY AND ENERGY LOSS BY PHOTORESPIRATION

From the carboxylation/oxygenation ratio as given in Equation 7.1 the specificy factor, $V_c K_o/V_o K_c$, is shown to regulate photorespiratory energy loss. Although photorespiration was considered to be inevitable (1, 10), substantial differences in Rubisco specificity were subsequently found (4). The Rubisco specificity values ranged from 9 in the photosynthetic bacterium *Rhodospirillum rubrum,* to about 80 for all C$_4$ and most C$_4$ plants examined, with intermediate values of about 48 for cyanobacteria and 62 for green algae (4, 5). The Rubisco specificity in C$_3$ plants, as outlined in the previous section and shown in Fig. 7.6, gives a carboxylase: oxygenase ratio of 4:1. The specificity factor also determines the *CO$_2$ compensation point*—that CO$_2$ concentration where the rate of CO$_2$ fixation by phyotsynthesis

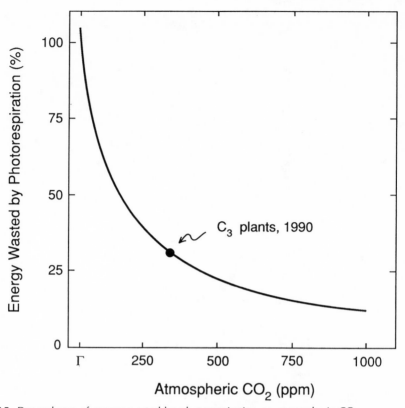

Fig. 7.5. Dependence of energy wasted by photorespiration on atmospheric CO_2.

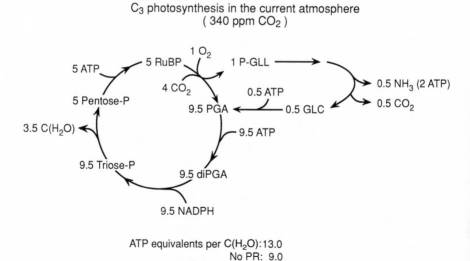

Fig. 7.6. Stoichiometry of C_3 photosynthesis and energy use in the current atmosphere (340 ppm CO_2, 21% O_2) at 25°C.

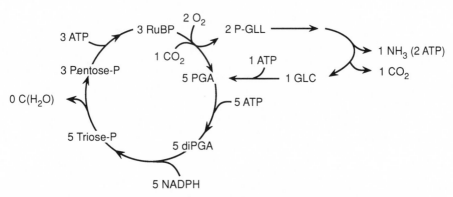

Fig. 7.7. Stoichiometry of C_3 photosynthesis and photorespiration at the CO_2 compensation point (25°C).

is equal to the rate of CO_2 evolved by photorespiration. In C_3 plants at 25°C in 21% O_2, the CO_2 compensation point is about 40 ppm, where the carboxylation/oxygenation ratio is 1:2 (Fig. 7.7). The 2 moles of P-glycolate synthesized by oxygenase activity leads to the release of 1 mol CO_2, which is refixed for a net CO_2 exchange rate of zero.

The specificity of *R. rubrum* Rubisco is 9. If this enzyme were present in the chloroplast of a C_3 plant, and the plant placed in air consisting of 340 ppm CO_2 and 21% O_2 at 25°C, the carboxylation/oxygenation ratio would be about 0.45:1 (Fig. 7.8). At this ratio, the rate of CO_2 evolution would exceed the rate of CO_2

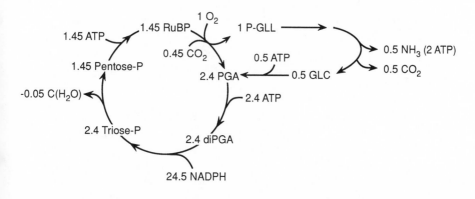

Fig. 7.8. Stoichiometry of photosynthesis in the current atmosphere (340 ppm CO_2, 21% O_2) at 25°C if higher plant Rubisco is replaced by *R. rubrum* Rubisco in the chloroplast.

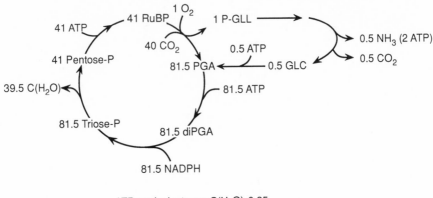

C₃ photosynthesis in the current atmosphere (340 ppm CO₂)
if another order-of-magnitude increase in Rubisco specificity

ATP equivalents per C(H₂O): 9.35
No PR: 9.0
Energy wasted by PR = 4%
Γ = 4 ppm CO₂

Fig. 7.9. Stoichiometry of photosynthesis and energy use in the current atmosphere (340 ppm CO_2, 21% O_2) at 25°C, if the CO_2/O_2 specificity of higher plant Rubisco is increased by an order of magnitude.

fixation, and thus the net CO_2 uptake would be negative and the plant would not survive. Thus the change in Rubisco specificity from 9 to 80, as has occurred in the evolution of photosynthetic organisms from photosynthetic bacteria to higher plants (4), has meant the difference between highly productive plants and plants that would not survive.

After it was determined that the interaction between photosynthesis and photorespiration was mediated by Rubisco (9), it was quickly realized that photorespiration could be reduced by appropriate alteration of the kinetic properties (11). If, for example, the Rubisco specificity were increased by another order of magnitude to support a carboxylation/oxygenation ratio of 40:1, the energy wasted by photorespiration would be reduced to a negligible 4% (Fig. 7.9). The CO_2 compensation point in such a plant would be about 4 ppm CO_2.

The evolution of Rubisco kinetics from an enzyme such as that found in *R. rubrum* to the C₃ higher plant enzyme, when compared at current atmospheric conditions, reduces photorespiratory energy loss from somewhat more than 100% to about 30% (Fig. 7.10). The intriguing question now is whether additional increases in specificity can be obtained. In one instance, the specificity of *Chlamydomonas reinhardtii* Rubisco was decreased and then increased by induced mutagenesis (3), demonstrating that a favorable change in Rubisco specificity might be obtained if the right mutation or series of mutations could be introduced into the enzyme.

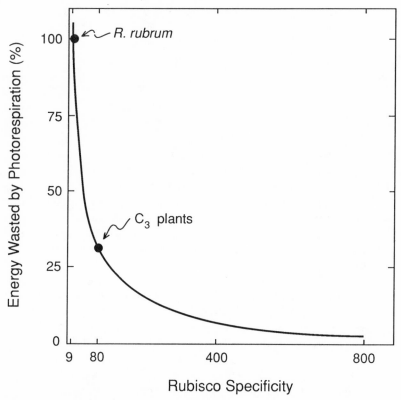

Fig. 7.10. Energy wasted by photorespiration at 340 ppm CO_2, 21% O_2, and 25°C as a function of Rubisco specificity.

CAN CONTEMPORARY PLANTS EFFICIENTLY UTILIZE MORE ENERGY?

Reduced loss of energy in photorespiration, accomplished either by an increase in Rubisco specificity or by the rising CO_2 concentration, increases the net carbon fixed by photosynthesis and available to the plant. It appears that many, if not most species, as presently constituted, cannot efficiently use much additional reduced carbon. In an extensive survey of the literature, Kimball (7) concluded that a doubling of atmospheric CO_2 concentration would lead to an average yield increase of about 33%. The plant response is highly species and environment specific, however; a wide range of responses in regard to both photosynthesis rate and growth were found in different plants (8).

A standard enzyme equation describes Rubisco carboxylation rate in the presence of the competitive inhibitor O_2. Namely:

$$v_c = \frac{V_c[C]}{[C] + K_c(1 + [O]/K_o)} \tag{7.5}$$

The increased rate of carboxylation at 680 ppm CO_2, compared to 340 ppm CO_2, should be capable of sustaining an increased rate of growth of about 50%, based on K_c and K_o values of 11 and 500 μM for CO_2 and O_2, respectively (6). Additionally, the reduction in photorespiration at 680 ppm CO_2 should lead to an additional 20% increase in net photosynthesis. Thus the photosynthesis rate of a C_3 plant at 680 ppm CO_2 should be about 70% greater than at 340 ppm.

There are at least two reasons why the plant does not realize the maximum benefit of increased CO_2. First, CO_2 enrichment leads to increased leaf starch accumulation (8). This is a poor "investment" decision by the plant, since starch is not capable of further increasing photoshnthetic capacity of the plant. If the starch were invested in more plant material, specifically during canopy development, the plant would increase photosynthetic area more rapidly and thus grow faster. The second reason photosynthesis input is not maximal at high CO_2 is that the plant recognizes, in an undetermined manner, that it not capable of properly utilizing all the additional carbon; thus RuBP carboxylase activity is reduced by deactivation of the enzyme (14). A similar deactivation response is observed when other environmental conditions occur that limit photosynthetic activity, such as reduced light intensity (13, 15).

These observations indicate that contemporary plants are not adapted to use CO_2 efficiently at double or triple the current atmospheric concentration. There is little question that, because the annual increase in CO_2 is relatively small, crop breeders and natural ecosystems will select those plants with development or investment strategies that optimize yield or are most competitive. However, plants that are adapted most efficiently to utilize twice or more CO_2 than is present in today's atmosphere do not presently exist, and thus we cannot accurately predict how tomorrow's plants will interact and alter the CO_2 content of tomorrow's atmosphere. If the answer to this question is truly important, it will be necessary to establish one or more regional or national growth facilities that will allow the selection of plants adapted to high CO_2 from large populations, using the standard techniques of crop breeding. Such a facility, which would comprise several acres under glass to permit control of temperature and atmospheric composition, would also be useful in compressing and evaluating the evolutionary time scale of natural ecosystems.

REFERENCES

1. Andrews, T. J., and Lorimer, G. H. 1978. Photorespiration—still unavoidable? *FEBS Lett.* **90**: 1–9.

2. Bassham, J. A., and Calvin, M. 1957. *The Path of Carbon in Photosynthesis.* Prentice-Hall, Englewood Cliffs, N.J.

3. Chen, X., and Spreitzer, R. J. 1989. Chloroplast intragenic suppression enhances the low CO_2/O_2 specificity of mutant ribulose-bisphosphate carboxylase/oxygenase. *J. Biol. Chem.* **264**: 3051–3.

4. Jordan, D. B., and Ogren, W. L. 1981. Species variation in the specificity of ribulose bisphosphate carboxylase/oxygenase. *Nature* **291**: 514–15.

5. Jordan, D. B., and Ogren, W. L. 1983. Species variation in kinetic properties of ribulose 1,5-bisphosophate carboxylase/oxygenase. *Arch. Biochem. Biophys.* **227**: 425–33.

6. Jordan, D. B., and Ogren, W. L. 1984. The CO_2/O_2 specificity of ribulose 1,5-bisphosphate carboxylase/oxygenase. *Planta* **161**: 308–13.

7. Kimball, B. A. 1983. Carbon dioxide and agricultural yield: An assemblage and analysis of 430 prior observatkons. *Agron. J.* **75**: 779–88.

8. Kramer, P. J. 1981. Carbon dioxide concentration, photosynthesis, and dry matter production. *BioScience* **31**: 29–33.

9. Laing, W. A., Ogren, W. L., and Hageman, R. H. 1974. Regulation of soybean net photosynthetic CO_2 fixation by the interaction of CO_2, O_2, and ribulose-1,5-diphosphate carboxylase. *Plant Physiol.* **54**: 678–85.

10. Lorimer, G. H., and Andrews, T. J. 1973. Plant photorespiration—an inevitable consequence of the existence of atmospheric oxygen. *Nature* **243**: 359–60.

11. Ogren, W. L. 1975. Control of photorespiration in soybean and maize. In *Environmental and Biological Control of Photosynthesis*, ed. R. Marcelle, pp. 45–52. Dr W. Junk b.v., The Hague.

12. Ogren, W. L. 1984. Photorespiration: Pathways, regulation, and modification. *Annu. Rev. Plant Physiol.* **35**: 415–42.

13. Perchorowicz, J. T., Raynes, D. A., and Jensen, R. G. 1981. Light limitation of photosynthesis and activation of ribulose bisphosphate carboxylase in wheat seedlings. *Proc. Natl. Acad. Sci. USA* **78**: 2985–89.

14. Sage, R., Sharkey, T. D., and Seemann, J. R. 1989. Acclimation of photosynthesis to elevated CO_2 in five C_3 species. *Plant Physiol.* **89**: 590–6.

15. Salvucci, M. E., Portis, A. R., Jr., and Ogren, W. L. 1986. Light and CO_2 response to ribulose-1,5-bisphosphate carboxylase/oxygenase in *Arabidopsis* leaves. *Plant Physiol.* **80**: 655–9.

16. Tolbert, N. E. 1980. Photorespiration. In *The Biochemistry of Plants*, ed. D. D. Davies, Vol. 2, pp. 487–523. Academic Press, New York.

8

C₂ Cycle in Green Algae and Algal Peroxisomes

HELMUT STABENAU

Algae possess ribulose-1,5-bisphosphate carboxylase/oxygenase with properties for synthesis of P-glycolate as in higher plants (16, 22, 29). They also contain P-glycolate phosphatase (15, 17, 24). Thus the pathway for the formation of glycolate is known and has been reported in the literature. The biosynthesis of glycolate can easily be demonstrated because algae are capable of specifically excreting this compound into their medium (5, 21, 23, 46, 48, 49). The amount of glycolate excreted differs with the kind of alga used, as reported in this chapter. It also depends on several external factors. High rates of glycolate excretion are obtained by aeration with high oxygen, by low CO_2 levels, and by use of high light intensities (31). Excretion of glycolate has been regarded as an indication that the conversion of this acid in algae is limited by a lack of, or at least an insufficiency of, specific enzyme activities. However, results are also available suggesting the presence of glycolate metabolism in algae.

GROWTH OF ALGAE AND EXCRETION OF GLYCOLATE AND GLYOXYLATE

Algae were grown photoautographically in continuous light as described previously (39, 40).

Algal cells were pelleted by centrifugation at $4000 \times g$ for 5 min. The resulting supernatant was acidified with H_2SO_4 (final concentration: 1.5 mM) and passed through a filter (pore size: 20 μm; Satorius, FRG). The eluate was analyzed by HPLC using an ion-exclusion column (Biorad, FRG). Separation of organic acids was performed with 1.3 mM H_2SO_4 at a temperature of 10°C and a flow rate of $0.6 \ ml \cdot min^{-1}$. Organic acids were detected by UV absorption at 210 nm.

PREPARATION OF CELL HOMOGENATES

Algal cells were pelleted from the nutrient medium by low-speed centrifugation and washed twice with distilled water. The resulting pellet was suspended in 10 ml grinding medium, containing 0.15 M Tricine buffer at pH 8, 15 mg KCl, 4 mg $MgCl_2$, 7.5 mg EDTA, and 100 mg Polyclar AT. Cells were homogenized with a Potter–Elvehjem homogenizer (*Eremosphaera*) or a Virtis homogenized (*Mougeotia*). Proteins were precipitated with ammonium sulfate (70% of saturation) and resuspended in grinding medium. Enzyme activity was assayed as described elsewhere (17, 51).

    Enteromorpha was collected from the German Wadden sea. Cells were washed with 3% NaCl. Cell homogenization, sucrose density centrifugation, and measurement of enzyme activities were performed as described previously (40).

RESULTS

Glycolate Oxidase and Glycolate Dehydrogenase: Two Different Enzymes for Oxidation of Glycolate in Algae

In higher plants oxidation of glycolate is catalayzed by a peroxisomal glycolate oxidase that uses oxygen as a direct electron acceptor, thereby forming H_2O_2. According to the literature, the same enzyme is present in several algae of different taxonomic groups, irrespective of whether they are unicellular or multicellular organisms (1, 45). Glycolate oxidase from *Mougeotia* has been isolated and characterized. Like the higher plant enzyme, it also oxidizes L-lactate and not D-lactate, but glycolate is the favored substrate (37).

    With regard to the group of green algae, the occurrence of glycolate oxidase is restricted to the Charophyceae, exclusively (6). Apparently all green algae other than Charophyceae contain a mitochondrial glycolate dehydrogenase that does not use oxygen as a direct electron acceptor that, in contrast to the glycolate oxidase, is tightly bound to membranes (4, 11, 24).

    The artificial electron acceptor dichlorophenolindophenol (DCPIP) is used mainly for testing the activity of the glycolate dehydrogenase. In vivo electrons may be transferred to a redox system of the respiratory chain which probably is a cytochrome (2, 26, 52). Then, in a final step, oxygen is also reduced, but to water and not to H_2O_2 (47). D-Lactate is an alternate substrate for the glycolate dehydrogenase, and the activity is generally higher with D-lactate than with glycolate (11).

    A preference for D-lactate over L-lactate has been regarded as a criterion for the presence of glycolate dehydrogenase in algae (11, 47). Inhibition by cyanide has also been considered a characteristic of glycolate dehydrobenase, while glycolate oxidase is not inhibited by cyanide (46). However, both criteria, though often used, may not be reliable because of the exceptions already known: In the marine diatom *Thalassiosira pseudonana* the glycolate-oxidizing enzyme was identified to be a dehydrogenase insensitive to cyanide and preferring L-lactate instead of the D isomer (27). The glycolate dehydrobenase in unicellular green *Eremosphaera* has been demonstrated to be only slightly inhibited (about 5%) in the presence of 2 mM KCN (38), whereas the L-lactate-converting glycolate oxidase from *Mesotaenium* is inhibited by 33% under the same conditions. Even

Table 8.1. Activity of Glycolate Dehydrogenase from *Fasciculochloris boldii* with Different Substrates.

Substrate	Activity (nkat·mg protein^{-1})[a]
Glycolate	0.22
D-Lactate	0.33
L-Lactate	0.08

Note: Tests were run with DCPIP. The enzyme did not use oxygen as electron acceptor, and no formation of H_2O_2 could be determined.

[a] nkat = nanokatal (an enzyme catalysis unit).

consumption of oxygen during oxidation of glycolate may be a questionable distinction marker for the two enzymes, because oxidation of glycolate in vivo is coupled to oxygen consumption even in a reaction catalyzed by glycolate dehydrogenase (38).

On the whole, it appears that the only reliable distinction marker for glycolate oxidase versus glycolate dehydrogenase, respectively, may be a test indicating the formation of H_2O_2 in the case of the oxidase. However, this test has been applied only in a few cases and so could explain some contradictory results in the literature.

It has been reported that *Chlorogonium* possesses glycolate oxidase (13), but this cannot be confirmed. All present results show that *Chlorogonium*, like the relaged alga *Chlamydomonas*, possesses glycolate dehydrogenase (32, 33). Also *Fasciculochloris* has been reported to possess glycolate oxidase (1). However, our data upon reinvestigation clearly demonstrate the presence of glycolate dehydrogenase in *Fasciculochloris* (Table 8.1).

Specific Activities of Glycolate Oxidizing Enzymes and Glycolate Excretion by Algae

In the alga *Mougeotia* grown at 2% CO_2 and 6000 lux, the specific activity of glycolate oxidase was 7.8 nmol·min^{-1}·mg protein^{-1}, which is relatively high, as is the activity of glycolate dehydrogenase in *Eremosphaera* (6.4 nmol·min^{-1}·mg protein^{-1}) (Fig. 8.1). However, the specific activity of glycolage dehydrogenase in *Chlamydomonas* grown under the same culture conditions was only 0.56 nmol·min^{-1}·mg protein^{-1} (unpublished); in higher plants the specific activity of glycolage oxidase is about 50 nmol·min^{-1}·mg protein^{-1} (53).

Increasing the light intensity to 20,000 lux and decreasing the CO_2 concentration to 0.03% resulted in an increase in the activity of glycolate oxidase in *Mougeotia* (factor 3.3) as well as of glycolate dehydrogenase in *Eremosphaera*, though in the latter case to a lower extent (factor 1.3) (Fig. 8.1). The specific activity of glycolate dehydrogenase in *Chlamydomonas* under these conditions was determined to be 1.15 nmol·min^{-1}·mg protein^{-1}.

The increase of the enzyme activities in *Mougeotia* and *Eremosphaera* was noticed not earlier than 16 h after the culture conditions were changed, except that a significant increase in P-glycolate phosphatase was measured in 3 h

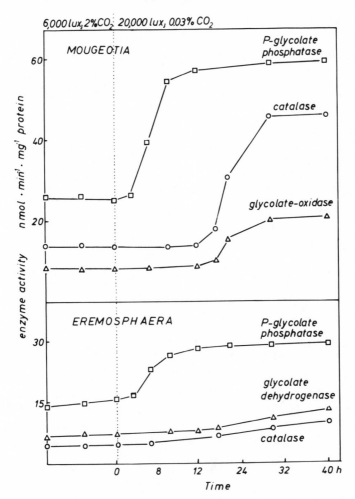

Fig. 8.1. Enzyme activities in *Mougeotia* and *Eremosphaera*. After changing the culture conditions by increasing the light intensity from 6,000 lux to 20,000 lux and decreasing the CO₂ concentration from 2% to 0.03%, an increase in enzyme activities was noticed.

Fig. 8.1). Concomitant with the increase in glycolate oxidase activity, a striking increase also of catalase activity was noticed in *Mougeotia* (Fig. 8.1).

Changing the culture conditions by reducing the CO_2 concentration and increasing the light intensity not only affects the enzyme activities but also induces the process of glycolate excretion which can be observed immediately (Fig. 8.2). Glycolate excretion stops after a few hours (Fig. 8.2). Because the activity of glycolate oxidase or glycolate dehydrogenase in neither of the algae is altered at this time, the cessation of glycolate excretion ought not be explainable by higher rates of glycolate oxidation. In this regard, other processes must be important (49).

Excretion of glycolate by *Chlamydomonas*, when compared with *Eremosphaera* and *Mougeotia*, is relatively high, possibly because glycolate dehydrogenase has extremely low activity in this alga. The specific activities of glycolate oxidase in

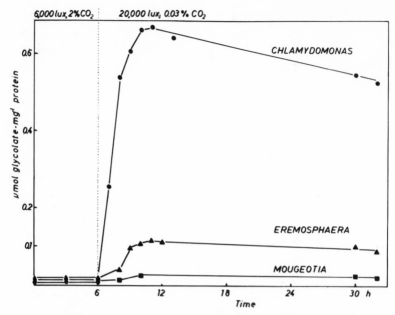

Fig. 8.2. Excretion of glycolate into the nutrient medium by *Chlamydomonas, Eremosphaera,* and *Mougeotia.* Excretion was noted only after the light intensity was increased and the CO_2 concentration was reduced.

Mougeotia is almost as high as the specific activity of glycolate dehydrogenase in *Eremosphaera.* Nevertheless, *Eremosphaera* excretes much more glycolate than *Mougeotia.* In view of this, it may be important that *Eremosphaera* is excreting not only glycolate but also glyoxylate (Fig. 8.3). No excretion of glyoxylate by *Mougeotia* and *Chlamydomonas* can be observed when the cells are suspended in a nutrient medium containing all compounds necessary for growth (Fig. 8.3). Excretion of glyoxylate was also reported in *Sphaerocystis* by Stewart and Codd (41).

Other C_2 Cycle Enzymes in Green Algae

In the C_2 cycle of higher plants, glyoxylage formed from glycolate is converted to glycerate by several steps, which require two aminotransferases (glyoxylate: glutamate and serine:glyoxylate), glycine oxidase, serine hydroxymethyltransferase, and a NADH:hydroxypyruvate reductase (7, 16), All of these enzymes have been demonstrated in algae (2, 9, 10, 14, 35, 40). The specific activities of the enzymes mentioned depend on the culture conditions (36). Increasing the light intensity and decreasing the CO_2 content in the aeration mixture, which favors the synthesis of glycolate, result in increasing enzyme activities (50, 52).

Compartmentation of C_2 Cycle Enzymes

Cellular distribution of C_2 cycle enzymes in the green algae has been studied most extensively with regard to *Mougeotia* and *Eremosphaera*, which may be regarded

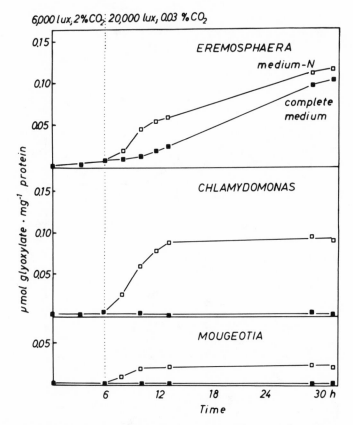

Fig. 8.3. Excretion of glyoxylate by *Chlamydomonas*, *Eremosphaera*, and *Mougeotia* after the light intensity was increased and the CO_2 concentration was reduced. Cells were cultured in a complete nutrient medium normally used for growth (filled symbols) or in the growth medium from which the nitrogen source was omitted (open symbols).

as representatives of two fundamentally different groups under the green algae. Both algae possess peroxisomes which, during gradient centrifugation, equilibrate between densities of 1.23 and 1.25 g·cm⁻³ (35, 36, 38). The peroxisomes from *Mougeotia* were found to contain catalase, glycolate, oxidase, aminotransferases, and hydroxypyruvate reductase, which are enzymes characteristic of the peroxisomes in green leaves (35). In addition, these peroxisomes possess enzymes of the fatty acid β-oxidation pathway (39, 50, 51). Thus they are very similar to leaf peroxisomes. Serine hydroxymethyltransferase is located in the mitochondria of *Mougeotia*, just as in higher plants (36). Different from *Mougeotia*, the peroxisomes from *Eremosphaera* do not contain enzymes involved in the metabolism of glycolate. Instead, glycolate dehydrogenase and hydroxypyruvate reductase are constituents of the mitochondria, exclusively. The mitochondria from *Eremosphaera* also contain serine hydroxymethyltransferase and aminotransferases for converting glyoxylate to glycine and serine to hydroxypyruvate (50). Therefore, the metabolism of glycolate should be possible in *Eremosphaera* as in higher plants, except that

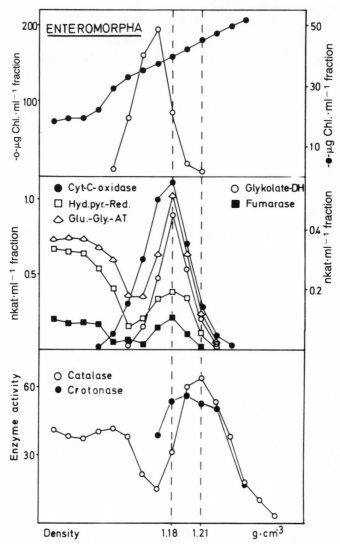

Fig. 8.4. Distribution of enzymes in a linear sucrose gradient after separation of cell organelles from *Enteromorpha*. Mitochondria are at a density of 1.18 g·cm^{-3}, and peroxisomes at a density of 1.21 g·cm^{-3}. Units of crotonase have to be multiplied by the factor of 0.02 to get the actual values.

in the algae the reactions would be associated with mitochondrial metabolism. In this case the peroxisomes are not involved in the metabolism of glycolate.

Peroxisomes were isolated also from several other algae of the group of green algae (32–5, 38, 40). Though there is less information on enzymes present in the organelles of these algae, all data available show that they are similar to the peroxisomes of either *Mougeotia* or *Eremosphaera*. Corresponding results with regard to *Enteromorpha* are presented in Figure 8.4.

Phylogenetic Aspects of Algal Peroxisomes

Peroxisomes of the leaf type demonstrated in *Mougeotia* seem to be characteristic of the Charophyceae, exclusively (30); these are also considered to be the only green algae containing glycolate oxidase. As far as known, the peroxisomes of all other green algae with glycolate dehydrogenase are not at all involved in the metabolism of glycolate (36). In this respect, mitochondria are rather expected to play a dominant role.

According to the literature, different evolutionary lines can be distinguished in the group of green algae, but only one of these leads toward higher plants (18, 25, 28, 42). It is postulated that all lines originate from ancestral flagellates which are as yet unknown. The Prasinophyceae are probably the phylogenetically oldest known green algae. Remarkably, in the phylogenetic scheme of Stewart and Mattox (42), algae possessing the peroxisomal glycolate oxidase are exclusively represented in the line toward higher plants (Fig. 8.5). In algae of the other two lines, mitochondria rather than peroxisomes are involved in the metabolism of glycolate. Obviously, C_2-cycle enzymes in these algae are constituents of mitochondria (Fig. 8.5).

As recently reported, the ancestral Prasinophycean algae (*Pyramimonas, Platymonas, Pedinomonas,* and *Heteromastix*) possess organelles with a structure similar to that of peroxisomes, but they do not contain catalase (40) and therefore should be named *microbodies*. Enzymes of the fatty acid β-oxidation pathway are the only enzymes detected in these microbodies so far, whereas enzymes expected to be involved in the metabolism of glycolate are constituents of mitochondria as in *Eremosphaera* or related algae (Fig. 8.5).

Consistent with the absence of catalase in the Prasinophycean microbodies, the enzyme that does oxidize the fatty acid acyl-CoA does not produce H_2O_2, in contrast to the enzyme in higher plant peroxisomes (8). The oxygen consumed during oxidation of acyl-CoA may rather be reduced to water as in *Mougeotia* and *Eremosphaera* (51). The peroxisomes from *Eremosphaera* contain catalase, but they also possess uricase, which produces H_2O_2. Uricase was not found in Prasinophycean algae (38).

According to the endosymbiontic theory, the ancestral flagellates are expected to have originated from prokaryotic cells. It seems to be an important fact that glycolate dehydrogenase and hydroxypyruvate reductase can be demonstrated in bacterial cells (12, 19). Therefore, the occurrence of these enzymes in the mitochondria of phylogenetically low-developed algae seems to be reasonable.

With regard to the evolutionary scheme of green algae by Stewart and Mattox, leaf-type peroxisomes should have developed from microbodies, similar to those in the Prasinophyceae. Also the peroxisomes found in *Eremosphaera* or *Enteromorpha* (Fig. 8.5) are expected to have originated from the simpler organelles. If this is the case, catalase and H_2O_2-producing oxidases are not originally constituents of peroxisomes. Subsequently, peroxisomes characterized by the possession of catalase and at least one H_2O_2-producing oxidase should have developed from organelles that according to their structure might be named microbodies.

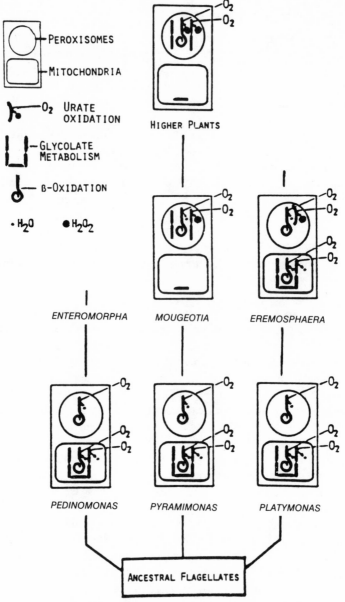

Fig. 8.5. Enzymes in peroxisomes/microbodies and in mitochondria of different green algae. In the modified-U symbol for glycolate metabolism, the right side symbolizes glycolate conversion to glycine (enzymes involved: glycolate oxidoreductase and glyoxylate:glutamate aminotransferase); the bottom line is for glycine conversion to serine (serine hydroxymethyltransferase); and the left side is for serine conversion to glycerate (serine:glyoxylate aminotransferase and hydroxypyruvate reductase).

According to Steward and Mattox (43), the algae used are representatives of three different evolutionary lines originating from ancestral flagellates which are unknown as yet.

Pedinomonas, Pyramimonas, and *Platymonas* are Prasinophycean algae, which are regarded

REFERENCES

1. Bullock, K. W., Deason, T. R., and O'Kelly, J. C. 1979. Occurrence of glycolate oxidase in some coccoid zoospore-producing green algae. *J. Phycol.* **15**: 142–6.

2. Collins, N., and Merrett, M. J. 1975. The localization of glycolate pathway enzymes in *Euglena*. *Biochem. J.* **148**: 321–8.

3. deVeau, E. J., and Burris, J. E. 1989. Glycolate metabolism in low and high CO_2-grown *Chlorella pyrenoidosa* and *Pavlova lutheri* as determined by O_2-labeling. *Plant Physiol.* **91**: 1085–93.

4. Floyd, G. L., and Salisbury, J. L. 1977. Glycolate dehydrogenase in primitive green algae. *Am. J. Bot.* **64**: 1294–6.

5. Fogg, G. E. 1976. Release of glycolate from tropical marine plants. *Austr. J. Plant Physiol.* **3**: 57–61.

6. Frederick, S. E., Gruber, P. J., and Tolbert, N. E. 1973. The occurrence of glycolate dehydrogenase and glycolate oxidase in green plants. *Plant Physiol.* **52**: 318–23.

7. Gerhardt, B. 1978. Microbodies/peroxisomes pflanzlicher zellen. In *Cell Biology Monographs*, Vol. 5, pp. 72–82. Springer-Verlag, New York.

8. Gerhardt, B. 1983. Localization of β-oxidation enzymes in peroxisomes isolated from nonfatty plant tissues. *Planta* **159**: 238–46.

9. Gross, W., and Beevers, H. 1989. Subcellular distribution of enzymes of glycolate metabolism in the alga *Cyanidium caldarium*. *Plant Physiol.* **90**: 759–805.

10. Gross, W., Winkler, U., and Stabenau, H. 1985. Characterization of peroxisomes from the alga *Bumilleriopsis filiformis*. *Plant Physiol.* **77**: 296–99.

---

as phylogenetically the most primitive algae. Peroxisomes of these organisms possess no catalase but only enzymes of the fatty acid β-oxidation pathway. The acyl-CoA oxidizing enzyme is an oxidase that produces H_2O. Enzymes of the glycolate metabolism are constituents of the mitochondria from the Prasinophycean algae, exclusively. During oxidation of glycolate by glycolate dehydrogenase, oxygen is reduced to H_2O. The mitochondria also contain enzymes of the fatty acid β-oxidation pathway. During oxidation of acyl-CoA, oxygen is also reduced to water.

In the evolutionary line toward higher plants, it seems that enzymes of the glycolate pathway (glycolate oxidoreductase, glyoxylate:glutamate aminotransferase, serine:glyoxylate aminotransferase, and hydroxypyruvate reductase) were transferred to the peroxisomes. Serine hydroxymethyltransferase, which remains mitochondrial in all organisms, is the only exception. Concomitant with the transfer, the glycolate-oxidizing enzyme, which is a dehydrogenase in the Prasinophycean algae, apparently became an oxidase that produces H_2O_2. In accordance with this, the peroxisomes also possess catalase. The peroxisomes in *Mougeotia* differ from those in higher plants in that the fatty acid acyl-CoA oxidase is an enzyme that produces H_2O as in the Prasinophycean algae. In higher plant peroxisomes, H_2O_2 is formed during oxidation of acyl-CoA. The mitochondrial fatty acid β-oxidation pathway apparently was lost during evolutionary development toward higher plants.

In the peroxisomes of higher developed algae of the other two lines [e.g., *Eremosphaera* and *Enteromorpha* (compare Fig. 8.4)], catalase has been demonstrated in addition to enzymes of the fatty acid β-oxidation pathway. Uricase is the H_2O_2-producing oxidase in the peroxisomes from *Eremosphaera*.

In general, it can be assumed that peroxisomes have developed from simpler organelles (microbodies) by receiving more enzymes. Thus they seem to have become more and more important. It also appears that H_2O_2-producing oxidases have become dominant with the development of peroxisomes.

11. Gruber, P. J., Frederick, S. E., and Tolbert, N. E. 1974. Enzymes related to lactate metabolism in green algae and low land plants. *Plant Physiol.* **53**: 167–70.

12. Harder, W., Attwood, M. M., and Quale, J. R. 1973. Methanol assimilation by *Hyphomicrobium* sp. *J. Gen. Microbiol.* **78**: 155–63.

13. Huang, A. C. H., Trelease, R. N., and Moore, T. S., Jr. 1983. *Plant Peroxisomes*, pp. 136–42. Academic Press, New York.

14. Husic, D. W., and Tolbert, N. E. 1987. NADH hydroxypyruvate-reductase and NADPH glyoxylate reductase in algae: Partial purification and characterization from *Chlamydomonas reinhardtii*. *Arch. Biochem. Biophys.* **252**: 1343–52.

15. Husic, H. D., and Tolbert, N. E. 1985. Properties of phosphoglycolate phosphatase from *Chlamydomonas reinhardtii* and *Anacystis nidulans*. *Plant Physiol.* 379: 394–98.

16. Husic, D. W., Husic, H. D., and Tolbert, N. E. 1987. The oxidative photosynthetic carbon cycle or C_2 cycle. *CRC* **5**: 45–100.

17. James, L., and Schwartzbach, S. D. 1982. Differential regulation of phosphoglycolate and phosphoglycerage phosphatases in *Euglena*. *Plant Sci. Lett.* **27**: 223–32.

18. Leedale, G. F. 1974. How many are the kingdoms of organisms. *Taxon* **23**: 261–70.

19. Lord, J. M. 1972. Glycolate oxidoreductase in *Escherichia coli*. *Biochim. Biophys. Acta* **267**: 227–37.

20. Lord, J. M., and Merrett, M. J. 1970. The pathway of glycolate utilization in *Chlorella pyrenoidosa*. *Biochem. J.* **117**: 929–37.

21. Merrett, M. J., and Lord, J. M. 1973. Glycolate formation and metabolism by algae. *New Phytol.* **72**: 751–67.

22. Miziorko, H. M., and Lorimer, G. H. 1983. Ribulose-1,5-bisphosphate carboxylase. *Annu. Rev. Biochem.* **52**: 507–42.

23. Moroney, J. V., Wilson, B. J., and Tolbert, N. E. 1986. Glycolate metabolism and excretion by *Chlamydomonas reinhardtii*. *Plant Physill.* **82**: 821–6.

24. Nelson, E. B., and Tolbert, N. E. 1970. Glycolate dehydrogenase in green algae. *Arch. Biochem. Biophys.* **141**: 102–10.

25. O'Kelly, C. J., and Floyd, G. L. 1984. Flagellar apparatus, absolute orientation and the phylogeny of the green algae. *Biosystems* **16**: 227–51.

26. Paul, J. S., and Volcani, B. E. 1974. Photorespiration in diatoms. I. The oxidation of glycolic acid in *Thallassiosira pseudodonana*. *Arch. Microbiol.* **101**: 115–20.

27. Paul, J. S., Sullivan, C. W., and Volcani, B. E. 1975. Photorespiration in diatoms: Mitochondrial glycolate dehydrogenase in *Cylindrotheca fusiformis*. *Arch. Biochem. Biophys.* **169**: 152–9.

28. Pickett-Heaps, J. D., and Marchant, H. 1972. The phylogeny of the green algae: A new proposal. *Cytobios* **6**: 255–64.

29. Satoh, H., Okada, M., Nokayama, K., and Miygji, K. 1984. Purification and further characterization of pyrenoid proteins and ribulose-1,5-bisphosphate carboxylase–oxygenase from the green alga *Bryopsis maxima*. *Plant Cell Physiol.* **25**: 1025–8.

30. Silverberg, B. A. 1975. An ultrastructural and cytochemical characterization of microbodies in the green algae. *Protoplasma* **83**: 269–95.

31. Stabenau, H. 1972. Ueber die Ausscheidung von Glykolsäure bei *Chlorogonium alongatum* Dangeard. *Biochem. Physiol. Pflanzen* **163**: 42–51.

32. Stabenau, H. 1974. Localization of enzymes of glycolate metabolism in the alga *Chlorogonium elongatum*. *Plant Physiol.* **54**: 921–4.

33. Stabenau, H. 1974. Verteilung von Microbody-Enzymen aus *Chlamydomonas* im Dichtegradienten. *Planta* **118**: 35–42.

34. Stabenau, H. 1976. Microbodies from *Spirogyra*: Organelles of a filamentous alga similar to leaf peroxisomes. *Plant Physiol.* **58**: 693–5.

35. Stabenau, H. 1980. Enzymes of glycolate metabolism in the alga *Mougeotia*. *Plant Physiol.* **65S**: 74.

36. Stabenau, H. 1984. Microbodies in different algae. In *Compartments in Algal Cells and Their Interaction*, ed. W. Wiesner, D. Robinson, and R. C. Starr, pp. 184–90. Springer-Verlag, Berlin.

37. Stabenau, H., and Säftel, W. 1982. A peroxisomal glycolate oxidase in the alga *Mougeotia*. *Planta* **152**: 165–7.

38. Stabenau, H., Winkler, U., and Säftel, W. 1984. Mitochondrial metabolism of glycolate in the alga *Eremosphaera viridis*. *Z. Pflanzenphysiol.* **114**: 413–20.

39. Stabenau, H., Winkler, U., and Säftel, W. 1984. Enzymes of β-oxidation in different types of algal microbodies. *Plant Physiol.* **75**: 531–3.

40. Stabenau, H., Winkler, U., and Säftel, W. 1989. Compartmentation of peroxisomal enzymes in algae of the group of Prasinophyceae. *Plant Physiol.* **90**: 754–9.

41. Steward, R., and Codd, G. A. 1981. Glycolate and glyoxylate excretion by *Sphaerocystis schroeteri* (Chlorophyceae). *Br. Phycol. J.* **16**: 177–82.

42. Steward, K. D., and Mattox, K. R. 1975. Comparative cytology, evolution and classification of the green algae with some consideration of the origin of other organisms with chlorophylls a and b. *Bot. Rev.* **41**: 104–35.

43. Steward, K. D., and Mattox, K. R. 1978. Structural evolution in the flagellated cells of green algae and land plants. *Biosystems* **10**: 145–52.

44. Tolbert, N. E. 1974. Photorespiration. In *Algal Physiology and Biochemistry*, ed. W. D. P. Stewart, pp. 474–505. Blackwell Scientific Publications, Oxford.

45. Tolbert, N. E. 1976. Glycolate oxidase and glycolate dehydrogenase in marine algae and plants. *Austr. J. Plant Physiol.* **3**: 125–32.

46. Tolbert, N. E. 1979. Glycolate metabolism by higher plants and algae. *Encycloped. Plant Physiol.* **6**: 338–52.

47. Tolbert, N. E. 1980. Photorespiration. In *The Biochemistry of Plants*, Vol. 2, ed. P. K. Stumpf and E. E. Conn, pp. 487–523. Academic Press, New York.

48. Tolbert, N. E., and Zill, L. P. 1956. Excretion of glycolic acid by algae during photosynthesis. *J. Biol. Chem.* **222**: 895–906.

49. Tolbert, N. E., Husic, H. D., Husic, D. W., Moroney, J. V., and Wilson, B. J. 1985. Relationship of glycolate excretion to the DIC pool in microalgae. In *Inorganic Carbon Uptake by Aquatic Photosynthetic Organisms*, ed. W. J. Lucas and J. A. Berry, pp. 211–29. American Society of Plant Physiologists.

50. Winkler, U., Säftel, W., and Stabenau, H. 1981. Studies on aminotransferases participating in the glycolate metabolism of the algae *Mougeotia*. *Plant Physiol.* **70**: 340–43.

51. Winkler, U., Säftel, W., and Stabenau, H. 1988. β-Oxidation of fatty acids in algae: Localization of thiolase and acyl-CoA oxidizing enzymes in three different organisms. *Planta* **175**: 91–8.

52. Yokota, A., Nakamo, Y., and Kitaoka, S. 1978. Metabolism of glycolate in mitochondria of *Euglena gracilis*. *Agric. Biol. Chem.* **42**: 121–8.

53. Zelitch, I. 1971. *Photosynthesis, Photorespiration and Plant Productivity*. Academic Press, New York.

9

The Interaction of Photorespiratory Carbon Metabolism with Mitochondrial Pyruvate Metabolism

JOANNA GEMEL, JAN A. MIERNYK, and DOUGLAS D. RANDALL

Several chapters in this volume describe how increasing carbon dioxide in the atmosphere will certainly have a multitude of effects on the plant world. Our present state of knowledge suggests that C_3 plants would potentially have the most to gain as they should exhibit decreasing amounts of photorespiration. However, CO_2 enrichment studies have shown that increased net photosynthesis and increased biomass are not necessarily the result of long-term increases in carbon dioxide (e.g., Refs. 3, 18). As we consider our future with increasing carbon dioxide and global warming, we need to keep in mind the incredible flexibility as well as the redundancy in metabolism, particularly when one ponders the ability of plants to survive in spite of their sessile nature.

In the past, increases in atmospheric oxygen and increases in plant photo-respiration were pressures that led to metabolic adaptions, including the very obvious "trap and pump" mechanism of C_4 photosynthesis, which increases the CO_2 substrate around Rubisco, and the photorespiratory carbon cycle, to salvage as much carbon as possible from phosphoglycolate. Most certainly, there have been many other less visible and more subtle adjustments that have occurred in metabolism. Most of the predictions concerning the effects of increased atmospheric carbon dioxide are based on a reduction in photorespiration, but it seems prudent to keep in mind that we shall begin to reverse as well the less visible adaptions that will likely affect our predictions about biomass, increased net photosynthesis, and efficiency. If some of these adaptions are difficult to reverse or overcome, the increasing CO_2 could conceivably be detrimental to C_3 plants. In this chapter, we describe what we believe is a rather significant adaption that some plants have made to deal more effectively or more efficiently with photorespiration and thus to reduce its cost to the plant. Figure 9.1 summarizes photorespiratory carbon metabolism and its occurrence in three subcellular organelles. Photorespiration originates in the chloroplasts and passes through the peroxisomes and mito-chondria before returning to the chloroplast, with potentially 75% of the carbon as glycerate.

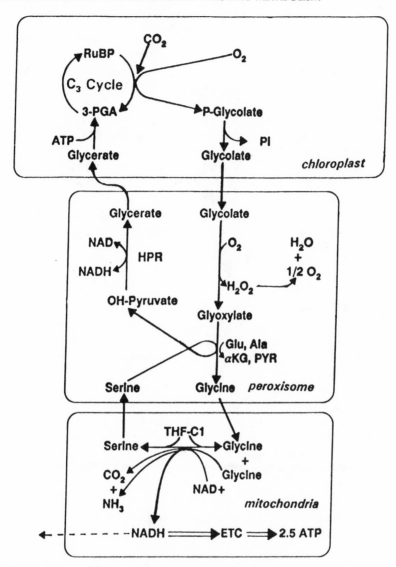

Fig. 9.1. Scheme summarizing the photorespiratory carbon cycle.

Any student of metabolism soon marvels at how various reactions or pathways are compartmentalized. This compartmentalization is critical to prevent futile cycling, to create unique reaction environments, to concengrate substrates and products, and to regulate specific reactions or pathways. As organisms evolve, they place new reactions in specific compartments or organelles to take advantage of organellar functions or to prevent reaction products from causing problems. The subcellular localization of the reactions of the photorespiratory ccarbon cycle illustrates this type of metabolic optimization very well. For example, phosphoglycolate, the initial product of photorespiration produced in the RuBP oxygenase reaction, can inhibit several reactions of the photosynthetic carbon reduction cycle and glycolysis. Thus, it must be dealt with quickly and specifically. Considering that both glycolysis and the photosynthetic carbon reduction cycle are full of phosphorylated intermediates, a general phosphatase in the chloroplast or cytosol would be diastrous, but waiting for the phosphoglycolate to make its way to the vacuole phosphatases would be equally disastrous. To prevent metabolic chaos, the plant has adapted by evolving a very substrate-specific and very fast phosphatase for phosphoglycolate in the chloroplast where phosphoglycolate is generated. The flavin oxidases (such as glycolate oxidase) that hield H_2O_2 are nearly surrounded with catalase inside the peroxisome, thus minimizing the oxidative effects of H_2O_2. The highly reactive glyoxylate produced in the glycolate oxidase reactions is immediately transaminated to the very unreactive and zwitterionic glycine by specific transaminases present in the peroxisome.

An examination of the remainder of the photorespiratory pathway shows that the remaining reactions yield products that are not "troublesome" metabolically. These reactions between glycine and glycerate could all just as well be in the peroxisome, where the NADH produced by glycine decarboxylase would be immediately available to the hydroxypyruvate reductase. However, the glycine-to-serine conversion occurs in the mitochondria. For this reason, elaborate shuttle mechanisms have been proposed to move the NADH out of the mitochondria and into the peroxisomes to satisfy the needs of the hydroxypyruvate reductase as well as to keep the pyridine nucleotide pool in the mitochondria in an oxidized state. If this is true, does it not seem peculiar that putting the glycine decarboxylase in the mitochondria causes an inordinate amount of shuttling of intermediates across membranes? Nature seldom misplaces a reaction. It suggests to us that the mitochondrial location of the glycine-to-serine conversion offers other advantages to the C_3 plant. This conclusion has support from a number of aspects: (a) The glycine decarboxylase complex can account for as much as 40% of the matrix protein in pea leaf mitochondria (2); (b) glycine is the preferred substrate over the Krebs cycle intermediates (6); (c) the peroxisomal and cytosolic NADH requirements can be easily met by the excess reducing equivalents from the chloroplast (10); and (d) maximal photosynthesis rates require mitochondrial ATP production (12). This latter point brings us to the long-standing controversy as to whether mitochondrial respiration is inhibited, curtailed, or unaffected during photosynthesis.

Numerous studies on this question have not resolved the issue definitively (9, 13). Recent studies by Krömer and colleagues (11, 12) have built a very strong case for a requirement for mitochondrial ATP production for maximal photosynthesis. They concluded that mitochondrial ATP would be necessary to meet the energy demands in the cytosol for such purposes as sucrose synthesis; and, if these demands were not met by mitochondrial ATP, the system would suffer from feedback inhibition, thus inhibiting photosynthesis. The question remaining, then, is what is the source of the reducing equivalents being utilized to drive oxidative phosphorylation? Is the Krebs cycle in full operation or partially active? If the Krebs cycle is not supplying the reducing equivalents, then what is the source? Can photorespiratory glycine oxidation provide the reducing equivalents in the mitochondria? If this is the case, then maybe photorespiration is not as negative as it was originally thought to be!

Studies where Krebs cycle intermediates were fed to leaves or leaf slices during photosynthesis suggest that the Krebs cycle was at least 80% operative (13). However, this was not an in vivo situation. An experiment more approximating the in vivo condition was run when Benson and Calvin were establishing the path of carbon in photosynthesis. They found that $^{14}CO_2$ fixation in the light labeled many metabolites, but the only Krebs cycle intermediate that was labeled was malate (1) (which can be labeled by the ubiquitous PEP carboxylase). When the tissue was put into darkness, a rapid labeling of the other Krebs cycle intermediates occurred. This suggests that in the light, when photosynthesis is occurring, the Krebs cycle is significantly curtailed.

This brings us to the question that our research group asked many years ago: Where is the most likely point or enzyme to control carbon entry into the Krebs cycle? An examination of Figure 9.2 shows that the cycle cannot turn without the product of the pyruvate dehydrogenase complex, acetyl-CoA. Thus, the pyruvate dehydrobenase complex (PDHC) became the target of our studies to determine how mitochondrial metabolism—particarly the Krebs cycle—was affected by photosynthesis and photorespiration. Actually the original goal was to find some positive function for photorespiratory carbon metabolism in addition to simply salvaging the carbon in phosphoglycolate.

The PDHC catalyzes the oxidative decarboxylation of pyruvate to acetyl-CoA, and concomitantly reduces NAD^+. This multienzyme complex is regulated by product feedback inhibition by acetyl-CoA and NADH (14, 16). It is particularly sensitive to NADH, with a K_i that is about one tenth of the K_m for NAD^+. PDHC is also regulated by covalent modification (phosphorylation–dephosphorylation) of the decarboxylase component of the complex. Phosphorylation acts like a switch to inactivate the complex, and dephosphorylation switches the complex back to the active form. The phosphorylation and dephosphorylation reactions are catalyzed, respectively, by an ATP-dependent kinase and a specific pyruvate dehydrogenase phosphate phosphatase (17). PDHC and its subtrate, pyruvate, connect glycolysis to the Krebs cycle, providing acetyl-CoA for citrate synthase and probably for numerous other acetyl-CoA requiring processes. The decarboxylation of pyruvate commits the remaining carbon atoms of this compound to a limited number of fates and is metabolically a rather costly step. For these reasons, we feel that there are multiple layers of regulation acting on this enzyme complex. Plants are unique in that they have a second isoform of PDHC in the plastids (7, 20); however, the plastid PDHC is not regulated by phosphorylation–dephosphorylation. Its function in the plastids is most probably to supply substrate for fatty acid and isoprenoid biosynthesis (5).

In situ studies using highly purified pea leaf mitochondria have established that when mitochondria are oxidizing various Krebs cycle intermediates and especially glycine (which is preferred over other substrates by leaf mitochondria) in State 3 and in the absence of any ATP-utilizing system, the PDHC is reversibly inactivated (phosphorylated) (Table 9.1). High levels (0.5–1.0 mM) of pyruvate can prevent this reversible inactivation (Table 9.1). Uncouplers and oligomycin prevent the inactivation of PDHC when intact leaf mitochondria are oxidizing substrates other than pyruvate; and when mitochondria oxidize extramitochondrial NADH under State 3 conditions, they also reversibly inactivate PDHC (15).

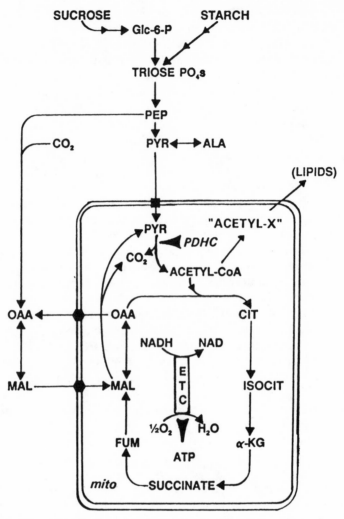

Fig. 9.2. Scheme illustrating the flow of carbon to the Krebs cycle in a plant cell.

Results from our in situ studies suggested that we had a potential mechanism to regulate carbon flux into the Krebs cycle, and that this mechanism could be operative during photosynthesis and photorespiration. Our initial hypothesis (ca. 1973) was that during photosynthesis, PDHC would be at least partially inactivated (phosphorylated) and the carbon flow to the Krebs cycle restricted. This hypothesis assumed that the chloroplast would supply the ATP and reducing equivalents to the cytosol. The work of Krömer and colleagues (11, 12), indicating that mitochondrial ATP was necessary for maximal photosynthesis, along with work which showed that the chloroplast adenylate transporter prefers to take up ATP (10), caused us to alter our hypothesis. Furthermore, with the discovery that Stage 3 mitochondrial glycine oxidation inactivated PDHC, we altered our hypothesis to suggest that photorespiratory carbon metabolism would cause a restriction in carbon flux to the Krebs cycle.

Table 9.1. Regulation of Pyruvate Dehydrogenase Complex (PDHC) Activation Status In Situ During Oxidation of Different Substrates in Pea Mitochondria.

Expt. I	V_0	PDHC[a]	Expt. II	V_0	PDHC[a]	Expt. III	V_0	PDHC[a]
CoA, NAD	0	100	CoA, NAD, Glu	0	100	CoA, NAD	0	100
Succinate	32	112	Succinate	38	100	Glycine[c]	32	111
ADP			ADP			ADP		
State 3	89	61	State 3	101	39	State 3	44	49
State 4	51	23	State 4	64	0	State 4	38	37
Pyruvate	51	61	Pyruvate	101	89	Malate	44	6
TPP[b]	114	112	TPP[b]	114	100	Pyruvate	76	100
						TPP[b]	76	100

Note: Experiments I–III represent typical assays in which subsequent additions into polarographic vessel were made—for example, in Expt. I, CoA, NAD, succinate, ADP, etc. Respiration rate and PDHC activity (spectrophotometric assay) were measured after each addition. The reaction mixture for oxygen uptake assays contained 0.1 mg·ml^{-1} of mitochondrial protein. For PDHC estimation, 20 µg mitochondrial protein was used. The following compounds were used at the concentration indicated: 1 mM NAD, 0.1 mM CoA, 0.1 mM thiamine pyrophosphate (TPP); 1 mM succinate, 0.1 mM glutamate, 1 mM glycine, 0.1 mM malate, 1 mM pyruvate, and 100 µM ADP. Respiration rate is expressed in nmol O_2 min^{-1}·mg^{-1} protein. Initial PDHC activity was 0.14 μmol·min^{-1}·mg^{-1} mitochondrial protein for Expts. I and II and 0.12 μmol·min^{-1}·mg^{-1} mitochondrial protein for Expt. III.

[a] The value expressed is the percentage of initial enzyme activity remaining.

[b] TPP = thiamine pyrophosphate.

[c] Tetrahydrofolate-limited rate.

Support for this last hypothesis was developed when our research group was able to make a reasonable estimate of the in vivo activation status of the PDHC. This in vivo assay of PDHC utilizes $^{14}CO_2$ release from [1-^{14}C]pyruvate under conditions that prevent both further phosphorylation and dephosphorylation and thereby fixing the in vivo steady-state phosphorylation or activation status of PDHC (4). We used this assay to determine the level of PDHC activity in the dark, in the light, and under a variety of atmospheric conditions as well as in the presence of photosynthesis and photorespiratory inhibitors. Our initial results (4) showed that PDHC activity was greatly reduced when the tissue was illuminated. This inactivation of PDHC was not due to a diurnal or circadian pattern, but was totally dependent upon illumination of the tissue (13). Figure 9.3A illustrates this light-dependent inactivation, and Figure 9.3B shows that the inactivation can be rapidly cycled, depending upon illumination and darkness. The initial inactivation (after a normal 14-h dark period) took about 45 min to reach a steady-state level of about 5–15% of dark steady-stage level of mitochondial PDHC activity. After this initial inactivation, the PDHC activity could be rapidly and repeatedly cycled through the inactive and active stages with light/dark intervals as short as 5 min.

We have concluded that the curtailment of the PDHC activity would significantly reduce the flux of carbon into the Krebs cycle during photosynthesis. To support this conclusion further, we inhibited photosynthesis with 3-(3,4-dichlorophenyl)-1,1-dimethylurea (DCMU), as shown in Figure 9.4, and observed

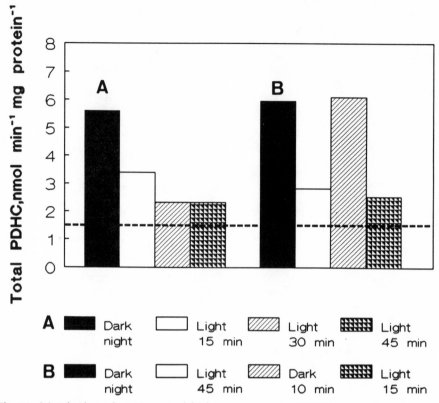

Fig. 9.3. (A) Light-dependent pyruvate dehydrogenase complex (PDHC) inactivation. Pea seedlings were dark-adapted overnight and then illuminated at 1000 $\mu E \cdot m^{-2} \cdot s^{-1}$ for 15, 30, or 45 min. (B) Reversibility of light/dark effect on PDHC activity. Pea seedlings, dark-adapted overnight, were illuminated for 45 min at 1000 $\mu E \cdot m^{-2} \cdot s^{-1}$. Subsequently, plants were transferred to darkness for 10 min and then reilluminated at the same light intensity for 15 min. Dashed line shows the level of estimated chloroplast PDHC activity.

that the light-dependent inactivation was almost completely inhibited. When etiolated seedlings were illuminated, there was no inactivation of PDHC (4); and when purified leaf mitochondria were illuminated with about half full sunlight, PDHC was not inactivated (8). When cell sap expressed from illuminated leaves was added to partially purified PDHC, the PDHC activity was unaffected, and the same result was obtained following gel filtration of the expressed cell sap from illuminated leaf tissue (8). This indicates that no substance(s) produced during photosynthesis was inhibiting PDHC.

Since glycine oxidation by purified leaf mictochondria leads to inactivation of the mitochondrial PDHC, it was reasonable to consider that photorespiratory carbon metabolism, particularly the mitochondrial glycine-to-serine conversion, could be the signal reaching the mitochondria and resulting in curtailment of carbon flux to the Krebs cycle. To test this hypothesis, we manipulated photo-respiratory carbon metabolism by changing the CO_2 and O_2 levels surrounding the tissue or by using inhibitors of reactions within the photorespiratory carbon

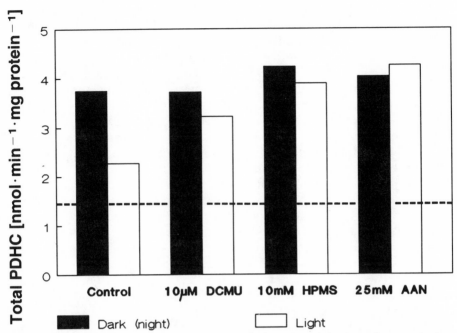

Fig. 9.4. Effect of DCMU and inhibitors of photorespiration on the light-dependent inactivation of pyruvate dehydrogenase complex (PDHC). Detached leaf pairs from pea seedlings were fed with DCMU, 2-hydroxypyridylmethane sulfonate (HPMS), aminoacetonitrile (AAN), or water (for control) via the transpiration stream in the dark for 1 h. PDHC activity was determined for dark treated tissue or after 1 h of illumination at 250 $\mu E \cdot m^{-2} \cdot s^{-1}$. Dashed line shows the level of estimated chloroplast PDHC activity.

pathway. Increasing the carbon dioxide or decreasing the oxygen or both increased the steady-state level of PDHC in the light and had no effect on the level of PDHC activity in the dark (8). Figure 9.4 shows that inhibitors (fed via the transpiration stream to detached leaf pairs) of either glycolate oxidase (2-hydroxypyridylmethane sulfonate) or glycine decarboxylase (aminoacetonitrile) prevented the in vivo light-dependent inactivation of PDHC. Furthermore, when glycine (2 mm) was fed to pea leaf slices *in the dark*, PDHC was inactivated, and this inactivation was prevented by inhibitors of glycine decarboxylase (8).

At least one logical conclusion to be drawn from these experiments is that products of the glycine decarboxylase reaction are providing conditions favoring the curtailment of carbon flow into the Krebs cycle through PDHC. The NH_4^+ produced in the oxidation of glycine can stimulate the PDH kinase that phosphorylates and inactivates the PDHC (19). Furthermore, if the pyridine nucleotide pool becomes reduced by 10%, PDHC is 40% inhibited by feedback. More importantly, the NADH produced by the glycine decarboxylase can be easily reoxidized by the electron-transport system to produce ATP, thus reducing the need to oxidize pyruvate and Krebs cycle intermediates. How much of the NADH needed to generate the mitochondrial ATP required during photosynthesis comes from glycine oxidation? The rate of photorespiration is roughly 30% of photosynthesis and thought to be 3–10 times the rate of dark respiration. For every turn of the Krebs cycle, with the input of 1 acetyl-CoA from pyruvate via PDHC, 3 CO_2 molecules are released (1 from pyruvate and 2 from the cycle), and 4 NADHs and 1 $FADH_2$ are generated, which should give rise to 12–15 ATPs. If CO_2 release is used as the basis for the minimum factor for comparing dark respiration and photorespiration and the lowest rate for photorespiration is considered to be 3 times the dark rate—then 9 glycine molecules are oxidized and 9 NADHs are produced. This yields 22–27 ATPs, or almost double the yield by 1 pyruvate into the Krebs cycle. This should certainly go a long way toward meeting the ATP requirements and also toward preserving the plant's carbon resources by not oxidizing the pyruvate.

If the gylcine oxidation produces more NADH than the electron transport system can reoxidize, the NADH can be reoxidized by the alternate oxidase, or the reducing equivalents can be shuttled out of the mitochondria using the malate–oxaloacetate exchange shuttle. This latter mechanism would also be disruptive to the Krebs cycle since it would require one of its reactions, catalyzed by malate dehydrogenase, to operate in the reverse direction simultaneously if the Krebs cycle were operational. Here again, our proposal that the Krebs cycle is restricted would help explain why the malate dehydrogenase could operate in reverse without causing problems to the cycle.

Our question, from several paragraphs ago, as to whether it is advantageous for the plant to have the photorespiratory glycine-to-serine reactions in the mitochondria can now be answered quite probably in the affirmative. All C_3 plants have glycine decarboxylase in their leaf mitochondria, and the oxidation of glycine can yield a large amount of ATP when linked to oxidative phosphorylation. To date, we have shown light-dependent inactivation of leaf PDHC in 8 C_3 plants (8), and it would seem metabolically economical to place the glycine decarboxylase in the mitochondria to provide a neat and efficient means of taking advantage of this oxidative reaction and conserve the pyruvate carbon at the same time. In Chapter 8, H. Stabenau points out that during algal evolution from the Chlorophyceae, which have all the enzymes for glycolate metabolism in the mitochondria, to the multicellular Charaphyceae and then to higher plants, the enzymes of the C_2 cycle moved from the mitochondria to the peroxisomes, except for glycine oxidase (Fig. 9.1). The close association between NADH generated during glycine oxidation and regulation of mitochondria respiration suggests a high essentiality for this coupling for regulating overall plant metabolism. Figure 9.5 summarizes the interaction among photorespiratory glycine metabolism, PDHC, the Krebs cycle, and the mitochondrial production of ATP.

One important point we wish to stress in this chapter is that plants are very frugal with their resources and have adapted their metabolism to get the most out of every situation—for example, even to deal with unavoidable reactions like that of Rubisco, which initiates photorespiration. Our example of how photosynthesis, photorespiration, and respiratory carbon metabolism have become intertwined, as well as our conclusions that plants regulate PDHC and thus limit the carbon obtained from glycolysis from unnecessarily entering the Krebs cycle, illustrate the difficulty in predicting the effects of increasing atmospheric carbon dioxide. If atmospheric carbon dioxide rises too quickly, there could be unpredicable effects

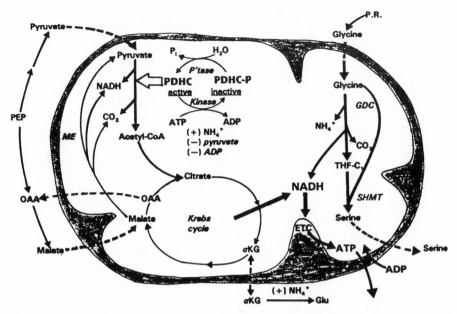

Fig. 9.5. Schematic illustrating pyruvate dehydrogenase complex (PDHC) as a primary control for carbon entry to the Krebs cycle when glycine oxidation occurs during photorespiration.

on metabolism, and initially at least, perhaps not all of these effects will be positive. However, we are confident that plants will adapt over time. Photosynthesis may require more glycolysis to support pyruvate oxidation for generation of mitochondrial ATP production, which in turn may alter the sucrose synthesis rates and/or capacities in some plants. It is obvious that we need to learn more about primary metabolism under the atmospheric conditions predicted for our future.

In summary, a question posed over 25 years ago while one of the authors (D.D.R.) was a student with Professor Tolbert—Does photorespiratory carbon metabolism carry any benefits other than to salvage part of the carbon of phosphoglycolate?—has been partially answered in the affirmative. We feel that photorespiratory carbon metabolism supports mitochondrial ATP production while conserving other carbon resources.

ACKNOWLEDGMENTS

This research was supported by the Missouri Agricultural Experiment Station and National Science Foundation Grants (DMB-8506473 and IBN-9201292). This is journal number 11,730 of the Missouri Agricultural Experiment Station.

This chapter is dedicated to Professor N. E. Tolbert and to his leadership in plant biochemistry.

REFERENCES

1. Benson, A. A., and Calvin, M. 1950. The path of carbon in photosynthesis. VII. Respiration and photosynthesis. *J. Exp. Bot.* **1**: 63–9.

2. Bourguignon, J., Neuburger, M., and Douce, R. 1988. Resolution and characterization of the glycine cleavage reaction in pea leaf mitochondria: Properties of the forward reaction catalyzed by glycine decarboxylase and serine hydroxymethyl transferase. *Biochem. J.* **255**: 169–78.

3. Bowes, G. 1991. Growth at elevated CO_2: Photosynthetic responses mediated through Rubisco. *Plant Cell Environ.* **14**: 795–806.

4. Budde, R. J. A., and Randall, D. D. 1990. Pea leaf mitochondrial pyruvate dehydrogenase complex is inactivated in vivo in a light-dependent manner. *Proc. Natl. Acad. Sci. USA* **87**: 673–6.

5. Camp, P. J., and Randall, D. D. 1985. Purification and characterization of the pea chloroplast pyruvate dehydrogenase complex: A source of acetyl-CoA and NADH for fatty acid biosynthesis. *Plant Physiol.* **77**: 571–7.

6. Dry, I. B., Day, D. A., and Wiskich, J. T. 1983. Preferential oxidation of glycine by the respiratory chain of pea leaf mitochondria. *FEBS Lett.* **158**: 154–8.

7. Elias, B. A., and Givan, C. V. 1979. Localization of pyruvate dehydrogenase complex in *Pisum sativum* chloroplasts. *Plant Sci. Lett.* **17**: 115–22.

8. Gemel, J., and Randall, D. D. 1992 Light regulation of leaf mitochondrial pyruvate dehydrogenase complex: The role of photorespiratory carbon metabolism. *Plant Physiol.* **100**: 908–14.

9. Graham, D. 1980. Effects of light on "dark respiration." In *The Biochemistry of Plants: A Comprehensive Treatise*, ed. D. D. Davis, Vol. 2, pp. 525–79. Academic Press, New York.

10. Heldt, H. W., and Flügge, U. I. 1987. Subcellular transport of metabolites in plant cells. In *The Biochemistry of Plants: A Comprehensive Treatise*, ed. D. D. Davis, Vol. 12, pp. 50–86. Academic Press, New York.

11. Krömer, S., and Heldt, W. 1991. On the role of mitochondrial oxidative phosphorylation in photosynthesis metabolism as studied by the effect of oligomycin on photosynthesis in protoplasts and leaves of barley (*Hordeum vulgare*). *Plant Physiol.* **95**: 1270–6.

12. Krömer, S., Stitt, M., and Heldt, W. 1988. Mitochondrial oxidative phosphorylation participating in photosynthetic metabolism of a leaf cell. *FEBS Lett.* **226**: 352–6.

13. McCashin, B. G., Cossins, E. A., and Canvin, D. T. 1988. Dark respiration during photosynthesis in wheat leaf slices. *Plant Physiol.* **87**: 155–61.

14. Miernyk, J. A., and Randall, D. D. 1987. Some kinetic and regulatory properties of pea mitochondrial pyruvate dehydrogenase complex. *Plant Physiol.* **83**: 306–10.

15. Moore, A. L., Gemel, J., and Randall, D. D. 1992. The activation state of mtPDC as a function of redox poise of Q pool, respiratory rate and respiratory substrate. *Plant Physiol.* **99S**: 101.

16. Randall, D. D., Miernyk, J. A., Fang, T. K., Budde, R. J. A., and Schuller, K. A. 1989. Regulation of the pyruvate dehydrogenase complexes in plants. *Ann. N.Y. Acad. Sci.* **573**: 192–205.

17. Randall, D. D., Williams, M., and Rapp, B. J. 1981. Phosphorylation–dephosphorylation of pyruvate dehydrogenase complex from leaf mitochondria. *Arch. Biochem. Biophys.* **207**: 437–44.

18. Ryle, G. J. A., Powell, C. E., and Tewson, V. 1992. Effect of elevated CO_2 on the photosynthesis, respiration and growth of perennial ryegrass. *J. Exp. Bot.* **43**: 811–18.

19. Schuller, K. A., and Randall, D. D. 1989. Regulation of pea mitochondrial pyruvate dehydrogenase complex: Does photorespiratory ammonium influence mitochondrial carbon metabolism? *Plant Phusiol.* **89**: 1207–12.

20. Williams, M., and Randall, D. D. 1979. Pyruvate dehydrogenase complex from chloroplasts of *Pisum sativum*. *Plant Physiol.* **64**: 1099–103.

10

Plant Respiration and the Concentration of Atmospheric CO_2

T. AP REES

This chapter has two aims. The first is simple: It is to pay tribute to the crucial contributions that Tolbert has made to our understanding of plant metabolism—in particular, to our knowledge of glycolate metabolism, photorespiration, and peroxisomes. The second aim is much more difficult: It is to follow Tolbert's instructions to adhere strictly to the theme of this volume—The Role of Photosynthesis in Regulating Atmospheric CO_2 and O_2—while examining the effects that plant respiration may have on any such regulation. Briefly, the answer must be very little.

Respiration produces CO_2 but is not directly affected by CO_2 concentration under natural conditions and is most unlikely to become so even if our worst fears of increased atmospheric CO_2 are realized. An increase in atmosopheric CO_2 will not affect respiration by a mass-action effect because the free-energy change during respiration is too large. We know that high concentrations of CO_2 can inhibit respiration, probably by interfering with the citric acid cycle (17), but the concentrations are far in excess of those envisaged for the atmosphere. For example, 5% CO_2 is required to reduce the oxygen uptake of bananas by 50% (22). Increasing CO_2 could affect respiration via an effect on cytosolic pH; again the predicted changes in CO_2 concentration are too small to produce a significant effect. Brown (7) has calculated that a doubling of the CO_2 concentration from 0.003% to 0.006% would increase intracellular pH only by 0.006 of a unit. Finally, the idea (23) that CO_2 effects metabolism via reactions between dissolved unhydrated CO_2 and cellular constituents, such as primary amines, has not really received any substantial support. In short, if plant respiration is going to play any role in regulating atmospheric concentrations of CO_2 as these rise, it will be indirect and not through any inhibitory effect on respiration per se. In order to assess the possibilities of any such effects, I shall follow my very specific brief from Tolbert and attempt to answer the questions he set forth.

THE MAGNITUDE OF PLANT RESPIRATION

The first question is how does the magnitude of dark respiration by plants compare to total biological respiration on Earth? In 1980, I discussed the methods available for measuring both the total consumption of substrates in plant respiration, and the contribution of the individual pathways to the complete process (4). I concluded that we had managed to make such measurements for only one plant tissue, the spadix of *Arum maculatum*. The situation has not improved significantly over the 1980s. Given that we have this much difficulty in making measurements on a single plant tissue, it therefore follows that even if we restrict ourselves to measurements of CO_2, any value for global plant respiration must be approximate in the extreme. The most convincing of the current estimates is probably that of Bolin (6), who suggests a value of 55 G tonnes of carbon per annum. This may be compared with values of 52 and 15 for decay of soil detritus and for wood decomposition, respectively. It is clear that plant respiration is appreciable in absolute terms, is comparable in magnitude to microbial decay, and is a significant component of the global carbon cycle.

RELATIONSHIP AMONG RESPIRATION, GROWTH, AND AVAILABILITY OF SUBSTRATE

The second question is, with increasing atmospheric CO_2 to produce more carbohydrate and overall growth, does dark respiration increases proportionally? What are the data on this, if any? The extent to which increased CO_2 actually will produce more carbohydrate and growth is dealt with by others. Here I consider the relationships between respiration and growth, and between respiration and substrate availability.

Growth requires energy, material, and information. In cells that are not photosynthesizing, the first two of these requirements are met by respiration. In a photosynthesizing cell, some of the products of the light reactions may be used directly to supply the energy for growth, but the respiratory pathways from triosephosphate onward will need to operate to produce the carbon skeletons needed for the polymer synthesis that occurs during growth. Thus, it follows that increased growth will require either increased respiration or an increased efficiency of existing respiration. We know too little about the latter to rule out the possibility of increased efficiency, but it is most unlikely that any increased growth resulting from higher concentrations of CO_2 will not be accompanied by increased rates of respiration.

There is ample evidence that growing tissues have high rates of respiration. In 1961, Beevers (5) wrote, "It is clear that the highest O_2 values (on a weight of tissue basis) are typically found in meristematic tissues, where cells are small and non-vacuolated, and lowest in those which have ceased active growth." This remains true today. For example, in pea roots (11) the rates of respiration of the apical 6 mm, of the stele 6–26 mm from the apex, and of the surrounding cortex were found to be, respectively, 565, 509 and 251 µl $CO_2 \cdot g^{-1}$ fresh wt$\cdot h^{-1}$. The first two regions sustain more biosynthesis than the cortical tissue. Freshly cut

Table 10.1. Relationship Between Respiration and Biosynthesis in Successive Segments of the Root of *Pisum sativum*.

Region of Root (mm from Apex)	Oxygen Uptake ($\mu l \cdot Segment^{-1} \cdot h^{-1}$)	^{14}C Recovered as Absorbed Insoluble Material %
0–0.4	0.26	9–12
0.4–1.0	3.50	10–19
1.0–1.6	3.60	15–24
1.6–3.0	5.30	13–25
3.0–6.0	2.60	8–21
6.0–16.0	4.90	9–20
16.0–26.0	4.60	9–21

Note: Samples of root sections were analyzed after incubation in 0.3 mM [6-^{14}C]glucose for 90 min.
Source: Data from Fowler and ap Rees (14).

disks of carrot storage tissue consume oxygen at 96 $\mu l \cdot g^{-1} \cdot h^{-1}$ and convert 14% of added [6-^{14}-C]glucose to polymeric material. The corresponding values for similar disks that have been aged and stimulated to grow are 221 $\mu l \cdot g^{-1} \cdot h^{-1}$ and 41% (6). The precise relationship between growth and respiration is difficult to measure because we lack quantitative information on the costs of growth, on the yield of respiration, and on the fraction of that yield that is devoted to growth. The data in Table 10.1 show how the respiration varies during the differentiation of the pea root. The incorporation of label from [6-^{14}C]glucose into insoluble material is an indication of the extent of polymer synthesis and growth. The correlation coefficient between this incorporation and the rate of respiration is 0.812. This provides clear evidence that increased respiration is associated with growth.

When the relationship between growth and respiration is considered at the level of the plant rather than on a cell or tissue basis, a more complex picture emerges. There is now appreciable evidence for considerable variation in the rates of respiratory gas exchange of mature tissues of quite closely related plants. Wilson (26) found that the oxygen uptake of the mature leaves of six genotypes of *Lolium perenne* ranged from 5.8 to 16.1 $\mu l \cdot min^{-1} \cdot g^{-1}$ dry weight. Of particular interest is the fact that these rates of respiration were found to be inversely proportional to the yield of dry matter during regrowth of plants that had been cut to a height of 5 cm. Further studies (10) revealed that the genotypes that differed widely in the respiration rates of their mature leaves did not do so with respect to the meristematic tissue at the base of the leaves, or to their mature roots. There is some evidence of other instances of a negative correlation between the rate of respiration and plant growth (25). Thus, present evidence suggests that increased growth will require increased respiration by the cells in which that growth is occurring, but that the resulting growth may produce mature tissue in which the respiration will vary quite widely from species to species. If increased CO$_2$ does lead to increased growth, the net effect will be an increase in plant respiration because respiration will be needed to support the growth and because there will be an increase in the amount of respiring material. The precise extent of this increase is not known, not only because we do not know the relationship between growth and respiration in growing cells, but more particularly because we do not understand the relationship between the respiration of mature tissues and the growth of the plant as a whole.

The implication from the studies of *Lolium* is that plants may differ in the efficiency with which they use their assimilates, and in the extent to which growth rate determines respiration. This most complex question brings us onto dangerous ground. First, it does not follow that because the yield of the slowly respiring plants is greater than that of the rapidly respiring plants, the latter are less efficient. The extent to which rapid respiration is associated with characteristics that are advantageous for survival and propagation has not been adequately explored. Second, even if we define efficiency in the relatively restricted sense of growth relative to respiratory substrate consumed, we cannot measure such efficiency. We need to know output and input. The requirements for growth are not established precisely; the details of the pathways—particularly those concerning intracellular compartmentation (3)—are frequently lacking. We are largely ignorant of the rates of turnover of the plant's constituents, and of transport costs in general. The yield of respiration is even more difficult to assess, as we have few estimates of flux, and again lack essential details of the pathway and their organization. A neglected but characteristic feature of plant respiration is that it appears to offer a wide range of alternative routes. Which alternative is taken in a particular instance is not known, but it is clear that the route taken affects yield. Consider the following as examples. Sucrose may be broken down by invertases or sucrose synthase. Starch may be degraded hydrolytically or phosphorolytically. The presence of pyrophosphate:fructose 6-phosphate 1-phosphotransferase provides an additional link between hexose monophosphates and triosephosphate (9). Carbon can leave glycolysis via pyruvate kinase or phosphoenolpyruvate carboxylase (8). $NADH_2$ may be oxidized through the conventional respiratory chain or via the cyanide-resistant alternative oxidase. These alternatives are not exclusive, and all may operate to varying degrees in any one tissue. It is particularly important to realize that not only do we not know the flux through these pathways, but we are also ignorant of their significance, or indeed, if they are significant in determining the plant's survival.

In general, I suggest that, although there is a complex relationship between plant growth and repiration, increased growth will entail increased respiration.

RESPIRATION AND SUBSTRATE AVAILABILITY

Closely related to the discussion in the previous section is the relationship between substrate availability and the rate of plant respiration—in particular, the extent to which an increase in substrate level will lead directly to an increase in respiration. This question is central to the current debate as to how plant respiration is controlled, and is relevant to the problem of how plant respiration might affect the concentration of atmospheric CO_2.

The conventional view of the control of plant respiration is that, like carbohydrate oxidation in other organisms, it is regulated primarily by the rate at which the products of respiration are used—that is, by the demand for ATP, NADH$_2$, NADPH$_2$, and carbon skeletons for biosynthesis. In the presence of adequate substrate, the rate of respiration is seen as being regulated by the demands made upon it. If this is so, then an increase in substrate would not affect the rate of respiration directly, but could do so indirectly through any effect that increased substrate had on growth. Evidence for this view is provided by correlations between respiration and the needs for energy and carbon skeletons, and by the evidence that the products of the respiratory pathways cause feedback inhibition of reactions that contribute to the control of the initial steps of carbohydrate oxidation.

Largely on the basis of estimates of differing respiratory efficiencies, relationships between substrate level and respiration, and activity of the alternative oxidase, Lambers and his colleagues have proposed a theory of "wasteful respiration." This holds that sugar in excess of that required for the production of carbon skeletons for growth, ATP production, osmoregulation, and storage as carbohydrate reserves is oxidized via a "wasteful respiration" proceeding through the cyanide-resistant pathway. If this is so, then an increase in atmospheric CO$_2$ might cause an increase in respiratory substrate, while wasteful respiration might result in the immediate return of most of that substrate to the atmosphere.

The evidence for wasteful respiration is given by Lambers (19, 20) and by Lambers and Rychter (21). I have discussed this evidence recently (2), and have argued that the case is not proven. As my arguments remain unchanged, I shall not repeat them. Whether or not one accepts the concept of wasteful respiration, we should bear in mind that the relationship between substrate availability and rate of respiration is not known. There are instances in which addition of substrate has been shown to increase respiration, but the tissues were stressed and the effect was to restore the rate to an initial level. Attempts to increase respiration above the rate in vivo by adding substrate have not produced the stimulation expected had there been a wasteful respiration (12). There are correlations between respiration and substrate content, and models in which carbon flux for biosynthesis and respirations are proportional to substrate concentration are well supported by experimental data (13). However, no casual relationships have been established. We do not know whether increased content of substrate causes wasteful respiration, or stimulates respiration indirectly by increasing growth, or whether substrate content can communicate in some way with feedback mechanisms responsible for respiratory control. In conclusion, it is likely that increased availability of substrate will increase plant respiration, not directly, but through increased demands made by increased growth.

RESPIRATION OF PLANTS GROWN AT HIGH CO$_2$

The only satisfactory way of determining the relationship between plant respiration and atmospheric CO$_2$ is by experiment. Sadly this is a much neglected topic (27). Few experiments have been reported; the results are disparate and are summarized in Table 10.2. Only two major conclusions may be drawn. First, the response of plant respiration to growth at elevated CO$_2$ varies with the species, the tissue, and

Table 10.2. Effects of Growth at High Concentrations of CO_2 on Rate of Respiration.

Plant	CO_2 Concentration During Growth $\mu l\ CO_2 \cdot L^{-1}$ air	Days Grown in High CO_2	Respiration Rate of Plant from High CO_2 as a Percentage of Plant Growth in Air	References
Soybean, leaf.	1000[a]	8	190	Hrubec et al. (16)
		13	123	
		15		Gifford et al. (15)
Wheat, roots	680[b]	15	98	
		22	78	
		29	97	
Mungbean, roots	680[b]	27	80	Poorter et al. (24)
Sunflower, roots	680[b]	24	126	
Plantain, shoots	700[c]	14	120	
		21	82	
		28	109	
Plantain, roots	700[c]	7	221	
		14	167	
		21	120	
		28	100	

[a] Plants transferred to high CO_2 after germination and growth in air for 10 days.
[b] Plants germinated and grown in high CO_2.
[c] Plants transferred to high CO_2 after germination and growth in air for 18 days..

the period of exposure to the high concentration of CO_2. If respiration responds to the demands made upon it, then such variation is perhaps not surprising. The second conclusion is that we do not have enough data, and our knowledge is too narrowly based.

RESPIRATION AND CONCENTRATION OF ATMOSPHERIC CO_2

The final question is will changes in plant respiration have any effect on net CO_2 uptake by photosynthesis that will alter the atmospheric CO_2? The answer to this question must almost certainly be no, as organic carbon is continuously recycled over a relatively short period of time. Plants can only respire what they photosynthesize. Much (30–50%) of the carbon fixed in a day's photosynthesis is promptly respired to CO_2 the following night (21). The remainder will be stored or built into the plant structure. By and large, the former will be respired eventually, and the latter will be broken down through respiration and fermentation by microorganisms. The net effect is that the carbon will be released as CO_2. The rate of this microbial release of CO_2, relative to that of photosynthesis, might change. For example, an increase in temperature resulting from elevated atmospheric CO_2 could have differential effects on the rate of photosynthesis, on the one hand, and the rates of microbial respiration and fermentation on the other. If this occurs, the amount of organic carbon locked up in the soil could be altered with a consequent effect on the concentration of atmospheric CO_2. However, the key factors in any such change are likely to be photosynthesis, and microbial respiration and fermentation, rather than the respiration of higher plants. Plant respiration cannot exceed photosynthesis for long and allow life to continue. Thus any increase in plant respiration is unlikely to affect the concentration of atmospheric CO$_2$ because the substrate must be obtained from that CO$_2$. In the short term, the changes in the rate of plant respiration may affect the amplitude of the annual downturn cycle of CO$_2$, but in the long term this effect is not likely to regulate the atmospheric concentration of CO$_2$. The only way plant respiration can mitigate the rise in atmospheric CO$_2$ is by stopping, and thus killing, the goose that lays the golden egg.

ACKNOWLEDGMENTS

I am grateful to Rachel Averill and Ian McKee for their critical reading of my manuscript.

REFERENCES

1. ap Rees, T., and Beevers, H. 1960. Pentose phosphate pathway as a major component of induced respiration of carrot and potato slices. *Plant Physiol.* **35**: 839–47.

2. ap Rees, T. 1988. Hexose phosphate metabolism by nonphotosynthetic tissues of higher plants. In *The Biochemistry of Plants: A Comprehensive Treatise*, Vol. 14, ed. J. Preiss, pp. 1–33. Academic Press, New York.

3. ap Rees, T. 1987. Compartmentation of plant metabolism. In *The Biochemistry of Plants: A Comprehensive Treatise*, Vol. 12, ed. D. D. Davies, pp. 87–115, Academic Press, New York.

4. ap Rees, T. 1980. Assessment of the contributions of metabolic pathways to plant respiration. In *The Biochemistry of Plants: A Comprehensive Treatise*, Vol. 12, ed. D. D. Davies, pp. 87–115. Academic Press, New York.

5. Beevers, H. 1961. *Respiratory Metabolism in Plants*. Harper, New York.

6. Bolin, B. 1986. Requirements for a satisfactory model of the global carbon cycle and current status of modeling efforts. In *The Changing Carbon Cycle: A Global Analysis*, ed. J. R. Trabalka and D. E. Reichle, pp. 403–24. Springer-Verlag, New York.

7. Brown, A. W. 1985. CO$_2$ and intracellular pH. *Plant Cell Environ.* **8**: 459–65.

8. Bryce, J. H., and ap Rees, T. 1985. Rapid decarboxylation of the products of dark fixation of CO$_2$ in roots of *Pisum* and *Plantago*. *Phytochemistry* **24**: 1635–8.

9. Dancer, J. E., and ap Rees, T. 1989. Relationship between pyrophosphate:fructose-6-phosphate 1-phosphotransferase, sucrose breakdown, and respiration. *J. Plant Physiol.* **135**: 197–206.

10. Day, D. A., DeVos, O. C., Wilson, D., and Lambers, H. 1985. Regulation of respiration in the leaves and roots of two *Lolium perenne* populations with contrasting mature leaf respiration and yield. *Plant Physiol.* **78**: 678–83.

11. Dick, P. S., and ap Rees, T. 1976. Sucrose metabolism by roots of *Pisum sativum*. *Phytochemistry* **15**: 255–9.

12. Farrar, J. F. 1985. The respiratory source of CO$_2$. *Plant Cell Environ.* **8**: 427–38.

13. Farrar, J. F. 1990. The carbon balance of fast growing and slow-growing species. In *Causes and Consequences of Variation in Growth Rate and Productivity of Higher Plants*, ed. H. Lambers, M. L. Cambridge, H. Konings, and T. L. Pons, pp. 241–56. SPB Academic Publishing bv, The Hague.

14. Fowler, M. W., and ap Rees, T. 1990. Carbohydrate oxidation during differentiation in roots of *Pisum sativum. Biochim. Biophys. Acta* **201**: 33–44.

15. Gifford, R. M., Lambers, H., and Morison, J. I. L. 1985. Respiration of crop series under CO_2 enrichment. *Physiol. Plant* **63**: 351–5.

16. Hrubec, T. C., Robinson, J. M., and Donaldson, R. P. 1985. Effects of CO_2 enrichment and carbohydrate content on the dark respiration of soybean. *Plant Physiol.* **79**: 684–9.

17. Knee, M. 1973. Effects of controlled atmosphere storage on respiratory metabolism of apple fruit tissue. *J. Sci. Fd. Agric.* **24**: 1289–98.

18. Kraus, E., Wilson, D., Robson, M. J., and Pilbeam, C. J. 1990. Respiration: Correlation with growth rate and its quantitative significance for the net assimilation rate and biomass production. In *Causes and Consequences of Variation in Growth Rate and Productivity of Higher Plants*, ed. H. Lambers, M. L. Cambridge, H. Konings, and T. L. Pons, pp. 187–98. SPB Academic Publishing bv, The Hague.

19. Lambers, H. 1982. Cyanide-resistant respiration: A non-phosphorylating electron transport pathway acting as an overflow. *Physiol. Plant* **55**: 478–85.

20. Lambers, H. 1985. Respiration in intact plants and tissues: Its regulation and dependence on environmental factors, metabolism and invaded organisms. In *Encyclopedia of Plant Phhsiology* (*New Series*), Vol. 18, ed. R. Douce, D. A. Day, pp. 418–73. Springer-Verlag, Berlin.

21. Lambers, H., and Rychter, A. M. 1990. The biochemical background of variation in respiration rate: Respiratory pathways and chemical composition. In *Causes and Consequence of Variation in Growth Rate and Productivity of Higher Plants*, pp. 199–225. SPB Academic Publishing bv, The Hague.

22. McGlasson, W. B., and Wells, R. B. H. 1972. Effects of O_2 and CO_2 on respiration, storage life and organic acids of green bananas. *Aust. J. Biol. Sci.* **25**: 34–42.

23. Mitz, M. A. 1979. CO_2 biodynamics: A new concept of cellular control. *J. Theor. Biol.* **80**: 537–51.

24. Poorter, S., Pot, S., and Lambers, H. 1988. The effect of an elevated atmospheric CO_2 concentration on growth, photosynthesis and respiration of *Plantago major. Physiol. Plant* **73**: 533–9.

25. Van Der Werf, A., Hirose, T., and Lambers, H. 1990. Variation in root respiration: Causes and consequences for growth. In *Causes and Consequences of Variation in Growth Rate and Productivity of Higher Plants*, ed. H. Lambers, M. L. Cambridge, H. Konings, and T. L. Pons, pp. 228–40. SPB Academic Publishing bv, The Hague.

26. Wilson, D. 1975. Variation in leaf respiration in relation to growth and photosynthesis in *Lolium. Ann. Appl. Biol.* **80**: 323–38.

27. Woodward, F. I., Thompson, G. B., and McKee, I. F. 1991. The effects of elevated concentrations of carbon dioxide on individual plants, populations, communities and ecosystems. *Ann. Bot.* **67**: 23–38.

IV
CO₂-CONCENTRATING PROCESSES

11

Roles of C_4 Photosynthetic Plants During Global Atmospheric CO_2 Changes

CLANTON C. BLACK

C_4 photosynthesis likely developed from two environmental pressures—namely, a lowering of the Earth's atmospheric CO_2 levels such that CO_2 was limiting for C_3 photosynthesis and a concurrent warming of the Earth's atmosphere. Atmospheric CO_2 levels and temperatures are predicted to increase through this century and into the next. Predictably C_4 plants will respond to and modify these environmental changes because C_4 plants have developed very effective mechanisms for modifying their internal CO_2 levels in both the gas and aqueous micro-environments, and warmer weather favors them. The C_4 leaf lowers its CO_2 level of the internal gas phase, well below atmospheric, thus forming a substantial CO_2 diffusion gradient; but simultaneously the aqueous CO_2 level in bundle sheath cells is raised well above atmospheric CO_2 gas/water equilibrium values such that CO_2 is not rate-limiting for C_4 photosynthesis. Therefore, C_4 plants have a strong control over the microenvironments of CO_2 within their leaves which, in turn, strongly influences global atmospheric CO_2 levels because these plants can decrease atmospheric CO_2 levels (a) faster and (b) lower than can other photo-synthetic plants and (c) without a loss of CO_2 in the light—that is, without external expression of leaf photorespiration.

As the predicted increases in CO_2 and temperature occur into the twenty-first century, the roles and responses of C_4 plants will be as follows:

1. The stomatal conductance of C_4 plants will decrease, and water use efficiency will increase.
2. The deleterious influences of low temperatures on C_4 plants will be partially relieved.
3. C_4 crops and weeds will rapidly migrate to warmer geographic regions.
4. C_4 weeds will become prevalent over a greater geographic area.
5. Global dry matter production from C_4 plants will increase slightly (ca. 10–15%) as a result of the increases in CO_2 and C_4 growth in more extensive geographic areas.
6. C_4 photosynthesis will continue to lower the global CO_2 levels each growing season.

7. In an emergency, more extensive growth of C$_4$ plants may serve to lower global atmospheric CO$_2$ levels rapidly.
8. In general, C$_4$ plants will be favored in environments with several stresses such as high temperatures, water deficits, and low soil fertility.

Following the discovery of C$_4$ photosynthesis, an overarching paradigm quickly developed about the environmental responsiveness of C$_4$ plants in that, when compared with the more established C$_3$ photosynthesis, unique response patterns to environmental changes emerged quite plainly with C$_4$ plants. Even though the early C$_4$ biochemical work occurred with plants grown in tropical environments, it was soon evident that C$_4$ plants possessed many traits that allowed them to be highly productive and competitive in a large variety of agricultural and natural ecological environments (5, 23). These environmentally related traits of C$_4$ plants formed distinct patterns, including:

1. Little effect from a lowering of atmospheric O$_2$ levels or a rise in CO$_2$ levels on growth or photosynthesis
2. High optimal temperatures for growth and photosynthesis
3. Low tolerance for cold temperatures
4. Efficient utilization of resources such as water, nitrogen, iron, CO$_2$, and other nutrients in productive growth.
5. Nonsaturating responses of leaf and canopy photosynthesis to increasing sunlight irradiances.

Today we know the mechanisms and basis for many of these environmental traits, and these allow us to predict how C$_4$ plants will respond to environmental changes and to predict roles for C$_4$ plants in the control of global atmospheric CO$_2$ levels.

Such distinct responses to a variety of environmental changes are integrated in C$_4$ plants with biochemical and structural traits to support strongly the environmental paradigm so firmly associated with the C$_4$ pathway of CO$_2$ assimilation (3, 10, 23, 30). How traits are integrated can even be artistically illustrated in the leaves of many C$_4$ plants, which exhibit a classical cell architecture, called *Kranz leaf anatomy* (19). A unique leaf anatomy is an integral part of C$_4$ photosynthesis because to be functional, C$_4$ plants require the distinct expression of specific genes and other biochemical activities within two green leaf cell types (11, 21, 41). When one considers global CO$_2$ changes, a crucial trait of C$_4$ plants, derived from combining structural and biochemical traits, is their ability to modify favorably their internal leaf CO$_2$ microenvironments such that current atmospheric CO$_2$ levels do not limit C$_4$ photosynthesis or growth! Indeed, C$_4$ photosynthesis apparently evolved in response to a lowering of global CO$_2$ levels, by C$_3$ photosynthesis, such that the CO$_2$ level of air became rate-limiting for C$_3$ photosynthesis (presented elsewhere in this volume).

Therefore, concentrating upon CO$_2$, we shall (a) recount some experiments on the environmental responsiveness of C$_4$ plants; (b) consider the cellular architecture of C$_4$ leaves in regard to the pathway CO$_2$ follows during C$_4$ metabolism; (c) integrate ideas about environmental traits with cellular gene expression and biochemistry within C$_4$ leaves to learn how they regulate their internal leaf CO_2 microenvironments; and (d) ask how C$_4$ plants regulate global atmospheric CO_2 levels, and predict their roles and responses as global CO_2 levels and temperatures increase, using as an example a doubling of atmospheric CO_2 and a 5°C rise in temperature.

ENVIRONMENTAL RESPONSIVENESS OF C$_4$ PLANTS

The biochemical discoveries that began uncovering the foundations of C$_4$ photosynthesis appeared in 1965–6 (24, 29). However plants with unusual or seemingly inexplicable environmental traits were studied much earlier, even though these plants were not known to be C$_4$! For example, shortly after World War I, a wide range of plant water-use efficiency values were found in a study with a number of higher plant species (42). But such variations among plant species made little intellectual sense until the detection of C$_4$ photosynthesis. With that discovery, it was soon realized that this range of water-use efficiency values could be subdivided into two nonoverlapping groups that were formed by two discrete types of plants, namely, C$_4$ and C$_3$ species (5). Over the next couple of years, by comparing a host of environmental, structural, and biochemical traits in C$_4$ plants versus C$_3$ plants, and then versus Crassulacean acid metabolism (CAM) plants as well, an intellectual organization of these traits developed that literally engulfed and redirected the thinking of plant biologists throughout the world. C$_4$ plants include some of our most productive crops (e.g., corn, sugarcane, and sorghum) but also weeds considered to be the world's worst by agriculturists (e.g., purple nutsedge, bermuda grass, and pigweed) (3, 10, 23, 25). Because we wish to concentrate upon CO_2 in the context of the global environment and the roles of C$_4$ plants, it is vastly beyond the scope of this chapter to provide a thorough review of this renaissance period in plant biology; instead a list of specialized reviews is provided (e.g., Refs. 11, 21, 31, 41).

Regarding environmental CO_2, C$_4$ leaf photosynthesis is near saturation at current levels of atmospheric CO_2, and C$_4$ plants maintain a high rate of photosynthesis in air without a detectable loss of CO_2 in the light due to photorespiration. How were these conclusions reached? In the early work on the influence of CO_2 concentration upon corn, a C$_4$ plant, Moss et al. (36) demonstrated that a curvilinear response of photosynthesis occurred as CO_2 was varied from about 20 up to 300 μl $CO_2 \cdot L^{-1}$ air with little indication of saturation. However, after the discovery of C$_4$ photosynthesis, more detailed studies demonstrated that C$_4$ photosynthesis saturated slightly above current atmospheric CO_2 level (illustrated in Fig. 11.1). Also shown in Figure 11.1 is the lack of CO_2 release by C$_4$ leaves at low CO_2 levels. In other words, there is little loss of CO_2 from C$_4$ leaves in the light, showing that little photorespiratory CO_2 is lost from C$_4$ leaves (10, 23). Related to this concept was the idea, recognized three decades ago, that the light CO_2 compensation concentration for C$_4$ leaves is nearly zero (35); that is, a C$_4$ leaf will remove the CO_2 from a closed atmosphere to nearly zero.

In asking why so little CO_2 is lost from C$_4$ leaves, we soon learned that the C$_4$ mesophyll cell effectively traps the CO_2 derived either from the atmosphere or from internal leaf respiratory processes. Indeed a C$_4$ leaf can lower its internal

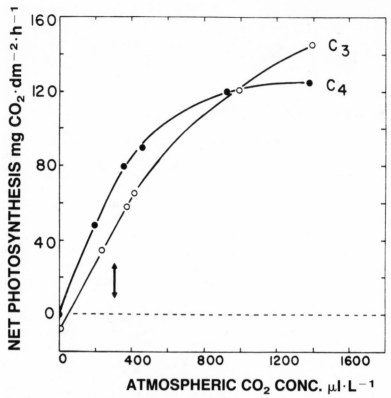

Fig. 11.1. The responses of C$_4$ and C$_3$ leaf photosynthesis to increasing atmospheric CO$_2$ concentrations. The vertical arrow indicates 350 μl CO$_2 \cdot$l^{-1} air. [Adapted from Brown (7).]

gaseous CO$_2$ much below atmospheric. As shown in Figure 11.2, the intercellular CO$_2$ concentration in C$_4$ leaves has to reach only about 100 μl CO$_2 °$L^{-1} air, whereas current air contains about 360 μl $\cdot$ L^{-1}. Thus by lowering the leaf airspace CO$_2$ level to about 1/3 that of air, C$_4$ leaves create a diffusion gradient for atmospheric CO$_2$. C$_4$ leaves maintain this ratio somewhat near 1:3 as atmospheric CO$_2$ increases several fold (31, 40, 48). Today we know that the combination of this gradient, internal leaf CO$_2$ in HCO$_3^-$ scavenging, and the movement of C$_4$ organic acids to the bundle sheath cells for decarboxylation, results in the release of CO$_2$ at well above saturation concentration for the carboxylation of RuBP.

Thus *a pivotal feature of C$_4$ plants is their ability to favorably modify their internal CO$_2$ microenvironments such that CO$_2$ is not rate-limiting.* It is also important to note that additional advantages accrue to the C$_4$ plant as a result of this highly efficient internal leaf CO$_2$ metabolism. Since intercellular CO$_2$ concentrations can fall to quite low levels without greatly decreasing the CO$_2$ fixation rate, the C$_4$ leaf can effectively conserve water by restricting the stomatal aperture. As cited earlier, the idea of efficient water use by C$_4$ plants was gained soon after the discovery of C$_4$ photosynthesis by reinterpreting the earlier literature (5, 12). In addition, more recent experiments with elevated CO$_2$ levels also show that C$_4$ plants actually increase their water use efficiency (i.e., stomatal conductance

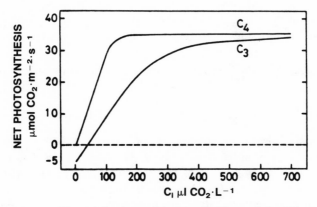

Fig. 11.2. The responses of C_4 and C_3 photosynthesis to leaf intercellular CO_2 concentrations (C_i). Data are from measurements on corn (C_4) and sunflower (C_3) leaves. [Adapted from Raschke (40).]

decrease) as atmospheric CO_2 levels rise (40, 48); the reason is that stomatal apertures decrease in size when a higher intercellular CO_2 level (Fig. 11.2) occurs from the higher atmospheric CO_2. Figure 11.3 demonstrates an almost linear increase in water use efficiency in two C_4 plants when the CO_2 level in which the plants were grown was increased about three fold. There was some increase in plant growth, but less than half of the water was needed to produce the same dry matter as in air (Fig. 11.3). Thus, the C_4 plant represents an adaptation of

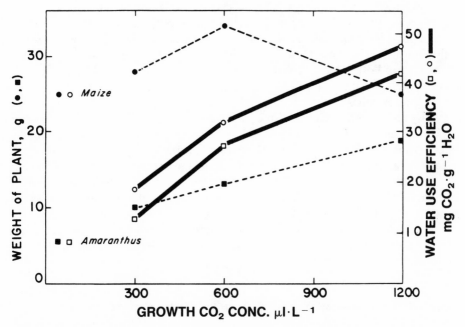

Fig. 11.3. The responses of water use efficiency and dry matter production to increasing CO_2 concentrations used for growing two C_4 plants, corn and *Amaranthus*. [Adapted from Black (4).].

photosynthesis for the effective utilization of water; and, at the projected increase in atmospheric CO$_2$, the number of plants required to produce a unit of dry matter will decrease. And, because photosynthesis is a linear function of leaf nitrogen content (6, 7), the amount of dry matter produced per unit of available N will increase, particularly with C$_4$ plants growing in soils with low N contents (48).

CELLULAR ARCHITECTURE OF C$_4$ LEAF PHOTOSYNTHESIS

The idea of a special cellular architecture in association with C$_4$ photosynthesis came from the work of at least four separate research groups late in 1967 (for a historical account, see Refs. 3, 23). Much earlier, about 1880 (19), the special cellular arrangement was called "Kranz leaf anatomy"; it was depicted as green bundle sheath cells centrifugally surrounding leaf veins that were themselves surrounded by green mesophyll cells. This apparent anatomical oddity of some plants unfortunately was not widely appreciated botanically nor understood functionally until the late 1960s when it was finally shown that Kranz leaf anatomy leaf plants were C$_4$ plants (23). Early in the 1970s, it was thought that all C$_4$ plants had Kranz leaf anatomy (23). Even though Kranz anatomy plants will be used here to illustrate C$_4$ photosynthesis, two caveats regarding C$_4$ leaf anatomy should be kept in mind: (a) Not all C$_4$ plants have Kranz leaf anatomy, and (b) not all plants with green bundle sheath cells are C$_4$ plants. For example, *Suaeda monoica* is a C$_4$ plant that has only rudimentary bundle sheath cell chloroplasts, but it has two layers of green cells under the epidermis of its semicylindrical leaves (43). In sharp contrast to *Suaeda* and to other Kranz anatomy plants, *Panicum milioides*, a C$_3$/C$_4$ intermediate plant (8), has well-developed bundle sheath chloroplasts, but it lacks the proper enzyme balance in the two leaf cell types to perform C$_4$ photosynthesis. Hence the mere presence of two types of green leaf cells is not sufficient to establish the presence of C$_4$ photosynthesis; the appropriate enzmymes and other biochemical processes—for example, the ability to strongly control internal leaf CO$_2$ levels—must be expressed by the two green cells in order for C$_4$ photosynthesis to occur (next section).

Early studies on C$_4$ biochemistry versus cellular ultrastructure noted also that C$_4$ bundle sheat cells often were coated with a suberized layer that was relatively impermeable to CO$_2$ (30). This suberized layer is critical in restricting CO$_2$ leakage, because it allows the C$_4$ bundle sheath cell to raise its CO$_2$ concentration well above equilibrium *both inter*cellularly and within mesophyll cells. Therefore, in all documented cases, the distilled essence of C$_4$ leaf cellular architecture can be stated simply: Two green cell types are always found in a C$_4$ plant. These cells do not have a fixed morphological or anatomical arrangement but rather, in an interdependent manner, share the overall tasks required to complete C$_4$ leaf photosynthesis.

CELLULAR GENE EXPRESSION AND C$_4$ LEAF BIOCHEMISTRY

Indeed the requirement for gene expression in two separate cells in order to complete various biochemical pathways and cycles is likely the greatest advance

Table 11.1. Enzymes Expressed Primarily in Mesophyll Cells or in Bundle Sheath Cells/Strands of C$_4$ Leaves.

Enzymes	Mesophyll	Bundle Sheath[a]
	% in Cell Type	
Carbon assimilation		
Carbonic anhydrase	99	<1
PEP carboxylase	98	<2
Malic dehydrogenase (NADP)	99	<1
Pyruvate P$_i$ dikinase	99	<1
Rubisco	<2	98
Ribulose 5-P kinase	<1	99
Malic enzyme (NAD$^+$ and NADP$^+$)	<1	99
Nitrogen assimilation		
Nitrate reductase	100	0
Nitrite reductase	100	0
Sulfur assimilation		
ATP sulfurylase	<2	98

[a] Each of these cell types was isolated as whole cells, protoplasts, or bundle sheath stands, and was used to assay for the enzymes (13, 14, 18, 22, 33, 39).

in C$_4$ plants! In other words, C$_4$ metabolism is much more than the sum of the carboxylation/decarboxylation reactions, of photosynthetic CO$_2$ metabolism since the division of biochemical activities in two cell types extends into essentially all of the C$_4$ leaf's biochemistry. Table 11.1 gives a condensed list of enzymes that are expressed in mesophyll versus bundle sheath cells for carbon, nitrogen, and sulfur metabolism. Today we know that complete C$_4$ metabolic pathways for the assimilation of carbon, nitrogen, and sulfur all require the spatial placement of essential functions into each of these cooperating cells.

An overall integrated outline of CO$_2$, nitrogen, and sulfur photoassimilation is given in Figure 11.4 for a C$_4$ leaf. A division of labor between the two green cell types is evident in the photoassimilation of these essential elements; illustrated is the fact that the assimilation of carbon and nitrogen is initiated in the C$_4$ mesophyll cell while the assimilation of sulfur is initiated in the bundle sheath cell. Clearly the compartmentation of C$_4$ metabolism into two types of green cells shows that nature has evolved ways to use each green cell type to carry out individual functions in an efficient cooperative manner for the assimilation of essential elements by C$_4$ plants.

It is evident immediately that each cell type does not need to synthesize a full complement of enzymes—for example, Rubisco, PEP carboxylase, nitrate reductase, ATP sulfurylase, and so on—for the whole organism to photoassimilate its essential nutrients (Figure 11.4, Table 11.1). Therefore, the whole C$_4$ plant is quite efficient in utilizing its nutritional resources (e.g., nitrogen, sulfur, iron, CO$_2$, etc.). This efficient utilization of resources allows C$_4$ plants to play a major role in many ecosystems (e.g., in periods of environmental changes, as in times of increasing

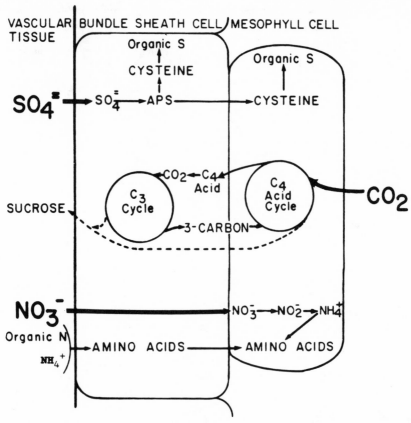

Fig. 11.4. Cellular pathways for the photoassimilation of carbon, nitrogen, and sulfur in mesophyll and bundle sheat cells of C$_4$ plants. [Adapted from Black (4), and Campbell and Black (11).]

atmospheric CO$_2$ levels, when more carbon can be reduced with the same amount of nutrient investment). Note that C$_4$ plants are treated as a group here, even though three subtypes of C$_4$ plants are known to exist on the basis of the decarboxylation of 4-carbon acids (10, 21). In outline, each of the three C$_4$ subtypes carries out a similar set of processes in regard to CO$_2$ and HCO$_3^-$, but each use different decarboxylases. Their environmental responses to such parameters as temperature increases and essential nutrient are likely similar, although these areas have not been thoroughly investigated.

A more detailed pathway for CO$_2$ assimilation within C$_4$ leaves, shown in Figure 11.5, demonstrates how C$_4$ plants regulate the levels of CO$_2$ and HCO$_3^-$ within their leaf microenvironments. Since PEP carboxylase in the mesophyll cell is known to use HCO$_3^-$ as its substrate (22), atmospheric CO$_2$ first must be converted to HCO$_3^-$. It was widely assumed that carbonic anhydrase equilibrates CO$_2$ and HCO$_3^-$ in leaf tissues (23). That question in C$_4$ plants was addressed quantitatively recently when it was shown that the C$_4$ carbonic anhydrase is operating close to its K_m within C$_4$ mesophyll cells (22). Indeed carbonic anhydrase

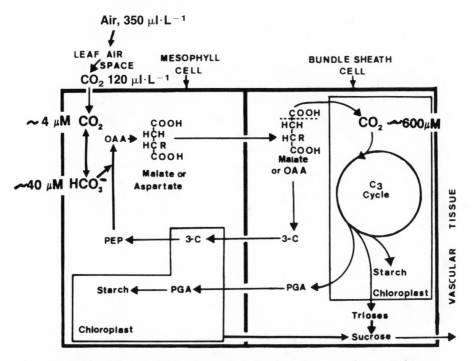

Fig. 11.5. The C$_4$ pathways of CO$_2$ photoassimilation from the atmosphere to Rubisco, showing the intercellular CO$_2$ gas concentration; the CO$_2$ concentration in mesophyll cells in equilibrium with HCO$_3^-$ catalyzed by carbonic anhydrase; PEP carboxylation; then bundle sheath cell decarboxylation; and the CO$_2$ concentration available for Rubisco. The assimilation pathways and cycles are completed in outline through to the synthesis of sucrose for translocation and starch storage. [Adapted from Ray and Black (41) and Hatch and Burnell (22).]

is the first enzyme of CO$_2$ metabolism within C$_4$ plants, and apparently its activity may limit C$_4$ photosynthesis (9, 17, 22, 26).

Thus C$_4$ photosynthesis is a cooperative activity between two green cells: (a) mesophyll cells, which trap CO$_2$ as HCO$_3^-$ and 4-carbon acids; and (b) bundle sheath cells, which decarboxylate the C$_4$ acids, maintain elevated CO$_2$ concentrations, and refix the CO$_2$ via C$_3$ photosynthesis (Fig. 11.5). This character of interdependence among leaf cells in which each cell type has several required roles in order to complete a metabolic cycle is not characteristic of any other photosynthetic organism. Hence the thesis will be developed that C$_4$ *plants play a unique role in the global atmospheric CO$_2$ cycle because of their strong ability to control their internal leaf microenvironments.*

HOW DO C$_4$ LEAVES REGULATE THEIR INTERNAL GAS AND AQUEOUS CO$_2$ MICROENVIRONMENTS?

Looking in more detail at CO$_2$ metabolism within C$_4$ leaves, we have seen that the gas phase within the leaf is reduced to levels perhaps 1/3 that of our current air (Fig. 11.2). The gaseous level of CO_2 in equilibrium with the aqueous portion of mesophyll cells is about 4 µM CO_2 in solution (values given in Fig. 11.5). The first enzyme utilized in CO_2 metabolism is the mesophyll carbonic anhydrase; it converts dissolved CO_2 to HCO_3^-. PEP carboxylase rapidly fixes the HCO_3^- as a C_4 acid carboxyl group. C_4 acids then are transported to bundle sheath cells for decarboxylation, where the Rubisco substrate, CO_2, is released and reached levels that essentially saturate Rubisco (about 600 µM, Fig. 11.5). The suberized bundle sheath cell, discussed earlier, prevents the leakage of CO_2 in and out of the bundle sheath cell, and the low carbonic anhydrase there (Table 11.1) is insufficient to convert it to HCO_3^- (16, 17, 20, 22, 26). Therefore the bundle sheath cell aqueous microenvironment of the chloroplast has a CO_2 concentration nearly 100 times higher than intercellular air in equilibrium with water. Also this result supports the conclusion that C_4 photosynthesis is limited by either the amount or the activation state of Rubisco.

This ability to regulate strongly the CO_2 and HCO_3^- levels within the C_4 leaf in a stepwise, cooperative fashion means that C_4 photosynthesis is not limited by current atmospheric CO_2 levels. In an overview, in the complete C_4 pathway CO_2 is converted to HCO_3^-, HCO_3^- is carboxylated and released as CO_2, the CO_2 is carboxylated, and finally the carboxyl of 3-phosphoglycerate (3-PGA) is reduced via glyceraldehyde 3-P dehydrogenase. Both green cell types reduce trioses, store starch, and synthesize sucrose for export throughout the C_4 plant (Fig. 11.5). Of course, at any step, CO_2 can be subject to diffusion; but the unique two-cell-architecture, combined with the required spatial enzyme expressions in each cell, results in a very efficient mechanism for C_4 plants to regulate their internal CO_2 microenvironments such that a highly efficient fixation of atmospheric CO_2 occurs, and essentially no CO_2 is lost from the leaves of illuminated C_4 plants.

Note that C_4 plants are insensitive, relative to C_3 plants, to changes in atmospheric O_2 between 2% and our current 21%. O_2 insensitivity is reflected in the lack of influence on the action spectrum of photosynthesis on the photosynthesis quantum requirement, on the rate of photosynthesis, or on growth (7, 10, 11, 23, 31).

HOW WILL C$_4$ PHOTOSYNTHESIS RESPOND TO A DOUBLING OF ATMOSPHERIC CO$_2$ LEVELS AND TO A 5°C RISE IN TEMPERATURE?

This question is asked because of current predictions that such environmental changes will occur through the middle of the twenty-first century. The understanding of C_4 leaf mechanisms just presented can be used to predict the responses of C_4 plants to global climate changes. If Figure 11.5 is used as the model, the respective values would be 700 µl $CO_2 \cdot L^{-1}$ of air in the atmosphere; about 230 µl $CO_2 \cdot L^{-1}$ in the leaf intercellular spaces; and about 9 µM CO_2 in solution in mesophyll cells. There carbonic anhydrase would equilibriate 9 µM CO_2 with about 90 µM HCO_3^- for PEP carboxylase; and finally C_4 acids would move to bundle sheath cells to release CO_2 at about 600 µM. Hence C_4 plants still would form a strong CO_2 gradient from the atmosphere and maintain a saturated CO_2 level for photosynthesis.

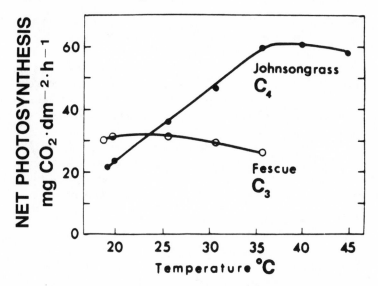

Fig. 11.6. Temperature response curves of a C_4 and C_3 photosynthetic plant (12).

Environmentally, the "Achilles heel" or weakest triat of C_4 plants is their failure to cope with low temperatures. As the temperature falls below about 10–15°C, C_4 plants do not function well; the reasons for this deleterious response are not clear, though enzyme lability has been implicated. But a 5°C rise in global temperatures would favor C_4 plants because of the more linear response of photosynthesis to temperature increases and a higher temperature optimum. In contrast, C_3 plants would derive little benefit from a 5°C temperature rise, except to lengthen their growing season, because they reach photosynthesis and growth optima at lower temperatures and have flatter temperature response curves than C_4 plants (illustrated in Fig. 11.6). It is widely known that C_4 plants often are annuals and that they grow well in warm/hot seasons. Hence a 5°C temperature increase would strongly favor C_4 plants.

Over the years of studying C_4 plants, three weak points have been identified. Because these are related to temperature and to CO_2 metabolism, they must be highlighted. The deleterious influences of low temperatures were just noted. And on the basis of their kinetic properties, two key enzymes of CO_2 metabolism—carbonic anhydrase and Rubisco—seem to operate near maximum capacity. How C_4 plants may regulate the expression of these proteins as global climates change is unknown. If one wishes to improve C_4 plants, however, all three aspects deserve more intensive research.

A "BELL JAR MODEL" OF HOW C₄ PLANTS MODIFY GLOBAL ATMOSPHERIC CO₂ LEVELS

In considering how C_4 plants help regulate global CO_2 levels, one can imagine a miniaturized model of our world and some terrestrial plant-related atmospheric

CO$_2$ control mechanisms from a bell jar experiment. An experiment directly showing the competition for atmospheric CO$_2$ between C$_4$ and C$_3$ plants was done several decades ago by placing both C$_3$ and C$_4$ plants in a sealed, continuously illuminated, clear container (37, 47). With the limited CO$_2$ available in these "closed bell jar-type studies," C$_4$ plants competed so effectively for CO$_2$ that the C$_3$ plants soon died. The C$_4$ plant essentially lowered the jar's CO$_2$ level, removing even respiratory CO$_2$, until the C$_3$ plants exhausted their own reserves and died; the jar's atmospheric CO$_2$ level decreased to near zero.

Several years ago we repeated those bell jar studies, (a) adding CAM plants and (b) switching to a normal day/night cycle. Under these conditions of direct competition for CO$_2$, C$_4$ plants survived the competition for CO$_2$ by C$_3$ plants, but the CAM plants survived *both* the C$_4$ and the C$_3$ plants because:

1. The CAM plants fix CO$_2$ at night.
2. CAM plant stomata seal their leaves to eliminate daytime gas losses.
3. CAM plant leaves raise the CO$_2$/O$_2$ ratio within their photosynthetic tissues such that O$_2$ does not inhibit their daytime photosynthesis (10, 23, 28). These results illustrate not only that photosynthesis reduces atmospheric CO$_2$ but also that both C$_4$ and CAM plants can reduce atmospheric CO$_2$ to lower levels that can C$_3$ plants. CAM plants are not considered further in this chapter except to note that they seldom are the dominant plants in major masses of terrestrial vegetation; they are not extensively used as crops nor are highly competitive in many geographical regions.

These bell jar studies will be used along with our features of C$_4$ plants to make a large extrapolation regarding global CO$_2$ control mechanisms. It is hypothesized that these *bell jar studies are useful in developing an understanding of how global atmospheric CO$_2$ levels are regulated by C$_4$ photosynthesis*. To present this thesis, the following features are assumed: (a) C$_4$ plants have a CO$_2$ compensation point near zero (12, 35, 37, 47), and hence, they can lower atmospheric CO$_2$ more than can C$_3$ plants; (b) C$_4$ plants are mostly annuals growing predominantly in warm temperate and tropical terrestrial environments (2, 5, 45, 46); (c) C$_4$ plants show little photosynthetic fractionation of stable carbon isotopes in contrast to C$_3$ and CAM plants (1, 44); and (d) no submerged aquatic C$_4$ plants are known.

To test this thesis partially, some data literature on global CO$_2$ levels and seasonal photosynthetic activities were plotted (Fig. 11.7). In building global carbon cycle models, it is widely accepted that "photosynthesis" plays a balancing role in the maintenance of atmospheric CO$_2$ levels (15, 27, 32, 38). Primarily photosynthetic organisms play a major role in lowering atmospheric CO$_2$ levels. This influence is easily seen in the annual peaks and troughs in seasonal data on atmospheric CO$_2$ levels (15, 31, 38). The effect also can be observed in annual peaks and troughs in the δ^{13}C values of atmospheric CO$_2$ (27, 32) because some photosynthetic organisms, particularly C$_3$ plants, are known to preferentially fix ^{12}C over ^{13}C (44). The annual peaks and troughs in the atmospheric CO$_2$ levels and δ^{13}C values are coincident (27, 32). Accepting that these annual cycles are due broadly to "photosynthetic organisms," our thesis is that *the influences of C$_4$ plants are manifested within these annual global cycles in that C$_4$ photosynthesis partially determines the depth of the seasonal CO$_2$ trough and the δ^{13}C value.*

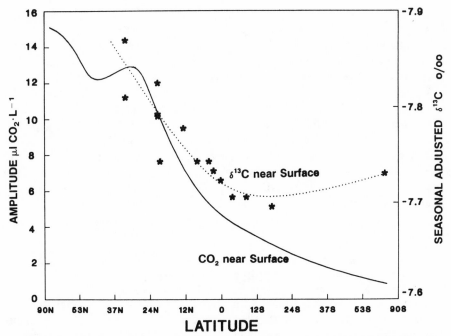

Fig. 11.7. Global latitudinal changes in the seasonal CO₂ concentration and in the δ¹³C value of atmospheric CO₂. *Left ordinate*: Latitudinal plot of the peak-to-peak amplitude in the annual atmospheric CO₂ concentrations near the Earth's surface. [Adapted from Fraser et al. (15) and Pearman et al. (38).] *Right ordinate*: Latitudinal plot of the seasonally adjusted atmospheric CO₂ δ¹³C values also near the earth's surface. [Adapted from Kealing et al. (27) and Mook et al. (32).]

First if we examine the annual amplitude of atmospheric CO₂ levels, it is strongly dependent upon latitude (Fig. 11.7). The amount of terrestrial plant photosynthesis is strongly dependent upon latitude, with the northern hemisphere containing about 65% of the terrestrial photosynthetic organisms. The height of the CO₂ amplitude clearly shows this concentration of terrestrial photosynthetic organism in the northern hemisphere (Fig. 11.7). C₄ plants are concentrated between the middle latitudes, particularly north of the Equator, and of course on land. The seasonal cycles in CO₂ levels (15, 31, 38) quickly reach a trough in warm seasons when C₄ plants are most active. Then in cooler seasons the levels rise slowly, indicating the inactivity of C₄ plants and that C₃ photosynthesis alone (e.g., evergreen forests, jungles, etc.) cannot reduce CO₂ faster than it is being released into the atmosphere. Thus *C₄ plants are the primary photosynthetic component in determininig the peak to amplitude in the CO₂ level within each annual global cycle.*

Next, we know C₄ plants don't strongly fractionate stable carbon isotopes (44). In Figure 11.7 we see a higher fractionation at northern latitudes that is moderated nearer the Equator with none in the southern hemisphere. Thus *the role of C₄ photosynthesis is to reduce carbon isotope fractionation.* Presumably in the southern hemisphere both the low annual amplitude in the CO₂ level and the low δ¹³C fractionation also are partially due to a large oceanic influence and to less terrestrial photosynthesis (15, 27, 32, 38).

In an overview on how C$_4$ plants regulate global CO$_2$ levels, evidence has already been presented that C$_4$ plants have a powerful ability within their leaves to regulate intercellular CO$_2$ levels well below atmospheric CO$_2$ (Fig. 11.5) to form a CO$_2$ gradient from the atmosphere to the mesophyll cell where dissolved CO$_2$ is converted to HCO$_3^-$. Hence a strong regulation of both the CO$_2$ and the HCO$_3^-$ pool exists within C$_4$ leaves such that C$_4$ photosynthesis is not rate-limited by atmospheric CO$_2$. C$_4$ plants form a steeper gradient for atmospheric CO$_2$ and more quickly remove atmospheric CO$_2$ to even lower levels than do C$_3$ plants. Therefore, a manifestation of the contributions of C$_4$ plant photosynthesis to the global carbon cycle is observed in both the amplitude of the seasonal cycle and in the fractionation of stable carbon isotopes (Fig. 11.7). On the basis of these considerations, it is hypothesized that *C$_4$ photosynthesis has a strong regulatory influence on global atmospheric CO$_2$ levels.*

CONCLUSIONS ON THE ROLES AND RESPONSES OF C$_4$ PLANTS TO INCREASES IN GLOBAL ATMOSPHERIC CO$_2$ LEVELS AND TEMPERATURES

The following conclusions, founded on the mechanisms and characteristics just presented for C$_4$ plants, illustrate the influences one can expect with C$_4$ plants from a global doubling of atmospheric CO$_2$ and a 5°C rise in temperature. One other ecological feature of C$_4$ plants—that they often are dominant weeds (2, 5)—must be included in order to understand the global responses of C$_4$ vegetation. In the top of the "world's worst weeds for agriculturists" (25), 14 of these are C$_4$ plants, whereas only 4 are C$_3$ plants (2, 5)! These weedy plants can migrate very rapidly, in a few years, because they already are spread over large geographical areas (45, 46). And they often are moved by common agricultural practices as with other seeds, tubers, or vegetative materials.

1. C$_4$ plant leaf stomatal conductance will decrease with increases in both atmospheric CO$_2$ and temperature; the water use efficiency of C$_4$ plants will thereby increase almost linearly with increases in CO$_2$.
2. Warmer temperatures will help C$_4$ plants, in both agricultural and natural ecosystems, to avoid the deleterious influences of cool low temperatures, and will lengthen their growing season. In brief, any increase in temperature— for example, 5°C in the twenty-first century—will favor C$_4$ plants, and a shift toward a higher proportion of C$_4$ plants relative to C$_3$ will occur in both agricultural and natural ecosystems.
3. Worldwide, C$_4$ plants will migrate to warmer geographical regions where they could not grow well previously, either because of unsuitable low temperatures of shortened growing seasons.
4. C$_4$ plants will migrate quickly to new geographical regions, in periods of a few years, to take advantage of short-term global climate changes.
5. On an annual basis, C$_4$ photosynthesis will continue to lower the global atmospheric CO$_2$ level markedly. Indeed a more extensive growth and cultivation of C$_4$ plants might be used in an emergency to lower global atmospheric CO$_2$ levels.

6. C$_4$ weeds will become more dominant problems in agricultural crops over larger beographical regions as temperatures rise.

7. The growth of C$_4$ plants will be favored in geographical regions characterized by arid climates and/or in regions of low soil fertility—for example, low nitrogen—because of the ability of C$_4$ plants to use their nutrient resources efficiently in dry matter production.

8. The dry matter production of individual C$_4$ crops will not increase appreciably—perhaps 10–15%—with a doubling of atmospheric CO$_2$. But C$_4$ plants will grow in more extensive geographical regions, and dry matter production from C$_4$ plants will thereby increase. They will lower CO$_2$ levels more than C$_3$ plants, particularly in warm seasons.

9. As CO$_2$ plants and temperatures rise, the nutrient resources available to terrestrial plants in most ecosystems will not change concurrently. Indeed, in most ecosystems the soil nutrient supply will be less—for example, as C$_3$ plant biomass increases. The competition for soil nutrients between C$_4$ and C$_3$ plants will increase, and C$_4$ plants will be favored. For example, increased nutritional requirements have been noted for C$_3$ crops grown under CO$_2$ enrichment conditions (34). C$_4$ plants will become more competitive with other plants at higher temperatures and CO$_2$ levels because C$_4$ plants are more efficient at utilizing nutrient resources. While the interactions between these organisms and their responses to limiting resources are extremely complex, C$_4$ plants will generally be favored in environments with multiple stresses such as higher temperatures, water deficits, and low soil mineral nutrition.

10. Research since the 1970s has shown that C$_4$ plants have three weaknesses— the deleterious effects of low temperatures and the limiting activities of carbonic anhydrase and Rubisco—all of which deserve more intensive study.

REFERENCES

1. Bender, M. M., Rouhani, I., Vines, H. M., and Black, C. C. 1973. ^{13}C/^{12}C ratio changes in crassulacean acid metabolism plants. *Plant Physiol.* **52**: 427–30.

2. Black, C. C. 1971. Ecological implications of dividing plants into groups with distinct photosynthetic production capacities. In *Advances in Ecological Research*, Vol. 7, ed. J. B. Cragg, pp. 87–114. Academic Press, New York.

3. Black, C. C. 1973. Photosynthetic carbon fixation in relation to net CO$_2$ uptake. *Annu. Rev. Plant Physiol.* **24**: 253–86.

4. Black, C. C. 1986. Effect of CO$_2$ concentration on photosynthesis and respiration of C$_4$ and CAM plants. In *Carbon Dioxide Enrichment of Greenhouse Crops*, ed. H. Z. Enoch and B. A. Kimball, chapter 3. CRC Press, Boca Raton, Fla.

5. Black, C. C., Chen, T. M., and Brown, R. H. 1969. Biochemical basis for plant competition. *Weed Sci.* **17**: 338–44.

6. Brown, R. H. 1978. A difference in N use efficiency in C$_3$ and C$_4$ crop plants and its implications in adaptation and evolution. *Crop Sci.* **18**: 93–8.

7. Brown, R. H. 1982. Response of terrestrial plants to light quality, light intensity, temperature, CO$_2$ and O$_2$. In *Handbook of Biosolar Resources*, Vol. 1, ed. A. Mitsui and C. C. Black, Part 2, pp. 185–212.

8. Brown, R. H., and Brown, W. W. (1975) Photosynthetic characteristics of *Panicum milioides*, a species with reduced photorespiration. *Crop Sci.* **15**: 681–5.

9. Burnell, J. N., and Hatch, M. D. 1988. Low bundle sheath carbonic anhydrase is apparently essential for effective C_4 pathway operation. *Plant Physiol.* **86**: 1252–6.

10. Burris, R. H., and Black, C. C. 1976. *CO₂ Metabolism and Productivity.* University Park Press, Baltimore.

11. Campbell, W. H., and Black, C. C. 1979. Cellular aspects of C_4 leaf metabolism. *Adv. Phytochem.* **16**: 223–48.

12. Chen, T. M., Brown, R. H., and Black, C. C. 1970. CO_2 compensation concentration, rates of photosynthesis, and carbonic anhydrase activity of plants. *Weed Sci.* **18**: 399–403.

13. Chen, T. M., Dittrich, P., Campbell, W. H., and Black, C. C. 1974. Metabolism of epidermal tissues, mesophyll cells and bundle sheath strands resolved from mature nutsedge leaves. *Arch. Biochem. Biophys.* **163**: 246–62.

14. Edwards, G. E., Lee, S. S., Chen, T. M., and Black, C. C. 1970. Carboxylation reactions and photosynthesis of carbon compounds in isolated mesophyll and bundle sheath cells of *Digitaria sanguinalis* (L,) Scop. *Biochem. Biophys. Res. Commun.* **39**: 389–95.

15. Fraser, P. J., Pearman, G. I., and Hyson, P. 1983. The global distribution of atmospheric CO_2. A review of provisional background observations, 1978–1980. *J. Geophys. Res.* **88C**: 3591–8.

16. Furbank, R. T., and Hatch, M. D. 1987. Mechanism of C_4 photosynthesis: The size and composition of the inorganic carbon pool in bundle sheath cells. *Plant Physiol.* **85**: 958–64.

17. Furbank, R. T., Jenkins, C. L. D., and Hatch, M. D. 1989. CO_2-concentrating mechanism of C_4 photosynthesis: Permeability of isolated bundles sheath cells to inorganic carbon. *Plant Physiol.* **91**: 1364–71.

18. Gerwick, B. C., Ku, S. B., and Black, C. C. 1980. Initiation of sulfate activation: A variation in C_4 photosynthesis plants. *Science* **209**: 513–15.

19. Haberland, G. 1914. *Physiological Plant Anatomy*, transl. of 4th edition, pp. 261–301. Macmillan, London.

20. Hatch, M. D. 1971. The C_4 pathway of photosynthesis: Evidence for an intermediate pool of carbon dioxide and the identity of the donor C_4 acid. *Biochem. J.* **125**: 425–32.

21. Hatch, M. D. 1987. C_4 photosynthesis: A unique blend of modified biochemistry, anatomy and ultrastructure. *Biochim. Biophys. Acta* **895**: 81–106.

22. Hatch, M. D., and Burnell, J. N. 1990. Carbonic anhydrase activity in leaves and its role in the first step of C_4 photosynthesis. *Plant Physiol.* **93**: 825–8.

23. Hatch, M. D., Osmond, C. B., and Slatyer, R. O. 1971. *Phytosynthesis and Photorespiration.* Wiley-Interscience, New York.

24. Hatch, M. D., and Slack, C. R. 1966. Photosynthesis by sugarcane leaves. A new carboxylation reaction and the pathway of sugar formation. *Biochem. J.* **101**: 103–11.

25. Holm, G., Plucknett, D. L., Pancho, J. V., and Herberger, J. P. 1977. *The World's Worst Weeds, Distribution and Biology.* Univ. of Hawaii Press, Honolulu.

26. Jenkins, C. L. D., Furbank, R. T., and Hatch, M. D. 1989. Mechanism of C_4 photosynthesis: A model describing the inorganic carbon pool in bundle sheath cell. *Plant Physiol.* **91**: 1372–81.

27. Keeling, C. D., Carter, A. F., and Mook, W. G. 1984. Seasonal, latitudinal, and secular variations in the abundance and isotopic ratios of atmospheric CO_2. 2. Results from oceanographic cruises in the tropical Pacific Ocean. *J. Geophys. Res.* **89D**: 4615–28.

28. Kenyon, W. H., Holaday, A. S., and Black, C. C. 1981. Diurnal changes in metabolite levels and Crassulacean acid metabolism in *Kalanchoe diagremontiana* leaves. *Plant Physiol.* **68**: 1002–1007.

29. Kortschak, H. P., Hartt, C. E., and Burr, G. O. 1965. Carbon dioxide fixation in sugarcane leaves. *Plant Physiol.* **40**: 209–13.

30. Laetsch, W. M. 1968. Chloroplast specialization in dicotyledons possessing the C$_4$ dicarboxylic acid pathway of photosynthetic CO$_2$ fixation. *Am. J. Bot.* **55**: 875–83.

31. Lemon, E. R., ed. 1983. *C$_4$CO$_2$ and Plants: The Response of Plants to Rising Levels of Atmospheric Carbon Dioxide.* Westview Press, Boulder, Colo.

32. Mook, W. G., Koopmans, M., Carter, A. F., and Keeling, C. D. 1983. Seasonal, latitudinal, and secular variations in the abundance and isotopic ratios of atmospheric carbon dioxide. 1. Results from land stations. *J. Biophys. Res.* **88C**: 10915–33.

33. Moore, R., and Black, C. C. 1979. Nitrogen assimilation pathway in leaf mesophyll and bundle sheath cells of C$_4$ photosynthesis plants formulated from comparative studies with *Digitaria sanguinalis* (L,) Scop. *Plant Physiol.* **64**: 309–13.

34. Mortensen, L. M. 1987. Review: CO$_2$ enrichment in greenhouses. Crop responses. *Scient. Horticult.* **33**: 1–25.

35. Moss, D. N. 1962. The limiting carbon dioxide concentration for photosynthesis. *Nature* **193**: 587.

36. Moss, D. N., Musgrave, R. B., and Lemon, E. R. 1961. Photosynthesis under field conditions. III. *Crop Sci.* **1**: 83–7.

37. Moss, D. N., Krenzer, E. G., and Brun, W. A. 1969. Carbon dioxide compensation points in related plant species. *Science* **164**: 187–8.

38. Pearman, G. I., Hyson, P., and Fraser, P. J. 1983. The global distribution of atmospheric carbon dioxide: 1. Aspects of observations and modeling. *J. Geophys. Res.* **88C**: 3581–90.

39. Potter, J. W., and Black, C. C. 1982. Differential protein composition and gene expression in leaf mesophyll cells and bundle sheath cells of *Digitaria sanguinalis* (L.) Scop. *Plant Physiol.* **70**: 590–7.

40. Raschke, K. 1986. The influence of the CO$_2$ content of the ambient air on stomatal conductance and the CO$_2$ concentration in leaves. In *Carbon Dioxide Enrichment of Greenhouse Crops*, ed. H. Z. Enoch and B. A. Kimball, chapter 7, CRC Press, Boca Raton, Fla.

41. Ray, T. B., and Black, C. C. 1979. The C$_4$ pathway and its regulation. In *Photosynthesis II: Photosynthetic Carbon Metabolism and Related Processes*, ed. M. Gibbs and E. Latzko, Vol. 6: *Encyclopedia of Plant Physiology*, New Series, pp. 77–101. Springer-Verlag, Berlin.

42. Shantz, H. L., and Piemeisel, L. N. 1927. The water requirement of plants at Akron, Colorado. *J. Agric. Res.* **34**: 1189–93.

43. Shomer-Ilan, A., Beer, S., and Waisel, Y. 1975. *Suaeda monoica*, a C$_4$ plant without typical bundle sheaths. *Plant Physiol.* **56**: 676–79.

44. Smith, B. N., and Epstein, S. 1971. Two categories of ^{13}C/^{12}C ratios for higher plants. *Plant Physiol.* **47**: 380–4.

45. Stowe, L. G., and Teeri, J. A. 1978. The geographic distribution of C$_4$ species of the dicotyledonae in relation to climate. *Am. Nat.* **112**: 609–23.

46. Teeri, J. A., and Stowe, L. G. 1976. Climatic patterns and the distribution of C$_4$ grasses in North America. *Oecologia* **23**: 1–12.

47. Widholm, S. G., and Ogren, W. L. 1969. Photorespiratory-induced senescence of plants under conditions of low carbon dioxide. *Proc. Natl. Acad. Sci. USA* **63**: 668–73.

48. Wong, S. C., Cowan, I. R., and Farquhar, G. D. 1985. Leaf conductance in relation to rate of CO$_2$ simulation. I. Influence of nitrogen nutrition, phosphorus nutrition, photon flux density, and ambient partial pressure of CO$_2$ during ontogeny. *Plant Physiol.* **78**: 821–5.

12

CO_2 and Crassulacean Acid Metabolism Plants: A Review

IRWIN P. TING

This report assesses the impact of elevated atmospheric CO_2 concentrations on succulent plants that manifest Crassulacean acid metabolism (CAM) or CAM photosynthesis, and, in addition, speculates on whether or not CAM plants could have a significant effect on atmospheric CO_2 concentration. CAM has been reviewed recently (29), and Black (3) has discussed the effects of CO_2 on photosynthesis and respiration in both C_4 and CAM plants.

It has been hypothesized that CAM is a metabolic adaptation to drought or otherwise dry conditions (13). The adaptation comes about by stomatal closure during the day and opening at night when the evaporative demand is low. Thus, CAM plants lose significantly less water than most other plants (13, 19). An equally interesting hypothesis that has not been developed is that CAM is an adaptation to low daytime CO_2 since many, if not most, are tropical epiphytes living in forest canopies that experience daily reductions in atmospheric CO_2 below ambient levels. Thus, the stomata of CAM plants are open at night when CO_2 levels are elevated by respiration, frequently above ambient levels. And indeed, Keeley (11) has presented evidence that certain submerged aquatic plants have a CAM-like metabolism with nocturnal CO_2 fixation that occurs in response to reduced daytime aqueous CO_2/bicarbonate in their aquatic natural environment.

CO_2 EFFECTS ON BIOCHEMISTRY AND PHYSIOLOGY

Accompanying the reverse-phase stomatal opening of CAM is massive nocturnal uptake of CO_2 brought about by carboxylation catalyzed by phosphoenolpyruvate carboxylase (PEPCase) and a malate dehydrogenase resulting in the net synthesis of malate:

$$\text{Phosphoenolpyruvate} + HCO_3^- \rightarrow \text{Oxalacetate} \rightarrow \text{Malate} \qquad (12.1)$$

As noted in Equation 12.1, bicarbonate and *not* free CO_2 is the substrate for carboxylation (13). The malate accumulates as malic acid throughout the night period in large water-storage vacuoles. During the subsequent light period, the malic acid leaves the vacuoles and is decarboxylated to free CO_2 and a three-carbon fragment. Since the stomata are closed in the light, no water is lost, and the CO_2, which is trapped in the photosynthetic tissue, is reduced photosynthetically by the usual mechanism catalyzed first by ribulose bisphosphate carboxylase (Rubisco). The biochemistry of the CAM process has recently been reviewed (30).

There are thus several points along the diurnal metabolic cycle of CAM that could be impacted by CO_2 concentrations. Osmond (21) has divided this diurnal cycle into four phases, which are useful for discussion. Phase I is the night CO_2 fixation phase in which the synthesis of malate and subsequent accumulation of malic acid respond to increased CO_2 concentration by greater rates of acid synthesis (4). Thus, elevated atmospheric CO_2 concentrations are expected to increase the rate of dark CO_2 fixation in CAM and, hence, also increase the amount of malic acid accumulated. However, CAM plants, like C_4 plants, saturate close to ambient levels of atmospheric CO_2, and thus, only a minimal effect of elevated CO_2 on dark fixation has been measured in CAM species (8, 22). The PEPCase-catalyzed carboxylation is evidently nearly saturated at present atmospheric CO_2 concentrations. In addition, there is clearly an upper limit to malid acid accumulation since accumulation of acid stops during extended dark periods (13).

Elevated atmospheric CO_2 may tend to close stomata at night; however, the stomata of CAM plants do not seem to be as sensitive to CO_2 at night as they are during the day (9). This is interesting in itself since the atomata of CAM plants have been assumed to function similarly to those of other plants (14, 31).

Since PEPCase is sensitive to high levels of CO_2, there could be inhibition of carboxylation (13); however, inhibitory levels of bicarbonate are in the millimolar range (17), and elevated CO_2 probably would not be an important factor at levels two or three times the preindustrial level.

Although dark, nonphotosynthetic CO_2 fixation by CAM can extend into the light period during Phase II when stomata are open maximally, much CO_2 is fixed photosynthetically largely by the C_3 pathway (22). Here, we expect elevated CO_2 concentrations to increase the rate of CO_2 fixation, but the latter also may tend to close stomata, compensating for the higher levels of CO_2 (9). If Phase II is completely analogous to C_3 photosynthesis, then the effects of high CO_2 on CAM plant Phase II would probably be similar to the effects on C_3 species. The actual effect of elevated atmospheric CO_2 during this phase would depend upon the extent to which C_3 photosynthesis and dark CO_2 fixation by CAM occur at this time.

Like C_3 plants, CAM plants show a postillumination CO_2 burst in which CO_2 is given off. According to the analysis with pineapple by Crews et al. (7), there is a primary first peak or burst, which is influenced by O_2 and CO_2 in a manner suggestive of photorespiration. A second peak follows which is insensitive to CO_2 and O_2 is most likely the result of decarboxylation of malate. The extent of the burst varies depending upon physiological state and, in particular, the amount of acid present.

During the major part of the light period, deacidification occurs as malate is decarboxylated, and the stomata are closed:

$$\text{Malate} \rightarrow \text{3-Carbon fragment} + CO_2 \qquad (12.2)$$

During Phase III, internal levels of CO_2 may reach concentrations of 1% or more (6, 12, 25). Decarboxylation is rapid, creating free CO_2 faster than can be assimilated by Rubisco-catalyzed carboxylation. Elevated atmospheric CO$_2$ resulting in more nocturnal malate synthesis would result in a greater rate and longer duration of internal CO$_2$ production by decarboxylation. It is doubtful if atmospheric CO$_2$ would have any significant direct effect on metabolism during Phase III since the stomata are closed. Any effects of high CO$_2$ will be indirect related to increased malate synthesis.

It is clear from published data that CAM plants have an active photorespiration system (7, 22). It is equally evident that the massive production and accumulation of CO$_2$ during decarboxylation would severely inhibit the oxygenase reaction of Rubisco, significantly reducing photorespiration during Phase III.

After malate depletion, photosynthesis eventually depletes the internal CO$_2$ which decreases to subambient levels. This is accompanied by stomatal opening, most likely in a cause-and-effect manner (31). When stomata are open toward the end of the light period, CO$_2$ is fixed by C$_3$ photosynthesis (Phase IV). Here, we expect that elevated atmospheric CO$_2$ concentrations would result in greater Phase IV CO$_2$ fixation. And indeed, several studies have shown enhanced late daytime CO$_2$ fixation in CAM plants under elevated CO$_2$ conditions. In *Kalanchoe blossfeldiana*, Osmond and Björkman (22) showed a steady increase in the rate of CO$_2$ fixation with increasing intracellular CO$_2$ up to about 600 μl $\cdot$ L^{-1} air, and Allaway et al. (1) showed an increase in daytime CO$_2$ fixation with increasing CO$_2$ in *Kalanchoe daigremontiana*. In another study, it was shown that pineapple responded to elevated CO$_2$ up to about 900 μl $\cdot$ L^{-1} toward the end of the light period, during Phase IV (7), but not at other times. These and other data also tend to support active photorespiration in CAM plants.

Thus, elevated atmospheric CO$_2$ concentrations would be expected to have an effect on all phases of CAM photosynthesis. I would predict from the analysis above that increased CO$_2$ would result in somewhat more growth of CAM succulent plants, but because of the limitations imposed by the less efficient dark CO$_2$ fixation process of CAM and the biochemical limitation of malate transport into the vacuole, coupled with the physical limitation imposed by the size of the storage vacuole, increased CO$_2$ assimilation and growth may be minimal. A few studies of CAM plant growth at elevated CO$_2$ have been conducted that serve to support these tentative conclusions (see below).

EXTRAORDINARY HIGH CO$_2$

Many studies on the effect of very high CO$_2$ concentrations have been conducted with CAM plants (4, 27). Bonner and Bonner (4) showed that enriched CO$_2$ concentrations up to an optimal 2% enhanced the rate of dark CO$_2$ fixation and malic acid accumulation. Much higher concentrations tended to be inhibitory, perhaps by stomatal closure (31) or inhibition of PEPcarboxylase (17). Experiments by Wood (35) showed that enhanced dark CO$_2$ fixation up to 10% CO$_2$ is temperature dependent, with higher fixation occurring at lower temperatures. High CO$_2$ in the range of 5% or more inhibits deacidification in the light (5, 18), most likely by continued acid synthesis either by the PEPCase reaction or reversal of the decarboxylation step. Leaves of *Bryophyllum* demonstrate a circadian rhythm of CO_2 fixation when kept at constant temperature and continuous light. This rhythm can be entrained with 5% CO_2 (2). It is not clear how these effects of extraordinarily high CO_2 on the metabolism of CAM plants relate to questions of enhanced atmospheric CO_2 only two to three times greater than preindustrial levels. Few studied are available that have used CO_2 concentrations intermediate between $1,000 \mu l \cdot L^{-1}$ and $10,000 \mu l \cdot L^{-1}$ (i.e., 1%).

EFFECT OF ELEVATED CO₂ ON PLANTS THAT SHIFT FROM C₃ TO CAM

Many succulent plants shift from C_3 to CAM, depending upon development and/or environmental conditions, particularly drought and/or salt stress (29, 32, 34). Thus, the effect of elevated CO_2 concentrations on succulents will depend in large part on whether they are functioning as C_3 or CAM plants. And those succulents that go through a developmentally dependent shift from C_3 to CAM will be more affected by high CO_2 during young stages when they are C_3 than during older, more mature stages when they are CAM. It is possible that some of the positive growth effects of elevated CO_2 on CAM plants may have been the result of plants functioning as C_3 plants rather than CAM plants.

Many succulent plants go through a developmental stage in which there is only daytime gas exchange, but still a diurnal fluctuation of organic acids through the CAM pathway (32). This phase, termed *CAM cycling*, results from refixation of respiratory CO_2 (23). Although we can only speculate, it would appear that elevated CO_2 would increase CO_2 fixation similarly to C_3 plants. Since CO_2 is available from internal decarboxylation of malate and diffusion from outside the leaf, the photosynthetic carboxylation system may be saturated. In this case, CAM cycling tends to reduce photorespiration, and I would predict little or no effects of elevated CO_2.

When *Portulacaria afra* is well watered, it shows C_3 photosynthesis but tends to have high organic acid levels. When water stressed, it shifts to CAM. When in C_3, neither reduced atmospheric CO_2 ($40 \mu l \cdot L^{-1}$) nor enhanced CO_2 ($950 \mu l \cdot L^{-1}$) will induce CAM (9). In a similar study with *Mesembryanthemum crystallinum*, which can be induced into CAM by salt or water stress, neither high nor low atmospheric CO_2 has an effect on the induction of CAM (34). The annual *Mesembryanthemum* differs from the perennial *Portulacaria* in that when in C_3, the former has low acid levels (34) whereas the latter has high acid levels (9). In well-watered C_3 *Portulcaria*, both high and low levels of CO_2 tended to reduce acid levels but, as stated, with no induction of CAM (9). Acids are evidently reduced at low CO_2 because of the lack of a bicarbonate substrate for carboxylation. It is less clear why high CO_2 should reduce acid levels except that stomata may close, restricting CO_2 entrance (9). Thus the shift from C_3 to CAM in these plants is largely independent of CO_2 concentrations.

EFFECTS OF ELEVATED CO₂ ON GROWTH

Kalanchoe blossfeldiana was shown to increase rates of dry matter production when grown in glasshouses with enriched CO_2 up to $900 \mu l \cdot L^{-1}$ (16). *Kalanchoe blossfeldiana* is an example of a plant that shows both C$_3$ photosynthesis and CAM photosynthesis, depending upon the environmental growth conditions. When grown under flower-inducing short days, it tends to be CAM (24), whereas on long days it is C$_3$. In addition, at lower temperatures, it tends to be more C$_3$-like than CAM-like (33). In the Mortensen experiments (16) with enriched CO$_2$, *K. blossfeldiana* was grown on 21-h days and presumably was functioning as a C$_3$ plant. Thus, the increased dry matter production may have been the result of a CO$_2$ effect on C$_3$ photosynthesis rather than on CAM.

Agave vilmoriniana plants were grown outside in Phoenix, Arizona in open air chambers under four CO$_2$ concentrations from ambient to 885 μl$\cdot$L^{-1} and under wet and dry conditions (10). Only the dry-treated plants responded positively to CO$_2$ enrichment. In another study, the same species was grown at Riverside, California, in open air chambers and at two levels of CO$_2$: ambient and about 750 μl$\cdot$L^{-1} (26). There were only slight effects on stomatal conductance and CO$_2$ assimilation as a result of the enriched CO$_2$. Thus, these *Agave* experiments are difficult to assess.

Nobel (19) predicts, on the basis of his growth experiments with *Agave deserti* and *Ferocactus acanthodes*, that these plants will experience a 1% increase in growth with each 10-ppm increase in atmospheric CO$_2$. In these studies, a corresponding increase in CO$_2$ fixation of about 30% was measured after a period of adjustment to elevated CO$_2$ concentrations raised from 350 to 650 ppm (20).

ASSESSMENT OF ELEVATED CO$_2$ AND CAM PLANTS

It seems clear that elevated CO$_2$ will mostly affect CAM plants in a manner similar to C$_4$ plants. When in CAM, the carboxylating system (PEPCase) is largely saturated at present ambient concentrations of CO$_2$. Thus, we can predict that there will be a minimal increase in dark CO$_2$ fixation and little impact on CAM. During the late period, when CAM plants fix CO$_2$ through the C$_3$ pathway similarly to C$_3$ plants, the influence of increased atmospheric CO$_2$ will be greater inasmuch as Rubisco is not CO$_2$ saturated. The extent to which there is increased CO$_2$ assimilation and concomitant increased growth should depend largely on the degree to which the CAM plant performs C$_3$ photosynthesis. Thus, I predict that elevated CO$_2$ concentrations would have a greater positive impact on those succulents that can function in either C$_3$ photosynthesis or CAM.

Under most conditions, CAM plants grow relatively slowly, and their physiology and biochemistry seem to be largely adapted to sustained production. However, under certain conditions, CAM plants are known to compete with other plants very well. As an example, in the southwestern United States in overgrazed regions, cacti tend to become dominant. Introduced cacti in Australia and southern Africa also point to extreme success of CAM succulents (see Ref. 13 for a discussion of CAM plant growth in exotic environments). In the Caribbean, two South African succulents, *Sansevieria trifasciata* and *Bryophyllum pinnatum*, are among the most aggressive weeds (28). In part, the success of these plants can be attributed to CAM. Although they do not compete well with native grasses or other vegetation during the growing season, during periods of drought, they tend to survive and indeed may even grow. Over a period of years, they frequently become extremely competitive with native flora and may even become dominant, as did *Opuntia inermis* in Australia (14). It seems reasonable to conclude that CAM plants are most aggressive when they grow as C$_3$ plants during the wet seasons. Elevated atmospheric CO$_2$ would be expected to give CAM plants, under these circumstances, an even greater competitive advantage.

The observation that somata of CAM plants are not as sensitive to CO$_2$ at night as they are during the day (9) leads to interesting speculation. Since stomata of all plants tend to close at elevated CO$_2$ concentrations (15), there should be less of a positive effect on daytime CO$_2$ fixation than would be predicted on the basis of increased CO$_2$ as a substrate for photosynthesis. I predict that CAM plants will not be affected to the same degree as other plants if their stomata are less sensitive to CO$_2$ at night when they fix the majority of exogenous CO$_2$. This reasoning, if correct, would tend to support the notion that CAM is, in part, an adaptation to diurnal fluctuation of CO$_2$ in forest canopies.

The impact of CAM plants on elevated atmospheric CO$_2$ is much more difficult to assess. Although the CAM species make up about 10% of the total of flowering plants, nowhere are they very dominant (13). Thus, the total biomass of CAM plants is relatively small. Therefore, given (a) the lower rates of CO$_2$ fixation attributed to CAM plants in comparison with C$_3$ and C$_4$ plants, (b) the relativelh low biomass of CAM plants the world over, and (c) the minimal increase in CO$_2$ fixation and subsequent growth as a result of elevated CO$_2$, it is reasonable to conclude that CAM plants will not be a significant sequestering source for CO$_2$. A possible exception might be in areas where CAM plants become dominant.

ACKNOWLEDGMENTS

This research was supported in part by National Science Foundation grants DCB-8807860 and CMB-8426981.

REFERENCES

1. Allaway, W. G., Austin, B., and Slayter, R. O. 1974. Carbon dioxide and water vapour exchange parameters of photosynthesis in a Crassulacean plant *Kalanchoe daigremontiana*. *Aust. J. Plant Physiol.* **1**: 397–405.

2. Anderson, C. M., and Wilkins, M. B. 1989. Phase resetting of the circadian rhythm of carbon dioxide assimilation in *Bryophyllum* leaves in relation to their malate content following brief exposure to high and low temperatures, darkness and 5% carbon dioxide. *Planta* **180**: 61–73.

3. Black, C. C. 1986. Effects of CO$_2$ concentration on photosynthesis and respiration of C$_4$ and CAM plants. In *Carbon Dioxide Enrichment of Greenhouse Crops*, ed. H. Z. Enoch and B. A. Kimball, Vol. II: *Physiology, Yield, and Economics*. pp. 29–40. CRC Press, Boca Raton, Fla.

4. Bonner W., and Bonner, J. 1948. The role of carbon dioxide in acid formation by succulent plants. *Am. J. Bot.* **35**: 113–17.

5. Bruinsma, J. 1958. Studies on the Crassulacean acid metabolism. *Acta Bot. Neerl.* **7**: 531–88.

6. Cockburn, W., Ting, I. P., and Sternberg, L. O. 1979. Relationships between stomatal behavior and internal carbon dioxide concentration in Crassluacean acid metabolism plants. *Plant Physiol.* **63**: 1029–32.

7. Crews, C. E., Vines, H. M., and Black, C. C. 1975. Postillumination burst of carbon dioxide in Crassulacean acid metabolism plants. *Plant Physiol.* **55**: 652–7.

8. Holtum, J. A. M., O'Leary, M. H., and Osmond, C. B. 1983. Effect of varying CO_2 partial pressure on photosynthesis and on carbon isotope composition of carbon-4 of malate from the Crassulacean acid metabolism plant *Kalanchoe daigremontiana* Hamet et Perr. *Plant Physiol.* **71**: 602–9.

9. Huerta, A. J., and Ting, I. P. 1988. Effects of various levels of CO_2 on the induction of Crassulean acid metabolism in *Portulacaria afra* (L.) Jacq. *Plant Physiol.* **88**: 183–8.

10. Idso, S. B., Kimball, B. A., Anderson, M. G., and Szarek, S. R. 1986. Growth response of a succulent plant, *Agave vilmoriniana* to elevated CO_2. *Plant Physiol.* **80**: 796–7.

11. Keeley, J. E. 1983. Crasslacean acid metabolism in the seasonally submerged aquatic *Isoetes howelli*. *Oecologia* **58**: 57–62.

12. Enyon, W. H., Holiday, A. S., and Black, C. C. 1981. Diurnal changes in metabolite levels and Crassulacean acid metabolism in *Kalanchoe daigremontiana* leaves. *Plant Physiol.* **68**: 1002–1007.

13. Kluge, M. and Ting, I. P. 1978. Crassulacean acid metabolism. In *Analysis of an Ecological Adaptation*. Springer-Verlag, New York.

14. Mann, J. 1970. *Cacti Naturalised in Australia and Their Control*. Department of Lands, Queensland.

15. Morison, J. I. L. 1987. Intercullar CO_2 concentration and stomatal response to CO_2. In *Stomatal Function*, ed. E. Zeiger, G. D. Farquhar, and I. R. Cowan, pp. 229–51. Stanford University Press, Stanford, Cal.

16. Mortensen, L. M. 1983. Growth responses of some greenhouse plants to environment. XII. Effect of CO_2 on photosynthesis and growth of *Saintpaulia ionantha*, *Kalanchoe blossfeldiana*, and *Nephrolepis exaltata*. *Meld Norg LandbrHogsk* **62**: 1–16.

17. Mukerji, S. K., and I. P. Ting. 1968. Phosphoenolpyruvate carboxylase isoenzymes: Separation and properties of three forms from cotton leaf tissue. *Phytochemistry* **7**: 903–11.

18. Nishida, K. 1977. CO_2 fixation in leaves of a CAM plant without lower epidermis and the effect of CO_2 on their deacidification. *Physiol. Plant* **18**: 927–30.

19. Nobel, P. S. 1988. *Environmental Biology of Agaves and Cacti*. Cambridge University Press, New York.

20. Nobel, P. S., and Hartsock, T. L. 1986. Short-term and long-term responses of Crassulacean acid metabolism plants to elevated CO_2. *Plant Physiol.* **82**: 604–6.

21. Osmond, C. B. 1978. Crassulacean acid metabolism: A curosity in context. *Annu. Rev. Plant Physiol.* **29**: 379–414.

22. Osmond, C. B., and Björkman, O. 1975. Pathways of CO_2 fixation in the CAM plant *Kalanchoe daigremontiana* II. Effects of O_2 and CO_2 concentration on light and dark CO_2 fixation. *Aust J. Plant Physiol.* **2**: 115–62.

23. Patel, A., and Ting, I. P. 1987. Relationship between respiration and CAM-cycling in *Peperomia camptotricha*. *Plant Physiol.* **84**: 640–2.

24. Queiroz, O. 1974. Circadian rhythms and metabolic patterns. *Annu. Rev. Plant Physiol.* **24**: 115–34.

25. Spalding, M. H., Stumpf, D. K., Ku, M. S. B., Burris, R. H., and Edwards, G. E. 1979. Crassulacean acid metabolism and diurnal variations of internal CO_2 and O_2 concentrations in *Sedum praealtum* DC. *Aust J. Plant Physiol.* **6**: 557–67.

26. Szarek, S. R., Holthe, P. A., and Ting, I. P. 1987. Minor physiological response to elevated CO$_2$ by the CAM plant *Agave vilmoriniana*. *Plant Physiol.* **83**: 938–40.

27. Thomas, M., and Ranson, S. L. 1954. Physiological studies on acid metabolism in green plants. III. Further evidence of CO$_2$ fixation during dark acidification of plants showing Crassulacean acid metabolism. *New Physol.* **53**: 1–30.

28. Ting, I. P. 1989. Photosynthesis of arid and subtropical succulent plants. *Aliso* **12**: 387–406.

29. Ting, I. P. 1985. Crassulacean acid metabolism. *Annu. Rev. Plant Physiol.* **36**: 595–622.

30. Ting, L. P. 1987. Crassulacean acid metabolism. In *Model Building in Plant Physiology/Biochemistry*, ed. D. Newman and K. Wilson, pp. 45–7. CRC Press, Baton Rouge.

31. Ting, I. P. 1987. Stomata in plants with Crassulacean acid metabolism. In *Stomatal Function*, ed. E. Zeiger, G. D. Farquhar, and I. R. Cowan, pp. 353–66. Stanford University Press, Stanford.

32. Ting, I. P., and D. Sipes. 1985. Metabolic modifications of Crassulacean acid metabolism—CAM-idling and CAM-cycling. In *Nitrogen Fixation and CO$_2$ Metabolism*, ed. R. H. Burris, P. W. Ludden, and J. E. Burris, pp. 371–8. Steenbock Symp., Elsevier, New York.

33. Ting, I. P., Thompson, M. L., and Dugger, W. M. 1967. Leaf resistance to water vapor transfer in succulent plants: Effect of thermoperiod. *Am. J. Bot.* **54**: 245–51.

34. Winter, K. 1979. Effect of different CO$_2$ regimes on the induction of crassulacean acid metabolism in *Mesembryanthemum cystallinum* L. *Aust. J. Plant Physiol.* **6**: 589–94.

35. Wood, W. M. C. 1952. Organic acid metabolism in *Sedum praealtum*. *J. Exp. Bot.* **3**: 336–55.

13

Algal DIC Pumps and Atmospheric CO$_2$

J. A. RAVEN and A. M. JOHNSTON

Marine microalgae and cyanobacteria must play an important part in any attempt to model changes in, and regulation of, atmospheric CO$_2$. The bases for this assertion are two fold. One is that the oceans contain about 50 times as much inorganic C as does the atmosphere, so that any net production or consumption by the sum of land-based activities (vulcanism, photosynthesis, respiration, combustion) will tend to be buffered by the quantatively predominant inorganic C in the oceans (8). Ocean chemistry, including ocean biochemistry, must accordingly play a very important role in any long-term (hundreds of years or longer) account of atmospheric CO$_2$ content (8). The other underpinning is that the biogeochemistry of inorganic C in the surface waters of the oceans is dominated by the activities of microscopic photosynthetic O$_2$ evolvers. Thus, the CO$_2$ concentration at the sea surface and, consequently, the direction of the CO$_2$ flux across the atmosphere–ocean interface are controlled by the photosynthetic activities of phytoplankton, and by the processes that remove the resulting organic C from the ocean surface and those that regenerate CO$_2$ at, or return CO$_2$ regenerated elsewhere to, the ocean surface (8). The predominance of biological activity in controlling CO$_2$ concentration at the surface of the ocean is, despite the ocean phytoplankton, contributing less than half of global primary productivity—that is, substantially less than expected *pro rata* from the total areas of land and sea on our planet (Table 13.1; Refs. 12, 3). It is the generally nutrient-limited nature of oceanic primary productivity, and especially of the export to deeper waters of the products of phytoplankton productivity, that has led biogeochemists and geophysicists to relate the glacial–interglacial changes in atmospheric CO$_2$ content to nutrient supply to the euphotic zone, involving a complex interaction of large-scale ocean water movements and biogeochemistry (6, 7, 9, 30, 41, 71).

More immediately to the title of this chapter is a consideration of the extent of occurrence of dissolved inorganic carbon (DIC) pumps in marine phytoplankton organisms, and the signifcance of the presence or absence of a DIC pump on the response of the organisms to varying DIC availability, and interactions with the supply of light and of nutrients such as N, Fe, and Mn. Such considerations may be significant for the relation of ocean-surface CO$_2$ concentration to species composition, total primary productivity, and resource use efficiency (light, N, Fe, Mn) of growth and thus of the overall N, Fe, and Mn content of organic material. This latter value is a determinant of how much organic C can be removed from immediate recycling—that is, act as a C sink, per unit input of N, Fe, and Mn (8, for C/P ratios in sedimented material).

It is important to note that the *really* long term (i.e., thousands to millions of years) removal of C from the exogenic cycle through the atmo-, bio-, and hydrospheres involves burial and incorporation into the crust of particulate C derived from biological activity dependent on the input of "new" N, P, Fe, Mn, and so on. By "new" is meant a population of elements such as N, P, Fe, and Mn derived from sources other than recycling from prior productivity by mineralization, whether at the ocean surface or in the deep ocean followed by upwelling. In either case the recycled elements are associated with an increment of DIC in a ratio similar to that required for phytoplankton growth, and so involve input of atmospheric CO$_2$ during a further round of incorporation by primary productivity.

These considerations are of great potential significance for the regulation of atmosopheric CO$_2$ partial pressure, as well as of significance for the behavior of individual genotypes of marine phytoplankton, concerning both the species composition of the community and the phenotypic changes in the genotypes, as a function of atmospheric CO$_2$ composition. Here, then, there is an effect of the atmospheric CO$_2$ partial pressure on the organisms and, probably, vice versa. For other algal populations and, indeed, the aquatic higher plants in the same and adjacent benthic habitats, the influence of the organisms on atmospheric CO$_2$ levels are likely to be smaller than for oceanic phytoplankton (67) while the influence of atmospheric CO$_2$ on their behavior is frequently also smaller. The relatively small influence of photosynthesis by terrestrial (including lichenized), freshwater planktonic and benthic, and marine benthic algae on atmospheric CO$_2$ is a function of their use of DIC from pools small relative to the atmospheric content of CO$_2$, or the use of oceanic DIC, with fluxes small relative to those due to oceanic phytoplankton, by benthic marine algae (Table 13.1). The "small flux/small pool of DIC" argument is compounded from some freshwater and coastal communities by the production of CO$_2$ concentrations in solution in excess of the atmospheric equilibrium levels via a "CO$_2$ pump" from atmospheric CO$_2$, via terrestrial productivity, to groundwater and freshwater bodies. While these phototrophs contribute to energy flow in the usually phototrophic way, they influence atmospheric CO$_2$ levels only via reduction in the leak of CO$_2$ pumped into the freshwater or coastal water body by terrestrial photosynthesis. These organisms using "pumped" CO$_2$ (Table 13.2) are substantially buffered from changes in atmospheric CO$_2$ as far as their photosynthetic C assimilation is concerned. Algae not so supplied with a CO$_2$ subsidy are somewhat more influenced by variations in atmospheric CO$_2$; this is especially likely for terrestrial algae (including lichenized algae) (61).

The only algal community that has the potential to have a major influence on, as well as to be influenced by, the atmospheric CO$_2$ partial pressure is the oceanic phytoplankton community; this community will be the main topic of discussion in the rest of this chapter.

Table 13.1. Productivity per Unit Area of Habitat, Area of Habitat Worldwide, and Global Productivity, for Habitats Involving Algae.

Habitat	Total Area·m^{-2}	Organisms	Net Productivity		Reference
			g C·m^{-2}·yr^{-1}	10^{15} g C·yr^{-1} worldwide	
Marine (phytoplankton)	370×10^{12}	Microalgae cyanobacteria	81	30	12, 16
Marine benthic	$6.8 \times 10^{12\,a,b}$	Microalage	50	0.34	16
		Macrolagae	500	0.34	
Marine benthic	$0.35 \times 10^{2\,a}$	Angiospherms (saltmarshes plus seagrass beds)	1000^c	0.35^c	modified from 12
Inland waters	2×10^{12}	Phytoplankton, benthic algae and higher plants	290	0.58	16
Terrestrial	150×10^{12}	Mainly higher plants^d	400^d	60^d	table 5.5 of 3

[a] Area of marine benthic algae and marine benthic seagrass habitats are in series with the shallow-water phytoplankton habitat; their areas are included in the marine phytoplankton area.

[b] Marine benthic algal area includes coral reefs; following Charpy-Roubaud and Sournia (12), the net productivity of the benthic symbioses involving dinoflagellates and invertebrates such as foraminifera, coelengerates, and tridacnid clams is assumed to be zero; that is, all of the productivity of the zooxanthellae is respired by the invertebrate symbiont. While being of the opinion that these "carnivorous plants of the sea" (43) *do* show net primary productivity, no estimates of its global magnitude are known to the authors. The significance of free-living and symbiotic algae in reef ecosystems is discussed by Hillis-Colinvaux (24).

[c] Values between highest and lowest quoted by de Vooys (16).

[d] Includes terrestrial cyanobacteria and algae, both free-living and lichenized; no separate estimates of productivity by terrestrial algae and terrestrial higher plants are known to the authors.

OCCURRENCE OF DIC PUMPS IN MARINE PHYTOPLANKTON

Relative to the number of species involved and their importance in global productivity (Table 13.1) and, potentially, in the regulation of the atmospheric CO_2 partial pressure (see above), there are limited data on the occurrence of DIC pumps in marine phytoplankton. (a) The "hardest" criterion is data for a higher intracellular than extracellular CO_2 concentration during steady-state photosynthesis, and other evidence of a less direct nature that gives an indication of the occurrence or nonoccurrence of a DIC pump. These less direct lines of evidence include (b) half-saturation values for DIC in photosynthesis, expressed as CO_2, at pH values at or near 8.0; (c) the CO_2 compensation concentration; (d) and the capacity to use HCO_3^-, preferably by the most rigorous test of photosynthesis at high external pH values which occurs faster than the uncatalyzed conversion of HCO_3^- to CO_2. The data in Table 13.3 shows that most of the marine phytoplankton organisms tested using the four criteria enumerated above have a

Table 13.2. Supplementation of Air-Equilibrium CO$_2$ Concentrations as Taken Up from the Bulk Water Phase in the Light by Aquatic Phototrophs (Especially Those That Cannot Concentrate CO$_2$ by a Transmembrane Active DIC Flux).

Organisms	Supplementation of CO$_2$ Supply	Use of "CO$_2$ Pump" from Atmosphere/ Bulk Water	References
Freshwater rhodophyte macrophyte	Growth in rapidly flowing streams enriched in CO$_2$; turbulence generators; relatively high affinity for CO$_2$ by Rubisco.	Yes	27–29
Vascular plants of isoetid morphology	Predominant uptake of CO$_2$ via roots in sediment containing about 100 times the air-equilibrium CO$_2$ concentrations.	Yes	58
Some isoetids, a few other aquatics	Crassulacean acid metabolism: dark CO$_2$ incorporation into organic acids, light refixation by photosynthetic carbon reduction cycle.	Not necessarily	58
Aquatic vascular with parts contacting air	Use gas-phase CO$_2$.	No	40
Some phototrophic flagellates (Chrysophyceae, Dinophyceae, Cryptophyceae; also Dinophyceae symbiotic with marine invertebrates)	Phagotrophy; mainly a source of N, P, Fe, etc; organic C supply probably a secondary consideration	Not necessarily	43, 44

DIC concentrating mechanism (see also Refs. 1, 44, 48). A quantitatively important exception is, apparently, the coccollithophorid *Emiliania huxleyi* which can dominate large areas of the ocean. It is clear that further work is needed, especially in view of the three following theoretical or experimental points.

One of these points relates to data not shown in Table 13.3 on photosynthetic characteristics of two strains of *Synechococcus* spp., the oceanic "DC2" and the coastal "Syn" (21). The effect of O$_2$ on [^{14}C]DIC fixation at seawater DIC concentrations is consistent with the occurrence of photorespiration, although more data with varying DIC concentrations would be helpful in deciding on the origin of the O$_2$ effect. If its origin is in photorespiration, then these two strains of *Synechococcus* either lack a DIC pump which suppresses photorespiration, or have a relatively leaky DIC pump. This contrasts with data on DIC accumulation and gas exchange in other strains, of marine and freshwater origin, of *Synechococcus* (including *Anacystis*) and other cyanobacgeria (Table 13.3; Refs. 1, 44, 48).

Table 13.3. Occurrence of DIC Concentration Mechanisms in Marine Phytoplankton.

Higher Taxon	Genus, Species	$[CO_2]i >$ $[CO_2]o$ [a]	Low $K_{1/2}$ CO_2 [b]	Low CO_2 c.p. [c]	Use HCO_3^- [d]	References
Cyanobacteria	Synechococcus sp.[e]	+	+	N.T.	N.T.	5, 10, 31
	Coccochloris[e] peniocystis	+	+	+		
	Oscillatoria woronichinii	N.T.	+	N.T.	N.T.	
Chlorophyceae	Dunaliella salina	+	N.T.	N.T.	N.T.	10, 33, 75
	Dunaliella tertiolecta	+	+	+	+	
	Stichococcus minor[e]	N.T.	−	N.T.	(+)	
	Stichococcus bacillaris[e]	N.T.	+	N.T.	(+)	
Micromonadophyceae	Ω 48 23[e]	N.T.	+	N.T.	+	19
	Mantoniella squamata[e]	N.T.	+	N.T.	+	
Eustigmatophyceae	Nannochloropsis oculata[e]	N.T.	+	N.T.	N.T.	33
	Monallantus sp.[e]	N.T.	+	N.T.	N.T.	
Bacillariophyceae	Phaedactylum tricornutum	−	+	N.T.	N.T.	10, 18, 26, 36
		+	+	+	+	
		+	+	N.T.	+	
Prymnesiophyceae	Isochrysis galbana	+	+	+	N.T.	10
	Emiliania huxleyii	N.T.	−	N.T.	(−)	35, 36, 66, 69
Rhodophyceae	Porphyridium purpureum	N.T.	+	N.T.	N.T.	17
		+	+	N.T.	N.T.	10
	Porphyridium cruentum	N.T.	+	+	+	13
		N.T.	+	N.T.	N.T. >	2
Cryptophyceae	Chroomonas sp.	N.T.	+	+	N.T.	10

Note: Data on cyanobacteria and microalgae in, or freshly isolated from, symbioses with benthic fungi or coelenterates are not included (see Refs. 11, 61).

[a] + = a significantly higher intracellular $[CO_2]$, than extracellular $[CO_2]$ during steady-state photosynthesis, measured using silicone oil centrifugation. − = the silicon oil centrifugation technique did not show a higher internal than external CO_2 concentration during steady-state photosynthesis. N.T. = not tested.

[b] + = the light-saturated photosynthesis vs. DIC relationship at pH 8.0 (or the closest that the experimental media come to pH 8.0) shows saturation at DIC concentrations below the normal seawater concentration. Expressed as CO_2 concentration, the $K_{1/2}$ is at least 1/3 of the air-equilibrium concentration at pH 8. N.T. = not tested.

[c] + = a CO_2 compensation point concentration in solution that is slightly higher than or below that found in terrestrial C_4 plants (0.1 mmol·m^{-3}). − = a CO_2 compensation concentration closer to that found in terrestrial C_3 plants (∼ 1 mmol·m^{-3}). N.T. = not tested.

[d] + = inorganic C can be assimilated at high pH at a faster rate than uncatalysed HCO_3^- conversion to CO_2. − = inorganic C cannot be assimilated faster than CO_2 is produced from HCO_3^- at high pH. (+) = some less rigorous test shows that cells can use HCO_3^-. (−) = some less rigorous test shows that cells cannot use HCO_3^-. N.T. = not tested.

[e] The organisms indicated are of picoplankton size; the other organisms are larger (74).

It is, however, worth remembering a second point that very small phototrophic cells, either cyanobacterial or eukaryotic, have less of a problem with diffusive CO_2 supply to Rubisco than does a large cell, thanks to the relationship between photosynthetic rate per unit cell volume, which yields a lower area-based net flux of inorganic C into the smaller cell, and the thinner unstirred (boundary) layer around the cell (47). These two factors mean that the steady-state CO_2 and O_2 concentrations in the very small cells during steady-state photosynthesis with purely diffusive CO_2 and O_2 fluxes across the plasmalemma (and, for eukaryotes, plastid envelope membranes) are very similar to those in the bulk medium, so that the kinetics of Rubisco achieved in these cells is very similar to that achieved by Rubisco dispersed through an aqueous medium (47). Indeed, the aqueous-phase diffusion distance from the bulk phase to the site of Rubisco activity in a picoplankton cell without a CO_2-concentrating mechanism—that is, 1–2 μm (47)—is very similar to that in a terrestrial plant mesophyll cell (53). However, the aqueous-phase diffusion distance in the very small aquatic cell is a result solely of the size of the cell, while the corresponding distance in the terrestrial plant cell requires a lot of help from physics, from the composition of cell walls, and from cell-water relations parameters to maintain the air–water interface at a position near the outside of the cell wall, relative to Rubisco in the plastids. The aqueous level must be held constant within a few tens of nanometers, despite very large changes in leaf water potential values and leaf relative water content (see 42, 45). However, we must remember that Rubisco kinetics in cyanobacteria, and to a lesser extent in eukaryotic microalgae, are less favorable to net C fixation from air-equilibrium solutions relative to the Rubisco kinetics of higher C_3 plants, so that the similarity of aqueous-phase diffusion paths would yield less net C fixation per unit Rubisco in a given time in the picollankton cells than the terrestrial C_3 plants. Thus, diffusive CO_2 entry in picoplankton cells would not yield a high rate of CO_2 fixation (44, 47). Overall, the data on picoplankton and larger cells (Table 13.3) shows no trend to DIC pumping versus diffusive CO_2 entry as a function of cell size in marine phytoplankton.

The third point that apparently casts doubt on the effectiveness of the DIC pump in certain marine phytoplankton organisms is the $^{13}C/^{12}C$ ratio of organic material in their cells (48). The depletion in ^{13}C relative to the inorganic C in the seawater in which the organisms grew suggests that Rubisco is able to exert a substantial fraction of its ability to discriminate between $^{13}CO_2$ and $^{12}CO_2$ which, in turn, implies that a substantial fraction of the DIC pumped from the medium to the site of Rubisco activity is able to recycle to the medium. Both entry and exit of DIC appear to transport their DIC substrate without appreciable discrimination between ^{13}C and ^{12}C; this permits the discrimination against $^{13}CO_2$ by Rubisco to establish a $^{13}C/^{12}C$ in the intracellular DIC pool which exceeds that in the medium, with DIC efflux involving ^{13}C and ^{12}C in the ratio found in the intracellular steady-state pool (48).

If this efflux of DIC were "leakage" (44) through a different path from the energy-dependent DIC influx pump, it would impose a substantial energy penalty (48) on growth, as well (see below) as in terms of N, Fe, and Mn costs. However, data on two of the species (*Phaeodyactylum tricornutum* and *Isochrysis galbana*) whose $^{13}C/^{12}C$ ratios have been measured and which have a DIC pump (Table 13.3) show a high photon yield of photosynthesis in seawater under light-limiting conditions (48). The lack of an apparent energy penalty for DIC recycling across the pumping membrane can be rationalized by the DIC efflux, being mainly a result of some mode of operation of the DIC pump in a manner that permits exchange of DIC across the membrane without an increased energy cost of net DIC influx (48), a possibility that is currently being tested. Such exchange reactions occur in both primary active transport [e.g., the ATP-driven Cl^- influx at the plasmalemma of *Acetabularia* (23)], and a secondary active transport [e.g., the H^+ gradient-driven hexose influx at the plasmalemma of *Chlorella* (72)]. It should be noted that the large $^{15}N/^{14}N$ and $^{34}S/^{32}S$ discriminations noted for dissimilatory reduction of nitrate and of sulfate (64) also require that a porter-mediated influx of the anion coexists with an efflux of that anion. The reasoning is that, as with CO_2 assimilation, the reduction of nitrate and of sulfate is an intracellular process (20, 37) and that, in order for the discriminating potential of reduction to be manifest, the substrate pool must not be limited to what is supplied by a nondiscriminatory influx with no possibility for efflux. This argument is stronger for sulfate, where the complete sequence of reductive reactions from sulfate to sulfide is intracellular (37), than for nitrate, where the nitrate to nitrous oxide or dinitrogen reactions are extracellular (20) and exhibit $^{15}N/^{14}N$ discrimination (64). Where nitrate and sulfate efflux must be invoked to explain the isotope data, the argument of energetic efficiency requires, as for inorganic C, that the efflux is largely an exchange reaction not involving energy input but, rather, a mediated or nonmediated leak that must be balanced by additional, energy-requiring influx. Certainly $^{13}C/^{12}C$ ratios are very significant in a number of ways for interpreting the present and past functioning of marine phytoplankton organisms (Refs. 6–9, 41, 71). Alas, the possibility that algae with a DIC pump (which is energetically efficient in the sense of having a small efflux (=leak, to short-circuit it) can be distinguished paleontologically from those that lack a DIC pump and have a high conductance for CO_2 entry (60), is rendered less likely, if the possibility of a DIC exchange mode of operation of the pump is experimentally verified. While reiterating the need for more data, for example an open-ocean picoplanktonic prokaryotes and eukaryotes, it is likely that much of the primary productivity in the ocean is carried out by organisms that have a DIC-concentrating mechanism.

THE DIC-CONCENTRATING MECHANISM IN MARINE PHYTOPLANKTON: REGULATION BY, AND REGULATION OF, ATMOSPHERIC CO₂ PARTIAL PRESSURE?

Here we examine the influence of DIC-concentrating mechanisms on the effect of atmospheric CO_2 partial pressure on marine phytoplankton, and on the possible influence of the marine phytoplankton on atmospheric CO_2. These effects will be considered in relation to the glacial–interglacial changes in CO_2 partial pressures, and to the current increase in atmospheric CO_2.

The effects can be viewed in terms of both the phenotypic responses of organisms that pump DIC relative to those that do not, and the changes in species composition that alter the ratio of DIC-pumping and CO_2-diffusion organisms.

The most obvious response to changed CO_2 concentrations is the direct "substrate" effect. Most of the marine phytoplankton organisms tested have a DIC pump with a high DIC affinity, so that photosynthesis would still be saturated with DIC, even at CO_2 concentrations in seawater in equilibrium with a glacial atmosphere containing only 18–20 Pa CO_2, as well as in extant marine locations, which have the lowest dissolved CO_2 concentrations as a result of high temperatures (=lower CO_2 solubility) and/or biological reduction in the CO_2 concentration to a value lower than air equilibrium (see Table 13.3; Refs. 6–9, 41, 71). Furthermore, the highest CO_2 concentrations in surface seawater over the last few million years are unlikely to have led to "de-adaptation" of the DIC-pumping organisms to their CO_2-diffusion mode with the DIC pump repressed (see Ref. 1). It would thus appear that the marine phytolankton organisms with a DIC pump would have expressed the pump, and been saturated with DIC, over the entire range of spatial and temporal variations in surface seawater CO_2 and DIC concentrations found over the past few million years.

The residuum of marine phytoplankton organisms that have lower CO_2 (DIC) affinites and that appear to lack a DIC pump, are not saturated with present-day oceanic DIC concentrations in equilibrium with 35 Pa CO_2 except, perhaps, in polar oceans where the greater solubility of CO_2 combines with the higher CO_2 affinity of Rubisco at low temperatures (41, 57, 71). The glacial draw-down of atmospheric CO_2 partial pressures would have outweighed the lower mean sea-surface temperature in glacial times, so that the sea-surface CO_2 concentration would have been lower than today (or in the interglacials before anthropogenic CO_2 increase). This would have tended to favor the DIC-pumping genotypes over those that did not pump DIC, especially if they lacked extracellular carbonic anhydrate to produce CO_2 from HCO_3^-, all other environmental conditions being equal (see below).

This argument as to the increased relative fitness of DIC-pumping plants in glacial times, and their decreased relative fitness as anthropogenically induced atmospheric CO_2 accumulation continues, has been applied to terrestrial C_4 plants in comparison with C_3 plants to C_4- and C_3-like physiologies (greater or lesser involvement of DIC pumps) in cyanobacteria and eukaryotic algae using atmospheric CO_2—that is, terrestrial and (emersed) intertidal free-living and lichenized phototrophs (61, 70). The difference between DIC-pumping phototrophs and those that lack DIC pumps at low CO_2 levels is accentuated, if we consider the CO_2 compensation point concentrations for the two groups of organisms. Most data are available for terrestrial C_3 and C_4 plants, where CO_2 compensation partial pressures are some 5 and 0.5 Pa, respectively. At a glacial minimum CO_2 partial pressure of 18 Pa, the C_4 plant has a "working range" of CO_2 of $(18 - 0.5)$ or 17.5 Pa, while the C_3 plant has a range of $(18 - 5)$ or 13 Pa. This effect is somewhat attenuated by the (admittedly abruptly approached) CO_2 saturation of C_4 photosynthesis at 18–20 Pa; in contrast, C_3 photosynthesis is far from saturation and has an essentially linear relation to CO_2 partial pressure in the range of 5–20 Pa. Use of this CO_2 compensation concentration argument still favors the C_4 mode of C acquisition relative to the C_3 process to a greater extent than that of CO_2 affinity per se. It is not clear how far this argument can be applied to planktonic algae in the open ocean, since the predicted lower CO_2 compensation concentration for DIC-pumping organisms, relative to those that do not pump DIC, seems to have been tested only for organisms that pump DIC (10; see Table 13.3).

Less direct effects relate to the response of organisms with different photo-synthetic C-acquisition physiologies to resources other than DIC (e.g., light, N, Fe, Mn, and Zn). The general argument here is that DIC-pumping organisms are likely to have superior N, Fe, and Mn use efficiencies, and possibly a higher photon use efficiency, than does an organism lacking a DIC pump, when both are growing at air-equilibrium CO_2 concentrations at 15–20°C in seawater. These effects would be accentuated at lower CO_2 concentrations, such as occurred in glacial times, and attenuated or even reversed as CO_2 concentrations increase with the anthropogenic increase in atmospheric CO_2 partial pressure.

For light as a limiting resource, no substantial advances seem to have been made since the subject was reviewed by Raven and Lucas (54). A number of important papers have appeared and been reviewed by Raven (48) on the relative importance of supplementary ATP sources of varying photon requirement—that is, cyclic and pseudocyclic photo-phosphorylation—in powering the DIC pump in freshwater cyanobacteria and microalgae. They have added little to arguments as to the relative photon requirements of photosynthesis and growth with DIC pumping and with diffuse CO_2 entry. It is still possible to argue that a DIC-pumping microalga or cyanobacterium, which manages to suppress completely the oxygenase activity of Rubisco, and has little leakage of CO_2 to short-circuit the DIC pump, can have a lower photon requirement for growth than an alga or cyanobacterium with diffusive CO_2 entry, when both are growing in air-equilibrium seawater. This could be important if climatic change alters the mixing depth of the surface waters of the ocean (see Ref. 73). Furthermore, the lower CO_2 partial pressure in glacial times would accentuate any advantage in photon use efficiency of DIC pumping organisms, while the current increase in CO_2 partial pressure would reduce or reverse this advantage.

For chemical resources we again have the possibility of a higher resource use efficiency for a DIC-pumping organism than for an organism that does not pump DIC for photosynthesis or growth in air-equilibrium seawater. This is possible when resource use efficiency is defined as mole C assimilated per second per mole of resource in the organism, rather than mole C assimilated per mole photon absorbed in the case of light-limited growth. The arguments for a higher resource use efficiency of growth limited by N, Fe, or Mn in a DIC-pumping phytoplankton organism than in an organism relying on diffusive CO_2 entry are detailed by Raven (44, 49, 54–6, 70) for growth at air-equilibrium CO_2 levels. For N the argument hinges on the idea that achieved specific reaction rate of Rubisco is higher in a DIC-pumping organism. These algae also economize on enzymes of the photo-respiratory carbon oxidation cycle, which more than offsets the extra N costs associated with the DIC-pumping mechanism. Furthermore, the DIC-pumping organism in seawater has little "spare" light-harvesting, photochemical and downstream photosynthetic capacity over what is used in light-saturated photosynthesis. The organism relying on CO_2 diffusion tends to have substantial surplus photosynthetic capacity. If this were not the case, the stimulation of photosynthesis by DIC concentrations in excess of those in seawater would not be seen. This latter argument for N (46, 48, 49, 56, 61, 70) can also be applied to Fe and Mn. Here the greater quantity of Fe- and Mn-requiring thylakoid-associated catalysts needed for the achieved rate of DIC assimilation in an organism relying on CO_2 diffusion relative to one dependent on a DIC pump is accentuated by the additional phytosynthetic capacity of organisms relying on CO_2 diffusion, which is expressed only at DIC concentrations higher than those found in seawater (49). These effects on N, Fe, and Mn would be accentuated at lower atmospheric CO_2 partial pressures, where the N, Fe, or Mn use efficiency of the DIC-pumping organisms would be unaltered, while the efficiency of an organism relying on CO_2 diffusion would decrease owing to increased CO_2 partial pressures. The advantages associated with DIC-pumping should be reduced or even abolished (49). It must be emphasized that these effects are, at least for Fe and Mn, hypothetical, and experimental testing is required.

It is also possible that Zn might exhibit differential requirement in DIC-accumulating organisms and in organisms relying on CO$_2$ diffusion as a result of different requirements for the major Zn-requiring enzyme, carbonic anhydrase. At least in some DIC-pumping organisms, HCO$_3^-$ is delivered to the compartment that contains Rubisco. These cells might have very little intracellular carbonic anhydrase (38, 39) relative to the requirement for intracellular carbonic anhydrase for catalysis of DIC diffusion, when CO$_2$ is entering by diffusion from limiting external CO$_2$ concentrations (14, 53, 62, 63). To this possible differential effect must be added any differential carbonic anhydrase requirements for the interconversion extracellular CO$_2$ to HCO$_3^-$ (1, 70). Further theoretical and especially experimental data are needed on the requirement for Zn in carbonic anhydrase and in other catalysts (not Cu–Zn superoxide dismutase which occurs only in the Charophyceae among the green and other algae tested (4, 15). Clearly Zn, like Fe, Mn, and N, is a potentially growth-limiting resource in nature for marine phytoplankton organisms (30, 32, 68).

Overall, then, it is possible that DIC-pumping organism in seawater could have a higher N, Fe, and Mn use efficiency of photosynthesis and growth than an organism using CO$_2$ diffusion as the mechanism of DIC entry in what is for them a DIC-limited environment. Such a difference in resource use efficiency would mean that a phytoplankton community composed mainly of DIC-pumping organisms could produce more organic C for a given Fe, Mn, or N input than could a community-based or CO$_2$-diffusion organism in seawater equilibriated with today's atmosphere. This could lead to increased organic C sedimentation in the community dominated by the DIC-pumping cells, with a corresponding net CO$_2$ removal from the atmospheric per unit Fe, Mn, or N available for primary production and subsequently for sedimentation. Such effects could have been important in amplifying the effect of changed atmospheric CO$_2$ partial pressure. A decrease in atmospheric CO$_2$ partial pressure might favor DIC-pumping organisms, which could produce more organic C per unit Fe, Mn, and N, with the possibility of increased organic C sedimentation and CO$_2$ removal from the atmosphere. This could have been important at the interglacial–glacial transition, granted an initial "priming" by an increase in a limiting chemical resource from the Fe, Mn, and N resulting from changed ocean circulation or atmospheric transfer (6–9, 30).

The reverse argument applies if there is an increased CO$_2$ partial pressure. Here the CO$_2$ diffusion organisms might be favored by the increased DIC availability, which could reduce organic C production per unit Fe, Mn, or N available for assimilation *unless* the DIC availability increased to a level that saturates growth even of those organisms that rely on CO$_2$ diffusion. This would tend to increase the Fe, Mn, and N use efficiency of growth of these organisms and, thus, once again increase organic C production and precipitation and removal of CO$_2$ from the atmosphere. This argument could apply to the current and future anthropogenic increase in atmospheric CO$_2$, provided oceanic circulation patterns do not change so much as to grossly alter Fe, Mn, and N availability to phytoplankton.

Although Fe has been suggested, with considerable supporting evidence, as a causative agent of phytoplankton-mediated changes in atmospheric CO$_2$ partial pressure in glacial interglacial transitions (30), the major element involved in these hypotheses is P (6 to 9). If Fe, Mn, and N are *not* major nutrients involved in initiating or amplifying the changed DIC removal by phytoplankton in the glacial–interglacial transitions, then the role for DIC-pumping as opposed to CO$_2$ diffusion in planktonic primary producers relates to the changes in species balance and their different effectiveness in DIC assimilation as a function of atmospheric CO$_2$ partial pressure.

CONCLUSIONS

Different responses to potentially limiting resources (CO$_2$, photons, Fe, Mn, N) of the (predominant) DIC-pumping organisms, relative to the less common nonpumping planktonic primary producers, could have important influences on the removal of CO$_2$ from the atmosphere in interglacial–glacial transitions, and in the current anthropogenic CO$_2$ increase. However, both the experimental basis for the assumed differential growth responses of DIC-pumping and non–DIC-pumping cells under the different resources supply conditions in culture, and the occurrence of these resource supply conditions in nature, need much further work.

POSTSCRIPT

The text of this article is essentially as submitted for publication at the time (March 1990) of the symposium dedicated to Professor Tolbert, and as such it reflects the authors' perceptions at that time. Our subsequent data and views on this topic may be found in Raven (26, 50–2, 55).

ACKNOWLEDGMENTS

Work on DIC accumulation in aquatic organisms in the authors' laboratory is supported by N.E.R.C. The arguments on the "working range" of CO$_2$ partial pressures for C$_3$ and C$_4$ plants arose from a discussion between J.A.R. and Dr. J. V. Lake of the A.F.R.C. Central Office, Swindon, U.K. (now director of the European Environmental Research Organisation, Wageningen, The Netherlands) on a train journey from Aberystwyth to Shrewsbury. The arguments on the similarity of phenomena related to isotopic discrimination in dissimilatory nitrate and sulfate reduction to those of photosynthesis in DIC-pumping organisms arose from a discussion between J.A.R. and Dr. L. Handley in Invergowrie. This chapter is dedicated to N. E. Tolbert in recognition of his many achievements in plants, especially in algal biochemistry and physiology.

REFERENCES

1. Aizawa, K., and Miyachi, S. 1986. Carbonic anhydrase and CO$_2$ concentrating mechanisms in microalgae and cyanobacteria. *FEMS Microbiol. Rev.* **39**: 215–33.

2. Aizawa, K., Nakamura, Y., and Miyachi, S. 1985. Variation of PEPc activity in *Dunaliella* associated with changes in atmospheric CO_2 concentration. *Plant Cell Physiol.* **26**: 1199–203.

3. Ajtay, G. L., Ketner, P., and Duvigneaud, P. 1979. Terrestrial primary production and phytomass. In *The Global Carbon Cycle*, ed. B. Bolin, E. T. Degens, S. Kempe, and P. Ketner, pp. 129–81. John Wiley, Chichester.

4. Asada, K., Kanematsu, S., and Vehida, K. 1977. Superoxide dismutases in photosynthetic organisms: Absence of Cu–Zn enzyme in eukaryotic algae. *Arch. Biochem. Biophys.* **179**: 243–56.

5. Badger, M. R., and Andrews, T. J. 1982. Photosynthesis and inorganic carbon usage by the marine cyanobacterium *Synechococcus* sp. *Plant Physiol.* **70**: 517–23.

6. Boyle, E. A. 1988. Vertical oceanic nutrient fractionation and glacial/interglacial CO_2 cycles. *Nature* **331**: 55–6.

7. Boyle, E. A. 1988. Cadmium: Chemical tracer of deepwater paleoceanography. *Palaeoceanography* **3**: 471–89.

8. Broecker, W. S. 1982. Ocean chemistry during glacial times. *Geochim. Cosmochin. Acta* **46**: 1689–705.

9. Broecker, W. S., and Denton, G. H. 1989. The role of ocean–atmosphere reorganisations in glacial cycles. *Geochim. Cosmochim. Acta* **53**: 2465–501.

10. Burns, B. D., and Beardall, J. 1987. Utilization of inorganic carbon by marine microalgae. *J. Exp. Mar. Biol. Ecol.* **107**: 75–86.

11. Burris, J. E., Porter, J. W., and Laing, W. A. 1983. Effects of carbon dioxide concentration on coral photosynthesis. *Mar. Biol.* **75**: 113–16.

12. Charpy-Roubaud, C. and Sournia, A. 1990. The comparative estimation of phytoplanktonic, microphytobenthic and macrophytobenthic primary production in the oceans. *Mar. Microb. Food Webs* **4**: 31–57.

13. Colman, B., and Gehl, K. A. 1983. Physiological characteristics of photosynthesis in *Porphyridium cruentum*: Evidence for bicarbonate transport in a unicellular red alga. *J. Physol.* **19**: 216–19.

14. Cowan, I. R. 1986. Economics of carbon fixation in higher plants. In *On the Economy of Plant Form and Function*, ed. T. J. Givinsh, pp. 133–70. Cambridge University Press, Cambridge.

15. de Jesus, M. D., Tabatabai, F., and Chapman, D. J. 1989. Taxonomic distribution of copper–zinc superoxide dismutase in green algae and its phylogenetic importance. *J. Phycol.* **25**: 767–72.

16. de Vooys, C. G. N. 1979. Primary production in aquatic environments. In *The Global Carbon Cycle*, ed. B. Bolin, E. T. Degens, S. Kempe, and P. Ketner. pp. 259–92. John Wiley, Chichester.

17. Dixon, G. K., Patel, B. N., and Merrett, M. J. 1987. Role of intracellular carbonic anhydrase in inorganic carbon assimilation by *Porphyridium purpureum*. *Planta* **172**: 508–13.

18. Dixon, G. K., and Merrett, M. J. 1988. Bicarbonate utilization by the marine diatom *Phaeodactylum tricornutum* Bohlin. *New Physol.* **109**: 47–51.

19. Down, Kl, Raven, J. A., and Beardall, J. 1989. Inorganic carbon assimilation by picoplanktonic prasinophytes grown under a low PFD. *Br. Phycol. J.* **24**: 302.

20. Ferguson, S. J. 1988. The redox reactions of the nitrogen and sulphur cycles. *Symp. Soc. Gen. Microbiol.* **42**: 1–29.

21. Glover, H. E., and Morris, I. 1981. Photosynthetic characteristics of coccoid marine cyanobacteria. *Arch. Microbiol.* **129**: 42–6.

22. Goudriaan, J. 1987. The biosphere as a driving force in the global carbon cycle. *Neth. J. Agric. Sci.* **35**: 177–87.

23. Gradmann, D. 1989. ATP-driven chloride pump in giant alga *Acetabularia*. In *Methods of Enzymology*, Vol. 174: *Biomembranes*, Part U: *Cellular and Subcellular Transport: Eukaryotic (Nonepithelial) Cells*, ed. S. Fleischer and B. Fleischer, pp. 409–504. Academic Press, San Diego.

24. Hillis-Colinvaux, L. 1986. Historical perspectives on algae and reefs: Have reefs been misnamed? *Oceanus* **29**: 43–8.

25. Johnston, A. M., and Raven, J. A. 1991. The acquisition of inorganic carbon by *Phaeodacytlum tricornutum*. *Br. Phycol. J.* **26**: 89.

26. Johnston, A. M., and Raven, J. A. 1992. The acquisition of DIC by the marine diatom *Phaeodacylum tricornutum*. *Br. Phycol. J.* **27**: 26.

27. MacFarlane, J. J., and Raven, J. A. 1985. External and internal CO_2 transport in *Lemanea*: Interactions with the kinetics of ribulose bisphosphate carboxylase. *J. Exp. Bot.* **36**: 610–22.

28. MacFarlane, J. J., and Raven, J. A. 1989. Quantitative determination of unstirred layer permeability and kinetic parameters of Rubisco in *Lemanea mamillosa*. *J. Exp. Bot.* **40**: 321–27.

29. MacFarlane, J. J., and Raven, J. A. 1990. C, N and P nutrition of *Lemanea mamillosa* Kutz. (Batrachospermales, Rhodophyta) in the Dighty Burn, Angus, Scotland. *Plant Cell Environ.* **13**: 1–13.

30. Martin, J. H., and Fitzwater, S. E. 1988. Iron deficiency limits phytoplankton growth in the north-east Pacific subarctic. *Nature* **331**: 341–43.

31. Miller, A. G., and Colman, B. 1980. Evidence for HCO_3^- transport by the blue-green alga (Cyanobacterium) *Coccochloris peniocystis*. *Plant Physiol.* **65**: 397–402.

32. Morel, F. M. M., and Hudson, R. J. M. 1985. The geobiochemical cycle of trace elements in aquatic systems: Redfield revisited. In *Chemical Processes in Lakes*, ed. W. Stumm, pp. 251–81. John Wiley, New York.

33. Munoz, J., and Merrett, M. J. 1988. Inorganic-carbon uptake by a small celled strain of *Stichococcus bacillaris*. *Planta* **175**: 460–4.

34. Munoz, J., and Merrett, M. J. 1989. Inorganic-carbon transport in some marine eukaryotic microalgae. *Planta* **178**: 450–5.

35. Paasche, E. 1964. A tracer study of the inorganic C uptake during coccolith formation and photosynthesis in the coccolithophorid *Coccolithus huxleyi*. *Physiol. Plant. Suppl. III.*

36. Patel, B. N., and Merrett, M. J. 1986. Inorganic-carbon uptake by the marine diatom *Phaeodacylum tricornutum*. *Planta* **169**: 222–7.

37. Peck, H. D., Jr., and Lissolo, T. 1988. Assimilatory and dissimilatory sulphate reduction: Enzymology and bioenergetics. *Symp. Soc. Gen. Microbiol.* **42**: 99–132.

38. Price, G. D., and Badger, M. R. 1989a. Expression of human carbonic anhydrase in the cyanobacterium *Synechococcus* PCC 7942 creates a high CO_2-requiring phenotype. Evidence for a central role for caboxysomes in the CO_2-concentrating mechanism. *Plant Physiol.* **91**: 503–13.

39. Price, G. D., and Badger, M. R. 1989b. Isolation and characterization of high CO_2 requiring mutants of the cyanobacterium *Synechocccus* PCC 7942. Two phenotypes that accumulate inorganic carbon but are apparently unable to generate CO_2 within the carboxysome. *Plant Physiol.* **91**: 514–25.

40. Prins, H. B. A., and De Guia, M. B. 1986. Carbon source of the water soldier, *Stratiotes aloides*. L. *Aquat. Bot.* **26**: 225–34.

41. Rau, G. H., Takahashi, T., and Des Moines, D. J. 1989. Latitudinal variations in plankton $\delta^{13}C$: Implications for CO_2 and productivity in past oceans. *Nature* **341**: 516–18.

42. Raven, J. A. 1977. The evolution of land plants in relation to supracellular transport processes. *Adv. Bot. Res.* **5**: 153–219.

43. Raven, J. A. 1981. Nutritional strategies of submerged benthic plants: The acquisition of C, N and by rhizophytes and haptophytes. *New Phytol.* **88**: 1–30.

44. Raven, J. A. 1984. *Energetics and Transport in Aquatic Plants.* Alan R. Liss, New York.

45. Raven, J. A. 1984. Physiological correlates of the morphology of early vasuclar plants. *Bot. J. Linn. Soc.* **88**: 105–26.

46. Raven, J. A. 1984. A cost benefit analysis of photon absorption by photosynthetic unicells. *New Phytol.* **98**: 593–625.

47. Raven, J. A. 1986. Physiological consequences of extremely small size for dietotrophic organisms in the sea. In *Photosynthetic Picoplankton*, ed. T. Platt and W. K. W. Li, pp. 1–70. *Can. Bull. Fish. Aquat. Sci.* No. 214. Dept. Oceans and Fisheries Place, Ottawa, Ontario.

48. Raven, J. A. 1990a. Use of isotopes in estimating respiration and photorespiration in microalgae. *Mar. Microb. Food Webs* **4**: 59–86.

49. Raven, J. A. 1990b. Mn and Fe use efficiencies of phototrophy as a function of photon flux density for growth: Predictions and their relationship to photo- and bio-geochemistry. *New Phytol.* **116**: 1–18.

50. Raven, J. A. 1991. Impications of inorganic C utilization: Ecology, evolution and geochemistry. *Can. J. Bot.* **69**: 908–24.

51. Raven, J. A. 1991. Physiology of inorganic C acquisition and implications for resource use efficiency by marine phytoplankton: Relation to increased CO₂ and temperature. *Plant Cell Environ.* **14**: 779–84.

52. Raven, J. A. 1991. Plant responses to high O₂ concentrations: Relevance to previous high O₂ episodes. *Glob. Planet. Change* **5**: 19–38.

53. Raven, J. A., and Glidewell, S. M. 1981. Processes limiting photosynthetic conduct-ance. In *Processes Limiting Plant Productivity*, ed. C. B. Johnson, pp. 109–36. Butterworths, London.

54. Raven, J. A., and Lucas, W. J. 1985. The energetics of carbon acquisition. In *Inorganic Carbon Uptake by Aquatic Photosynthetic Organisms*, ed. W. J. Lucas and J. A. Berry, pp. 305–24. American Society of Plant Physiologisgs, Rockville, Md.

55. Raven, J. A., Osborne, B. A., and Johnston, A. M. 1985. Uptake of CO₂ by aquatic vegetation. *Plant Cell Environ.* **8**: 417–29.

56. Raven, J. A., Johnston, A. M., MacFarlane, J. J., Surif, M. B., and McInroy, S. 1987. Diffusion and active transport of inorganic carbon species in freshwater and marine macroalgae. In *Progress in Photosynthesis Research*, ed. J. Biggins, Vol. 4, pp. 33–340. Nijhoff, Dordrecht.

57. Raven, J. A., and Geider, R. 1988. Temperature and algal growth. *New Phytol.* **110**: 441–61.

58. Raven, J. A., Handley, L. L., MacFarlane, J. J., McInroy, S., McKenzie, L., Richards, J. H., and Samuelson, G. 1988. The role of root CO₂ uptake and CAM in inorganic C acquisition by plants of the isoetid life form: A review, with new data on *Eriocaulon decagulare*. *New Phytol.* **108**: 125–48.

59. Raven, J. A., and Ramsden, H. 1989. Similarity of stomatal index in the C₄ plant *Salsola kali* L. in material collected in 1843 and in 1987: Relevance to changes in atmospheric CO₂ content. *Trans. Bot. Soc. Edinb.* **45**(1988): 223–33.

60. Raven, J. A., and Sprent, J. I. 1989. Phototrophy, diazotrophy and palaeo-atmospheres: Biological catalysis and the H, C, N and O cycles. *J. Geol. Soc.* **146**: 161–70.

61. Raven, J. A., Johnston, A. M., Handley, L. L., and McInroy, S. G. 1990. Transport and assimilation of inorganic carbon by *Lichina pygmaea* under emersed and submersed conditions. *New Phytol.* **114**: 407–17.

62. Raven, J. A., and Johnston, A. M. 1991. Mechanisms of inorganic carbon acquisition in marine phytoplankton and their implications for the use of other resources. *Limnol. Oceanogr.* **36**: 1701–14.

63. Roeske, C. A., Widholm, J. M., and Ogren, W. L. 1989. Photosynthetic carbon metabolism in photoautotrophic suspension cultures grown at low and high CO_2. *Plant Physiol.* **91**: 1512–19. Cambridge University Press, New York.

64. Schidlowski, M., Hayes, J. M., and Kaplan, I. R. 1983. Isotopic inferences of ancient biochemistries: C, S, H and N. In *Earth's Earliest Biosphere: Its Origin and Evolution*, ed. J. W. Schopf, pp. 149–86. Princeton University Press, Princeton, N.J.

65. Sikes, C. S., Roer, R. D., and Wilbur, K. M. 1980. Photosynthesis and coccolith formation: Inorganic C sources and net reaction of deposition. *Limnol. Oceanogr.* **25**: 248–61.

66. Sikes, C. S., and Wilbur, K. M. 1982. Functions of coccolith formation. *Limnol. Oceanogr.* **27**: 18–26.

67. Smith, S. V. 1981. Marine macrophytes as a global carbon sink. *Science* **211**: 838–40.

68. Smith, S. V. 1985. Phosphorus versus nitrogen limitation in the marine environment. *Limnol. Oceanogr.* **29**: 1149–60.

69. Steeman Nielsen, E. 1966. The uptake of free CO_2 and HCO_3^- during photosynthesis of plankton algae with special reference to the coccolithophorid *Coccolithus huxleyii*. *Physiol. Plant* **19**: 232–40.

70. Surif, M. B., and Raven, J. A. 1990. Photosynthetic gas exchange under emersed conditions in eulittoral and normally submersed members of the Fucales and Laminariales: Interpretation in relation to C isotope ratio and N and water use efficiency. *Oecologia* **82**: 68–80.

71. Takahashi, T. 1989. The carbon dioxide puzzle. *Oceanus* **32**: 22–9.

72. Tanner, W. and Sauer, N. 1989. Uptake of sugars and amino acids by Chlorella. In *Methods in Enzymology*, Vol. 174: *Biomembranes*, Part U: *Cellular and Subcellular Transport: Eukaryotic (Nonepithelial) Cells*, ed. S. Fleischer and B. Fleischer, pp. 390–402. Academic Press, San Diego.

73. Tett, P. 1990. The photic zone. In *Light and Life in the Sea*, ed. P. J. Herring, A. K. Campbell, M. Whitfield, and C. Maddock, pp. 59–87. Cambridge University Press, Cambridge.

74. Thomsen, H. A. 1986. A survey of the smallest eukaryotic organisms of the marine phytoplankton. In *Photosynthetic Picoplankton*, ed. T. Platt and W. K. W. Li, pp. 121–58. Can. Bull. Fish. Aquat. Sci. No 214.

75. Zenvirth, D., and Kaplan, A. 1981. Uptake and efflux of inorganic carbon in *Dunaliella salina. Planta* **152**: 8–12.

14

Structure and Induction of Periplasmic Carbonic Anhydrase of *Chlamydomonas reinhardtii*

SHIGETOH MIYACHI and HIDEYA FUKUZAWA

Microalgae grown in 0.04% (air level) CO_2 show a higher affinity for inorganic carbon in photosynthesis than those grown at a high level of CO_2 (1–5%). This higher affinity has been accounted for by carbonic anhydrase (CA^{III}; carbonate hydrolyase, EC 4.2.1.1) and by putative CO_2-concentrating mechanisms that operate via active dissolved inorganic carbon (DIC) transporters (1). Until now, no component involved in the active DIC transporter has been identified in microalgae. On the other hand, the role and the structure of CA have been relatively well characterized. The enzyme is a widely distributed zinc metalloenzmye that catalyzes the reversible hydration of CO_2.

In photoautotrophically grown *Chlamydomonas reinhardtii* cells, CA activity is located mainly outside the plasma membrane, either in the periplasmic space or attached to cells wall (12, 24). Periplasmic CA has been identified as a glycoprotein with a subunit molecular mass estimated at 35–37 kDa (2, 23). The CA activity increases when *high-CO_2* cells [cells grown with the bubbling of 5% CO_2 enriched air] are transferred into the *low-CO_2* condition (grown with the bubbling of ordinary air). The induction of CA activity under the low-CO_2 condition was inhibited by the protein translation inhibitor cycloheximide (26), suggesting that the enzyme was synthesized de novo when the CO_2 concentration was lowered. Subsequently, evidence was provided confirming that the increase in CA activity paralleled the increase in CA protein (26). Isolation of poly(A)$^+$ RNA from *C. reinhardtii* cells and analysis of the in vitro translation products showed that the synthesis of the precursor polypeptide was greater with mRNA from low-CO_2 cells than from high-CO_2 cells, indicating that the induction of CA in *C. reinhardtii* is regulated at a level prior to translation (2, 22). The biosynthesis and intracellular processing of this enzyme were then studied (23). In the wall-less mutant of *C. reinhardtii* cw-15, which secretes CA into the culture medium, a 42-kDa precursor polypeptide was identified by in vivo pulse-chase experiments.

In *Chlorella vulgaris* 11h cells, the light intensity, which compensates respiration, saturated the induction of CA activity. Inhibition of photosynthesis by

3-(3,4-dichlorophenyl)-1,1-dimethylurea (DCMU) did not affect the induction, indicating that photosynthesis is not required in *Chlorella* (18). On the other hand, photosynthetic and nonphotosynthetic light were required for the induction of CA activity in *C. reinhardtii* (4). In the nonphotosynthetic light, blue light was effective for the induction (5).

In animals, there are seven isozymes of CA (CAI–CAVII), grouped according to their cellular distribution and biological function (21). The human isozymes— CAI, CAII, and CAIII—have been characterized at the nucleotide sequence level to have a M_r of about 30,000, with one zinc atom at the active sites (21). Three-dimensional structures of human CAI and CAII were determined by x-ray crystallography, and the amino acid residues essential for conformation of the enzyme have been pinpointed (17).

In this chapter, we summarize our recent biochemical and molecular biological studies on (a) subunit structure of the enzyme based on biochemical methods; (b) the biosynthetic processes and primary structure of the enzyme based on sequence analysis of cDNA; (c) the organization and expression of the genes encoding periplasmic CA.

SUBUNIT CONSTITUTION OF PERIPLASMIC CA IN *CHLAMYDOMONAS REINHARDTII*; PRESENCE OF 35-kDa and 4-kDa SUBUNITS

The enzyme was purified from low-CO$_2$ cells by inhibitor-conjugated affinity chromatography and then subjected to anion-exchange HPLC. The major peak of protein having highest enzyme activity (ca. 2000 units/mg of protein) was recovered, and this fraction was used to analyze the enzyme constitution of CA from *Chlamydomonas* (11). The purified enzyme showed a sharp symmetrical peak upon gel-filtration HPLC, and its molecular mass was estimated at 94 kDa. This value was different from the value (115 kDa) determined by Yagawa et al. (25). Therefore, the molecular mass of the holoenzyme was determined with a more sophisticated method, as follows.

One broad band with a molecular mass of 35 kDa was obtained by SDS–PAGE (lane 1, Fig. 14.1) as previously reported (23). Unexpectedly, analysis of the N-terminal amino acid sequence of the holoenzyme resulted in the release of two amino acids for each cycle of Edman degradation. The major cause for the double sequences might be heterogeneous subunit constitution of the enzme.

When analyzed by SDS–PAGE using 10–20% polyacrylamide gradient gel, the 35-kDa polypeptide was clearly separated into two bands, A$_1$ and A$_2$, with molecular masses of 35 and 36.5 kDa, respectively (lane 3, Fig. 14.1). These polypeptides were electroblotted onto a polyvinylidene difluoride filter, and their amino acid sequences were determined from the N-termini. Both polypeptides showed a single amino acid sequence (referred to as the A sequence; see below). Thus a polypeptide with another sequence was not observed from this SDS–PAGE gel. Instead, a polypeptide with a molecular mass of 4 kDa was detected when the quantity of the sample applied to SDS–PAGE was increased and when staining and destaining of the gel were carried out as quickly as possible. This polypeptide band became undetectable in the gel during the prolonged staining process—the

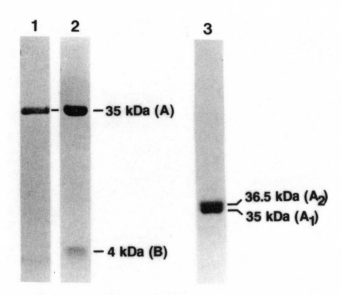

Fig. 14.1. SDS–PAGE analysis of carbonic anhydrase. The purified enzyme was electrophoresed in 12.5% gel for 2.5 h (1), in 15% gel for 2.5 h followed by glutaraldehyde fixation (2), and in 10–20% gradient gel for 4 h (3). The large subunit of carbonic anhydrase is indicated as 35 kDa (A) in lanes 1 and 2, and was separated into 35 kDa (A$_1$) and 36.5 kDa (A$_2$) in lane 3. The small subunit is shown as 4 kDa (B) in lane 2.

reason why subunits of *Chlamydomonas* CA had been presumed to be homogeneous in the previous report (25). To prevent the release of this polypeptide from polyacrylamide gel, the gel was fixed with glutaraldehyde. This resulted in a clear staining of the 4-kDa polypeptide (lane 2, Fig. 14.1). This 4-kDa polypeptide was electroblotted onto a polyvinylidene difluoride filter, and the N-terminal amino acid sequence was analyzed. The fact that the 4-kDa polypeptide had another sequence (referred as the B sequence, see below) indicates that CA is a hetero-oligomer.

ESTIMATION OF MOLECULAR MASS BY LALLS-HPLC

CA was incubated with varying concentrations of dithiothreitol, and the enzyme activity was measured at the indicated time (Fig. 14.2). Inactivation of the enzyme activity by dithiothreitol depends on its concentration and incubation time. Complete inactivation occurred after a 2-h incubation at 30°C with 10 mM dithiothreitol. The inactivated enzyme showed a single symmetrical peak at 38.9 kDa on gel-filtration HPLC (data not shown). This polypeptide had the A sequence and thus corresponded with the 35-kDa polypeptide obtained by SDS–PAGE. This observation suggests that reduction of disulfide bonds of the

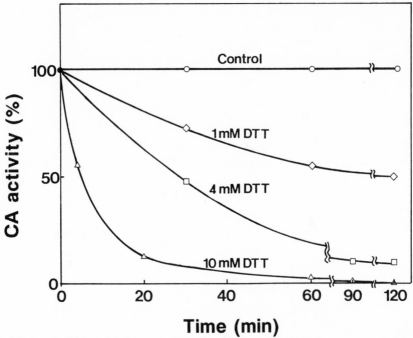

Fig. 14.2. Effect of dithiothreitol (DTT) on the enzyme activity of carbonic anhydrase (CA). The initial activity was 2000 units/mg protein. The enzyme was incubated at 30°C in 10 mM Tris-HCl (pH 8.0) with the indicated concentration of DTT. The enzyme activity was measured within 1 min after 100-fold dilution in 12.5 mM barbital-H₂SO₄ (pH 8.3).

enzyme causes dissociation of the holoenzyme into the 35-kDa and 4-kDa polypeptides. However, no polypeptide with the B sequence was detected because of interference with overlapping dithiothreitol.

Low-angle laser light high-performance liquid chromatography (LALLS-HPLC), which has recently been developed as a simple and precise method to estimate the molecular mass (14), was applied for the estimation of the molecular mass of CA. Figure 14.3 shows a typical elution profile of holoenzyme (a) and dithiothreitol-treated enzyme (b) from the same column, which was monitored with a light-scattering photometer and a differential refractometer connected in series. The molecular masses of the holoenzyme and the dithiothreitol-treated enzyme estimated with Equation 14.1 were 67 and 35 kDa, respectively.

$$M = \frac{k}{dn/dc} \times \frac{(\text{output})_{LS}}{(\text{output})_{RI}} \tag{14.1}$$

where k is a constant determined by the instrumental and experimental conditions, dn/dc is a specific refractive index increment, and $(\text{output})_{LS}$ and $(\text{outputs})_{RI}$ are outputs of the light-scattering photometer and the refractometer, respectively. The value of $k/(dn/dc)$ was calculated from the values of $(\text{output})_{LS}/(\text{output})_{RI}$ obtained using bovine serum albumin (molecular mass, 67 kDa) as standard.

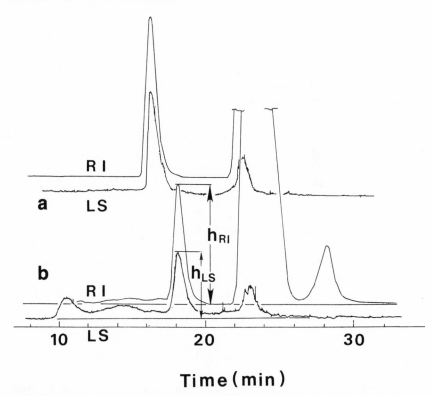

Fig. 14.3. Molecular weight estimation by LALLS-HPLC. RI (refractive index) indicates the output of the precision differential refractometry. LS (light scattering) indicates the output of the light-scattering photometry. (a) Holoenzyme; (b) dithiothreitol-treated enzyme.

The 4-kDa polypeptide did not show enough signal in the light-scattering for the determination of the molecular mass. Thus the mass ratio of the 35-kDa to the 4-kDa polypeptide in a holoenzyme was estimated from the respective peak areas of output of the refractometry (data not shown). The peak areas were 89% and 11% for the 35-kDa and 4-kDa polypeptides, respectively. The molar ratio obtained from the mass ratio by dividing the respective molecular mass was 1:1. On the basis of these results, we therefore concluded that the molar ratio of the holoenzyme to the 35-kDa and 4-kDa polypeptides is 1:2:2; namely that the holoenzyme consists of two 35-kDa (A) subunits and two 4-kDa (B) subunits.

AMINO ACID SEQUENCE DETERMINATION OF CA POLYPEPTIDES

Amino acid sequences of the pyridylethylated 35-kDa and 4-kDa polypeptides were analyzed. The N-terminal sequence of the 35-kDa polypeptide was as follows: Cys-Ile-Tyr-Lys-Phe-Gly-Thr-Ser-Pro-Asp-Ser-Lys-Ala-Thr-Val-Sel-Gly-(the A sequence). The complete sequence of the 4-kDa polypeptide was as follows: Ala-Glu-Ser-Ala-Asn-Pro-Asp-Ala-Tyr-Thr-Cys-Lys-Ala-Val-Ala-Phe-Gly-Gln-Asn-Phe-Arg-Asn-Pro-Gln-Tyr-Ala-Asn-Gly-Arg-Thr-Ile-Lys-Leu-Ala-Arg-Tyr-

His (the B sequence). The 4-kDa polypeptide consists of 37 amino acid residues and has a molecular mass of 4144. Only one cysteine was located at the 11th residue from the N-terminus of this polypeptide. Computer-assisted comparison of the amino acid sequence of the 4-kDa polypeptide to the data base library of protein sequences revealed that the 4-kDa polypeptide has a significant sequence similarity to C-terminal regions of mammalian CAs such as horse (21/37 including conservative amino acid substitutions), human (20/37), and *Rhesus macaque* (20/37). These mammalian CAs are monomeric polypeptides with a molecular mass of ca. 30 kDa (21).

To obtain more information about the amino acid sequence of the enzyme, products of a CNBr-cleavage of the carboxymethylated holoenzyme were separated by reverse-phase HPLC, and the sequence of the polypeptide at peak 7 was determined as follows: Arg-Pro-Asn-Asp-Ala-Ala-Arg-Val-Thr-Ala-Val-Pro-Thr-Gln-Phe-His-Phe-His-Ser-Thr-Thr-Glu. Computer-assisted comparison with the amino acid sequence data base indicates that the amino acid sequence contains an identical sequence (-Gln-Phe-His-Phe-His-), with the highly conserved region in animal CAs containing two zinc-liganded histidine residues. As the sequence of zinc-liganded region was not observed in the 4-kDa polypeptide, it was concluded that zinc exists only in the 35-kDa polypeptide.

ZINC CONTENT

The zinz content of the holoenzyme, analyzed by atomic absorption spectrophotometry, was 0.0345 µg zinc per 21 µg protein. If one assumes that the molecular mass of the holoenzyme is 76 kDa, one molecule of the holoenzyme contains two atoms of zinc, and accordingly the 35-kDa polypeptide contains one atom of zinc.

These biochemical studies revealed that (a) periplasmic CA of *C. reinhardtii* dissociates into the 35-kDa and the 4-kDa subunits upon treatment with dithiothreitol accompanied by a loss of the enzymatic activity; (b) the molar ratio of the 35-kDa subunit to the 4-kDa subunit is 1:1; (c) there is one cysteine residue in the 4-kDa subunit and at least one in the 35-kDa subunit; (d) one molecule of the holoenzyme contains two zinc atoms, and the 35-kDa subunit has the zinc-liganded site; (e) the 35-kDa subunit is further separated into two subcomponents (A_1 = 35 kDa, and A_2 = 36.5 kDa) with the same N-terminal amino acid sequences. Based on these results, a model was proposed as shown in Figure 14.4 (11). The enzyme is a heterotetramer; and, since the 35-kDa subunit was separated into two subcomponents, A_1 and A_2, three combinations of the A_1 and A_2 subcomponents were possible in the subunit constitution (A_1A_1BB, A_1A_2BB, and A_2A_2BB). Since the 4-kDa subunit has only one cysteine, which is required for the disulfide bond between the 35-kDa subunit and the 4-kDa subunit, it is apparent that no disulfide bond exists between the two 4-kDa subunits. One of cysteine residues is used for the disulfide bond between the 35-kDa and the 4-kDa subunits, and some of the rest between the 35-kDa subunits.

The heterogeneity of the 35-kDa subunit in SDS–PAGE may be attributable to substitution, addition, or deletion of amino acids of the polypeptide chain other

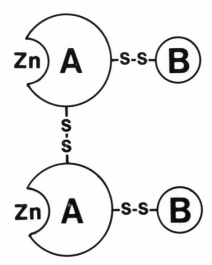

Fig. 14.4. A model for subunit constitution of *Chlamydomonas* carbonic anhydrase. A and B indicate the large and small subunits, respectively. The large subunit (A) is composed of two subcomponents, A_1 (35 kDa) and A_2 (36.5 kDa). Zinc atoms contained in the large subunits are shown as Zn. Disulfide bonds between subunits are shown by -S-S-.

than 17 amino acids from the N-terminus, or to modification in carbohydrate chains.

CLONING AND SEQUENCE ANALYSIS OF cDNA ENCODING PERIPLASMIC CA FROM *C. REINHARDTII*

Isolation and Characterization of CA cDNA

A λgt10 cDNA library was constructed using poly $(A)^+$ RNA from 2-h air-adapted *Chalmydomonas* cells (8). The CA activity was induced during this adaptation period (4). So that cDNA clones could be isolated two subunit-specific probes Pr68 and Pr56 were synthesized, based on the most probable codons in *Chlamydomonas* genes (16). The sequence of Pr68—5′-TCGGTGGTGG AGTGGAAGTGGAACTGGGTGGGGACGGCGGTGACGCGGTCGGCGG CGTGTTGGGGCG-3′—was complementary to a putative mRNA sequence encoding a part of the CNBr-cleaved peptide fragment—Arg-Pro-Asn-Asp-Ala-Ala-Asp-Arg-Val-Thr-Ala-Val-Pro-Thr-Gln-Phe-His-Phe-His-Ser-Thr-Thr-Glu—derived from the 35-kDa large subunit. The sequence of Pr56 was 5′-GTGCGGCCGTTGGCGTACTGGGGGGTTGCGGAAGTTCTGGCCGAAG GCCACGGCCTT-3′, which was completementary to a putative mRNA sequence encoding a part of the small subunit (Lys-Ala-Val-Ala-Phe-Gly-Gln-Asn-Phe-Arg-Asn-Pro-Gln-Tyr-Ala-Asn-Gly-Arg-Thr).

First, the cDNA library containing 5×10^5 recombinant phages was screened by ^{32}P-labeled oligonucleotide Pr68, which was specific to the 35-kDa subunit, and 11 independent cDNA clones were obtained. Second, the reprobing of the

```
                                                                    GAGTCATTACCTGCAACCCACTTGAACACC   -1

  +1  ATGGCGCTACTGGCGCTCTACTCCTGGTCGCGCTTGGCGCTGCGCCAGGCTTGCATCTACAAGTTCGGCACGTTCGCCGGACTCGCCGGCTGGCATCACTGG   120
   1   M  A  R  T  G  A  L  L  L  V  A  L  A  L  A  G  C  A  Q  A  C  I  Y  K  F  G  T  S  P  D  S  K  A  T  V  S  G  D  H  W    40

 121  GACCATGGCCTCAACGGCGAGAACTGGGAGGGCAAGGACGGCGCAGGCAACGCCTGGGTTTGCAAGACTGGCCGCAAGCAGTCGCCCATCAACGTGCCCCAGTACGTCCTGGACGGG   240
  41   D  H  G  L  N  G  E  N  W  E  G  K  D  G  A  G  N  A  W  V  C  K  T  G  R  K  Q  S  P  I  N  V  P  Q  Y  V  L  D  G    80

 241  AAGGGTTCCAAGAТTGCCAACGGCCTGCAAACCCAGTGGTCGTACCCTGACCTGATGTCCAACGGCACCTCAGTCATCAACAACGGCCACACCATCCAGGTGCAGTGGACTTAC   360
  81   K  G  S  K  I  A  N  G  L  Q  T  Q  W  S  Y  P  D  L  M  S  [N]  G  T  S  V  Q  V  I  N  N  G  H  T  I  Q  V  Q  W  T  Y   120

 361  AACTACGCCGGCCATGCCACCATCGCCATCCCTGCCATGCACAACCAGACCAACCGCATCGTGGACGTGCTGGAGATGCGCCCCAACGACGCCGCCGACCGCGTGACCGCTGCGTGCCACC   480
 121   N  Y  A  G  H  A  T  I  A  I  P  A  M  H  [N]  Q  T  N  R  I  V  D  V  L  E  M  R  P  N  D  A  A  D  R  V  T  A  V  P  T   160

 481  CAGTTCCACTTCCACTCCACCTCGGAGCACCTGCTGGCCGGCAAGATCTATCCCCTTGAGTTGCACATTGTGCACCAGGTGACTGAGAAGCTGGAGGCGTGCAAGGGCGGCTGCTTCAGC   600
 161   Q  F  F  H  S  T  S  E  H  L  L  A  G  K  I  Y  P  L  E  L  H  I  V  H  Q  V  T  E  K  L  E  A  C  K  G  G  C  F  S   200

 601  GTCACCGGCATCCTGTTCCAGCTGGACAACGGCCCCGATAACGAGCTGCTTGAGCCCATCTTTGCGAACATGCCCTCGCGGGAGGGCACCTTCAGCAACCTTCCAGCAGGCACCACCATC   720
 201   V  T  G  I  L  F  Q  L  D  N  G  P  D  N  E  L  L  E  P  I  F  A  N  M  P  S  R  E  G  T  F  S  N  L  P  A  G  T  T  I   240

 721  AAGCTGGGTGAGCTGCTGCCCAGCGACCGCGACTACGTAACGTACGAGGGCAGCCTGACCACCCCGCCCTGCTCTGTGGCAGTCATGACCCAGCCGCAGCGCATCAGC   840
 241   K  L  G  E  L  L  P  S  D  R  D  Y  V  T  Y  E  G  S  L  T  T  P  P  C  S  E  G  L  L  W  H  V  M  T  Q  P  Q  R  I  S   280

 841  TTCGGCCAGTGGAACCGCTACCGCCTGGCTGTGGGCCTGAAGGAGTGCAACTCCACGGAGACCGCCGCCGACGCCGGCCACCACCACCACCGCCGCCTGCTCCACAACCACGCCCAC   960
 281   F  G  Q  W  N  R  Y  R  L  A  V  G  L  K  E  C  [N]  S  T  E  T  A  A  D  A  G  H  H  H  H  R  R  L  L  H  N  H  A  H   320

 961  CTGGAGGAGGTGCCTGCCGCCACCTCCGAGCCCAAGCACTACTTCCGGCGCGTGATGCTGGCCGAGTCGGCGAACCCGATGCCTACACCGTCGGCACGGCTACCACCGTCGGCAGGCATTCCATTTTCCAGGCTT   1080
 321   L  E  E  V  P  A  A  T  S  E  P  K  H  Y  F  R  R  V  M  L  A  E  S  A  N  P  D  A  Y  T  C  K  A  V  A  F  G  Q  N  F   360

1081  CGCAACCCCAGTACGCCCACGGCCGCCACCATCAAGCTGGCCCGCTATCACTAAACTTCCCAGTAGTTAGTCACGCTACCACGTCGGCACGGCATTCCATTTTCCAGGCTT   1200
 361   R  N  P  Q  Y  A  N  G  R  T  I  K  L  A  R  Y  H                                                                      377

1201  TGCTTCACGGTTTGGTGTGTCATTCGATGGTGTTCTTGACGACCGGCCTTTGGCGGGCCTT   1320
1321  ACCGGACGGCGGCGGGGACCTCGTTTCTCCTGACTAGTAAAGAAGTAAGGAAGGTATGGAGTTGGTTCCACGA   1440
1441  CCGGTTCTCCTGATGAGTAGTGAGGCAAGTAAATACGTAGAGAGGCAATCGAGAGTGTCTCACGCGAGACATTTGCGTACAGGGGAGGCA   1560
1561  CCGGTTCTCCTGATGATCGTACTTATCGCAAGTTATATAAGGCTGGTGTAATAACGGAACTTGACAGCAATCGAGAGTGTCCACGGTATGGTTGCGCTCCGGTCTGGTCGGTTG   1680
1681  TTGGCGGGCTGGCCTCATGCGCGCTGCGCACATGCCACTATTCCGTCTCCAGTAGCTGCAATCCATGCCATATCCATAGTCGAATGTGAAGCATTGTTTCTT   1800
1801  GGAGATGGAGGACAGGAGACGCTGACCGGATGTTTTAAGACGTGCAGGATGTGGGAGGCGAGGTAGCTACAACGTGTACGAGAGTGCAGTTGAGGCAGAGACGTGTAAGACACATGCCATGGA   1920
1921  CAAAAAAAAAAAAAAAAAAAA   1942
```

same filters with ^{32}P-labeled probe Pr56 specific to 4-kDa subunit showed that all of these clones hybridized to Pr56. The 11 cDNA clones, therefore, were assumed to encode the large and small subunits. The insert *Eco*RI fragments of positive clones were 1.5–2.0 kb in length. Restriction analysis of these cDNA clones revealed that all of the isolated clones were derived from one kind of mRNA species. Two of the cDNA clones having the longest inserts, designated as gtCA1 and g5CA3, were used for further analysis.

To verify whether these cDNA clones encode CA polypeptides and to determine the complete amino acid sequence of CA, we sequenced internal cCNA fragments on both strands. Analysis of the nucleotide sequence of the cDNA insert in gtCA3 revealed that it consists of a 30-base-pair (bp) 5' untranslated region, a 1131-bp coding sequence encoding 377 amino acid residues, a 790-bp 3' untranslated region, and a 21-bp poly(A) segment (Fig. 14.5). The nucleotide sequence corresponding to the two synthetic probes were located in the open reading frame at positions 439–506 for PR68 and 1054–1109 for Pr56. The nucleotide sequence of the cDNA insert in gtCA1 was identical with that in gtCA3, except for the shorter 5' untranslated region (position from -21 to -1 in Fig. 14.5) and a 11-bp poly(A) tail in gtCA1.

The first ATG codon at the 31th nucleotide, which was numbered as $+1$, was assumed to be the translation initiation codon. This conclusion was based on the following observations: (a) No other in-frame ATG codon was found upstream from the NH$_2$-terminal cysteine residue of the large subunit polypeptide; (b) the nucleotide sequence ACACC upstream from the first triplet ATG codon matched with the consensus sequence CC(A/G)CC for eukaryotic initiation sites predicted by Kozak (13); and (c) the molecular mass of the in vitro transcription–translation product (41 kDa) from the gtCA3 cDNA fragment, which was immunoreactive with anti-deglycosylated CA antibody, was comparable to that of the in vitro translation product from *C. reinhardtii* poly (A)$^+$ RNA (data not shown).

Highly restricted codon usage in the coding sequence for *Chlamydomonas* CA and an extremely long 3' untranslated region of 790 bp were observed in the isolated cDNA fragment, as reported in many other *Chlamydomonas* genes (16). A conserved poly(A) signal sequence, TGTAA, was found 13 bases preceding the poly(A) site.

Amino Acid Sequence Analysis of CA Polypeptides

The amino acid sequence of *Chlamydomonas* CA polypeptides deduced from the nucleotide sequence is also shown in Figure 14.5. Amino acid sequences of peptide

---

Fig. 14.5. Nucleotide Sequence of 1972 bp cDNA fragment in gtCA3 and Deduced Amino Acid Sequence of *Chlamydomonas* Periplasmic CA. The sequences of *Eco*RI linkers (GGAATTCC) at both 5' and 3' ends are not included. The junction between the signal peptide and the large subunit and that between the large and the small subunit are indicated by arrowheads. Amino acid residues that are identical to those determined by peptide sequencing are underlined. A conserved poly(A) signal sequence (TGTAA) and translation termination codon (TAA) are indicated by thick bars. Potentially glycosylated asparagine residues are boxed.

Sequence alignment of *Chlamydomonas* carbonic anhydrase (CA) with human CAI, CAII, and CAIII.

```
Chlamydomonas CA   MARTGALLLVALALAGCAQACIYKFGTSPDSKATVSGDHWDHGLNGENWE-GKDGAGNAWV-----CKTGRKQSPINVPQYQVLDG   80
Human CAI                                            ASPDWGYDDKNGPEQWSKLYPIANGNNQSPVDIKTSETKHD   41
Human CAII                                           SHHWGYGKHNGPEHWHKDFPIAKGERQSPVDIDTHTAKYD   40
Human CAIII                                          AKEWGYASHNGPDHWHELFPNAKGENQSPVELHTKDIRHD   40

Chlamydomonas CA   KGSKIANGLQTQWSYPDLMSNGTSVQVTNNGHTIQVVQWTYNYAGHATIAIPAMHNQTNRIVDVLEMRPNDAADRVTAVPT      160
Human CAI          TSLK-----PISVSY-NP---ATAKEIINGHSFHVNFEDND-----NRIS----VLKGGPFSDSYRLF-----             91
Human CAII         PSLK-----PLSVSY-DQ---ATSLRLNNGHAFNVEFDDSQ-----DKA----VLKGGPLDGTYRLI-----             90
Human CAIII        PSLQ-----PWSVSY-DG---GSAKTILNNGKTCRVVFDDTY-----DRS----MLRGGPLPGPYRLR-----            90

Chlamydomonas CA   QFHFH---ST-----SEHLLAGKIYPLELHIVHQVTEKLEACKGGCFSVTGIL---FQLDNGPDNELLEPIFANMPSREGT     230
Human CAII         QFHFHWGSTNEHGSEHTVDGVKYSAELHVAHWNSAKYSSLAEAASKADGLAVIGVLMKVGEANPKLQKVLDALQAIK-T      169
Human CAII         QFHFHWGSLDGQGSEHTVDGVKKKYAAELHLVHWNT-KYGDFGKAVQQPDGLAVLGIFLKVGSAKPGLQKVVDVLDSIK-T    167
Human CAIII        QFHLHWGSSDDHGSEHTVDGVKYAAELHLVHWNP-KYNTFKEALKQRDGIAVIGIFLKIGHENGEFQIFLDALDKIK-T      167

Chlamydomonas CA   FSNLPAGTIKLGELLPSDRDYVTYEGSLTTPPCSEGLLWHVMTQPQRISFGQWNRYRLAVGLKECNSTETAADAGHHHH       310
Human CAI          KGKRAPFTNFDPSTLLPSSLDFWTYPGSLTHPPLYESVTWIICKESISVSSEQLAQFR-----                      227
Human CAII         KGKSADFTNFDPRGLLPESLDYWTYPGSLTTPPLLECVTWIVLKEPISVSSEQVLKFR-----                      225
Human CAIII        KGKEAPFTKFDPSLLFPACRDYWTYQGSFTTPPCEECIVWLLLKEPMTVSSDQMAKLR-----                      225

Chlamydomonas CA   HRRLLHNHAHLEEVPAATSEPKHYFRRVMLAESANPDAYTCKAVAFGQNFRNPQYANGRTIKLARYH                  377
Human CAI          -----------------------------SLLSNVEGDNAVPMQHNNRPTQPLKGRTVR-ASF                      260
Human CAII         -----------------------------KLNFNGEGEPEELMVDNWRPAQPLKNRQIK-ASFK                     259
Human CAIII        -----------------------------SLLSSAENEPPVPLVSNWRPPQPINNRVVR-ASFK                     259
```

fragments produced by CNBr cleavage (Arg-147 to Glu-169; based on the numbering for *Chlamydomonas* CA precursor polypeptide, see below) or proteinase Lys-C digestion (Ala-33 to Asn-45; Asp-53 to Cys-61; Gln-67 to Gln-76; Leu-242 to Leu 269) were identical with those parts of the predicted amino acid sequence of the open reading frame in the cDNA, as shown by the underlined amino acid residues except for Ser-168. In the peptide sequencing, Ser-168 was judged to be threonine. This discrepancy was caused by misreading weak signals for serine and 5-[(alkylthio)metnyl]-3-phenyl-2-thioxo-4-imidazolidiones on the HPLC chromatogram in the peptide sequencing. Additionally, the amino acid sequences of the NH_2-terminal regions of authentic enzyme subunits (Cys-21 to Asn-45 for the large subunit; Ala-341 to His-377 for the small subunit) were identical to the corresponding parts of the deduced polypeptide encoded by cDNA.

The fact that the NH_2-terminal residue (cysteine) of the large subunit was localized at the 21st residue in the putative CA precursor polypeptide and that 13 hydrophobic amino acid residues such as alanine and leucine were present in the NH_2-terminal leader peptide of 20 amino acids suggest that this peptide is the signal sequence that targets the premature enzyme to the endoplasmic reticulum. In the cDNA encoding *Chlamydomonas* periplasmic arylsulfatase, the hydrophobic signal sequence of 21 amino acid residues has also been reported (3).

Since the NH_2-terminal residue (alanine) of the small subunit is located at the 341st residue of the polypeptide encoded by the cDNA, the small subunit is produced by the proteolytic cleavage of a precursor polypeptide with a calculated mass of 41,626 Da. The cotranslational expression of the large and small subunits coincides with the model of subunit constitution that the holoenzyme is a heterotetramer consisting of two large and two small subunits. The mass of the large subunit (35,603 Da) predicted from the cDNA sequence was larger than that of the chemically deglycosylated enzyme (32,000 Da) (23). This difference may be accounted for by the possibility that some amino acid residues were removed from the COOH-terminal region of the predicted large-subunit polypeptide during maturation of the enzyme, or that the molecular mass reported by Toguri et al. was underestimated. Determination of the COOH-terminal residue of the native large subunit should clarify this discordance.

Three potential asparagine-linked glycosylation sites—Asn-Xaa-Thr/Ser (10)—were localized at Asn-101, Asn-135, and Asn-297 in the deduced amino acid sequence of the large subunit.

Chlamydomonas CA had sequence identity with human CA isozymes CAI, CAII, and CAIII of 20.4%, 21.8%, and 20.7%, respectively (Fig. 14.6). In particular, three zinc-liganded histidine residues (His-94, His-96, and His-119; based on CAI

Fig. 14.6. Amino Acid Sequence Comparison Between *Chlamydomonas* and Human CA Isozymes. Amino acid residues are shown by one-letter codes. Residues that are identical with those of *Chlamydomonas* CA are boxed. Filled triangles indicate possible asparagine-linked glycosylation sites. Putative zinc-liganded histidine residues and those forming the hydrogen-bond network to zinc-bound solvent molecules are depicted by filled and open circles, respectively. Arrows indicate junction sites between the putative signal peptide and the large subunit as well as between the large and the small subunits.

numbering) and those forming the hydrogen-bond network to zinc-bound solvent molecules predicted by x-ray crystallography of human CAI and CAII (Ser-29, His-64, Gln-92, Glu-106, His-107, Glu-117, Tyr-194, Thr-199, Trp-209, Asn-244, and Arg-246; based on CAI numbering) were conserved in *Chlamydomonas* CA, except for three or two amino acid residues found in human isozymes (Tyr-7, His-67, and His-200 in CAI; Tyr-6 and Asn-66 in CAII). Because the 4-kDa small subunit of *Chlamydomonas* CA has two amino acid residues (Asn-359 and Arg-361) essential for zinc folding, the small subunit is probably located near the active site in the holoenzyme. Aside from these amino acid residues, 39 other amino acid residues were also conserved in *Chlamydomonas* and human isozymes. It is possible that these conserved residues are important in catalytic activity of in forming the functional conformation of the enzyme.

These results showed that two subunits of *Chlamydomonas* CA shared significant sequence homology with human CA isozymes and that the two CA subunits are cotranslated from a single transcript, generating a single, large peptide that is processed posttranslationally to generate the two subunits. Thus, it appears that *Chlamydomonas* CA probably shares a common ancestry with the human CA isozymes but that the *Chlamydomonas* CA has acquired a posttranslational processing step to split the protein into two subunits during its evolution.

STRUCTURE AND EXPRESSION OF CA GENES

Identification of CA Genes on the *Chlamydomonas* Gene

Using the same oligonucleotide probes (Pr68 and Pr56) as used in cCNA cloning, two genes for periplasmic CAs have been cloned from *C. reinhardtii* C-9 cells (7, 9).

Genomic clones (F1 and F9) were isolated by screening the EMBL3 genomic DNA library with the ³²P-labeled oligonucleotide probes Pr68 and Pr56. Restriction mapping and partial nucleotide sequence analysis of the insert DNA fragments in the two phages revealed that the inserts in the two phages overlapped each other and that two CA genes were tandemly clustered in the *Chlamydomonas* genome as shown in Figure 14.7. The 5′ upstream gene for CA was designated as *CAH1*, and the 3′ downstream gene as *CAH2*, respectively. Positions that correspond to the two oligonucleotide probes Pr68 and Pr56 were located in the coding regions

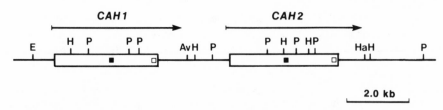

Fig. 14.7. Partial restriction map and the organization of two CA genes. Unfilled rectangles indicate coding regions including introns. The horizontal arrows indicate transcriptional orientations of the CA genes. Filled (■) and open (□) squares in the rectangles indicate the positions to which oligonucleotide probes Pr68 and Pr56, respectively, were hybridized. Av, *Ava*I; E, *Eco*RI; H, *Hinc*II; Ha, *Hae*III; P, *Pst*I.

of the two genes. The length of the *CAH1* coding region including 10 intervening sequences was 3.4 kb. Because the sequences of parts of *CAH1* were identical to those parts of isolated cDNA gtCA3, it was concluded that the cDNA isolated from low-CO_2 cells was derived from the 5' upstream gene *CAH1*. Another gene, *CAH2*, was located approximately 2.0 kb downstream from *CAH1*. The length of the predicted coding region of *CAH2* including 10 introns was 3.3 kb. The predicted amino acid sequence of the second gene *CAH2* showed 92% identity to that of the upstream gene *CAH1*, although both genes were interrupted by 10 introns at the identical positions. Therefore, the second gene *CAH2* was predicted to encode an isoenzyme of periplasmic CA (see below).

Differential Expression of the Two Genes, *CAH1* and *CAH2*, Regulated by Environmental CO_2

To clarify the expression pattern of the two genes, the changes in levels of transcripts in response to changes of CO_2 concentration during the cellular growth were examined by Northern blot hybridization using the gene-specific probes PrCAH1 and PrCAH2 (7). The nucleotide sequences of the two probes were complementary to those of parts of the mRNA transcripts from the two genes corresponding to the positions just after translation stop codons and had no significant homology with each other.

First, mRNA levels of the two genes *CAH1* and *CAH2* were assayed during the period when CA activity and protein accumulation increased (4)—that is, after the bubbling gas was changed from CO_2-enriched air (5%) to ordinary air (0.04%), under continuous illumination (Fig. 14.8A). The PrCAH1 probe specifically hybridized with the same length of mRNA of 2.0 kb, as was found with the ^{32}P-labeled cDNA probe (8). No accumulation of 2.0-kb transcript hybridized with the PrCAH1 was observed under the high-CO_2 condition (lane 1), even when the autoradiography was prolonged. The level of *CAH1* mRNA increased within 1 h and attained the maximum 2 h after the CO_2 concentration was lowered (lanes 1–5). In contrast, accumulation of the *CAH2* mRNA, which was detected as a 2.0-kb RNA at a much lower level than that of the *CAH1* in the fully induced conditions, was repressed within 1 h after lowering the CO_2 concentration (lanes 6–10). However, a small accumulation of the 2.0-kb mRNA was still observed 6 h after the air induction when the autoradiography was prolonged (data not shown).

Second, when low-CO_2 cells were aerated by CO_2-enriched air (5%) in light, the level of *CAH1* mRNA decreased to an undetectable level within 1 h (Fig. 14.8B, lanes 1–5), as was shown with the cDNA probe (8). On the other hand, the level of *CAH2* mRNA increased within 1 h and reached a maximum 2 h after the CO_2 concentration was increased (lanes 6–10). These results indicate that the expression patterns of the *CAH1* are essentially the same as observed when the cDNA probe is used, and that levels of the *CAH1* transcript are regulated in the reverse manner as compared with the case of the *CAH1*.

Addition of 20 μg per ml of actinomycin D, which preferably inhibits nuclear transcription, completely stopped the accumulation of both *CAH1* and *CAH2* mRNA transcripts (Fig. 14.8C, lanes 4 and 9). In contrast, addition of 80 μg/ml of refampicin, which is the transcriptional inhibitor specific for prokaryotic RNA

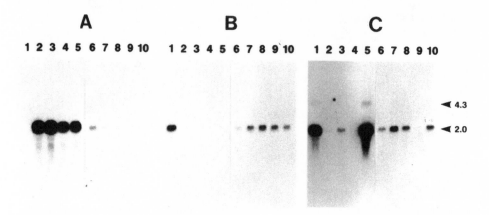

Fig. 14.8. Northern blot analysis of *Chlamydomonas* RNA using gene-specific probes, PrCAH1 (lanes 1–5), and PrCAH2 (lanes 6–10). (A) High-CO$_2$ cells (lanes 1, 6) were transferred to the low-CO$_2$ condition in light for 1 h (lanes 2, 7), 2 h (lanes 3, 8), 4 h (lanes 4, 9), and 6 h (lanes 5, 10). (B) Low-CO$_2$ cells (lanes 1, 6) were kept in high CO$_2$ for 1 h (lanes 2, 7), 2 h (lanes 3, 8), 4 h (lanes 4, 9), and 6 h (lanes 5, 10) in light. (C) Effects of light and inhibitors on CA transcript. Samples of high-CO$_2$ cells were all kept in low CO$_2$ for 4 h in light (lanes 1, 6), in the dark (lanes 2, 7), or in the presence of light with 10 μM DCMU (lanes 3, 8), 20 μg·ml^{-1} of actinomycin D (lanes 4, 9), or 80 μg·ml^{-1} of rifampicin (lanes 5, 10). kb, kilobases.

polymerases, did not have any effect on the accumulation of both CA mRNA species (lanes 5 and 10). These results confirmed that the two CA genes are both encoded by the nuclear gemome. In Figure 14.8C, lanes 1 and 5, additional bands of 4.3 kb were observed, since the contact period of autoradiography was longer than in panels A and B. The 4.3-kb band corresponded to the size of the precursor mRNA including introns predicted from the structure of the 5′ upstream gene *CAH1*.

Photoregulation of CA Expression

It has been shown that both photosynthetic and nonphotosynthetic light are required for the induction of CA activity in *C. reinhardtii* (4, 19). In the nonphotosynthetic light, blue light was effective for the induction (5). By use of the ^{32}P-labeled oligonucleotide probe Pr68, the effect of blue light on steady-state levels of CA mRNA was monitored (6). While illumination with only red light produced only a limited accumulation of the CA transcript, additional illumination with blue light at a low-energy fluence rate resulted in the induction of the 2.0-kb CA transcript to a level similar to that of white light-irradiated cells.

The *CAH1* mRNA did not accumulate when highCO$_2$ cells were transferred to the low-CO$_2$ condition in the dark (Fig. 14.8C, lane 2). Similarly, addition of 10 μM 3-(3,4-dichlorophenyl)-1,1-dimethylurea (DCMU) to air-adapting cells inhibited the accumulation of the *CAH1* mRNA (lane 3). The light requirement and inhibition by DCMU are consistent with the results obtained by Northern hybridization using a cDNA probe (8). Therefore, it was concluded that light is essential for the accumulation of the *CAH1* transcript as well as for lowering the CO_2 concentration, and that photosynthesis is involved in the transcriptional regulation of the *CAH1* expression. On the other hand, the level of the *CAH2* mRNA was higher when the CO_2 concentration was lowered in the dark (lane 7) than when it was lowered in light (lane 6). This result indicates that light has an inhibitory effect on *CAH2* expression. The fact that the level of the *CAH2* mRNA was not affected by the addition of $10 \, \mu M$ DCMU (lane 8) indicates that photosynthetic electron flow is not involved in the regulation of the *CAH2* mRNA accumulation. Although the activity and polypeptide accumulation of CA were low, these could be detected under the high-CO_2 condition (4, 6). In this condition, no *CAH1* mRNA was detected in the cells, and only the *CAH2* mRNA is translated into enzymatically active CA polypeptides under the high-CO_2 condition. Indeed, CA isozyme CA2, which is the *CAH2* gene product, was recently isolated from mutant cells that cannot produce *CAH1* product (15) and also from wild-type cells (20). The CA2 was shown to have heterotetrameric subunit structure, as in the case of *CAH1* product (20). The differential expression of the two genes may be important for photosynthetic adaptation to environmental changes. The *CAH2* product expecially may play a role just after the start of air-adaptation in light when cells have not yet accumulated enough *CAH1* products.

REFERENCES

1. Aizawa, K., and Miyachi, S. 1986. Carbonic anhydrase and CO_2 concentrating mechanisms in microalgae and cyanobacteria. *FEMS Microbiol. Rev.* **39**: 215–23.

2. Coleman, J. R., and Grossman, A. R. 1984. Biosynthesis of carbonic anhydrase in *Chlamydomonas reinhardtii* during adaptation to low CO_2. *Proc. Natl. Acad. Sci. USA* **81**: 6049–53.

3. DeHostos, E., Schilling, J., and Grossman, A. R. 1989. Structure and expression of the gene encoding the periplasmic arylsulfatase of *Chlamydomonas reinhardtii*. *Mol. Gen. Genet.* **218**: 229–39.

4. Dionisio, M. L., Tuszuki, M., and Miyachi, S. 1989a. Light requirement for carbonic anhydrase induction in *Chlamydomonas reinhardtii*. *Plant Cell Physiol.* **30**: 207–13.

5. Dionisio, M. L., Tsuzuki, M., and Miyachi, S. 1989b. Blue light induction of carbonic anhydrase activity in *Chlamydomonas reinhardtii*. *Plant Cell Physiol.* **30**: 215–19.

6. Dionisio-Sese, M. L., Fukuzawa, H., and Miyachi, S. 1990. Light-induced carbonic anhydrase expression in *Chlamydomonas reinhardtii*. *Plant Physiol.* **94**: 1103–10.

7. Fujiwara, S., Fukuzawa, H., Tachiki, A., and Miyachi, S. 1990. Structure and differential expression of two genes encoding carbonic anhydrase in *Chlamydomonas reinhardtii*. *Proc. Natl. Acad. Sci. USA* **87**: 9779–83.

8. Fukuzawa, H., Fujiwara, S., Yamamoto, Y., Dionisio-Sese, M. L., and Miyachi, S. 1990. cDNA cloning sequence, and expression of carbonic anhydrase in *Chlamydomonas reinhardtii*: Regulation by environmental CO_2 concentration. *Proc. Natl. Acad. Sci. USA* **87**: 4383–7.

9. Fukuzawa, H., Fujiwara, S., Tachiki, A., and Miyachi, S. 1990. Nucleotide sequences of two genes *CAH1* and *CAH2* which encode carbonic anhydrase polypeptides in *Chlamydomonas reinhardtii*. *Nucl. Acids Res.* **18**: 6441–2.

10. Hubbard, S. C., and Ivatt, R. J. 1981. Synthesis and processing of asparagine-linked oligosaccharides. *Annu. Rev. Biochem.* **50**: 555–83.

11. Kamo, T., Shimogawara, K., Fukuzawa, H., Muto, S., and Miyachi, S. 1990. Subunit constitution of carbonic anhydrase from *Chlamydomonas reinhardtii. Eur. J. Biochem.* **192**: 557–62.

12. Kimpel, D. L., Togasaki, R. K., and Miyachi, S. 1983. Carbonic anhydrase in *Chlamydomonas reinhardtii.* I. Localization. *Plant Cell Physiol.* **24**: 255–9.

13. Kozak, M. 1984. Possible role of flanking nucleotides in recognition of the ATG initiator codon by eukaryotic ribosomes. *Nuclei Acids Res.* **12**: 857–72.

14. Maezawa, S., and Takagi, T. 1983. Monitoring of the elution from a high-performance gel chromatography column by a spectrophotometer, a low-angle laser light scattering photometer and a precision differential refractometer as a versatile way to determine protein molecular weight. *J. Chromatogr.* **280**, 124–30.

15. Rawat, M., and Moroney, J. V. 1991. Partial characterization of a new isoenzyme of carbonic anhydrase isolated from *Chlamydomonas reinhardtii. J. Biol. Chem.* **256**: 9719–23.

16. Rochiax, J.-D. 1987. Molecular genetics of chloroplast and mitochondria in the unicellular green alge *Chlamydomonas. FEMS Microbiol. Rev.* **46**: 13–34.

17. Sheridan, R. P. and Allen, L. C. 1981. The active site electrostatics potential of human carbonic anhydrase. *J. Am. Chem. Soc.* **103**: 1544–50.

18. Shiraiwa, Y., and Miyachi, S. 1983. Factors controlling induction of carbonic anhydrase and efficiency of photosynthesis in *Chlorella vulgaris* 11h cells. *Plant Cell Physiol.* **24**: 919–23.

19. Spencer, K. G., Kimpel, D. L., Fisher, M. L., Togasaki, R. K., and Miyachi, S. 1983. Carbonic anhydrase in *Chlamydomonas reinhardtii.* II. Requirements for carbonic anhydrase induction. *Plant Cell Physiol.* **24**: 301–4.

20. Tachiki, A., Fukuzawa, H., and Miyachi, S. 1992. Characterization of carbonic anhydrase isozyme CA2, which is the *CAH2* gene product in *Chlamydomonas reinhardtii. Biosci. Biotechnol. Biochem.* **56**: 788–94.

21. Tashian, R. E. 1989. The carbonic anhydrases: Widening perspectives on their evolution, expression and function. *Bioessays* **6**: 186–92.

22. Toguri, T., Yang, S.-Y., Okabe, K., and Miyachi, S. 1984. Synthesis of carbonic anhydrase with messenger RNA isolated from the cells of *Chlamydomonas reinhardtii* Dangeard C-9 grown in high and low CO_2. *FEBS Lett.* **170**: 117–19.

23. Toguri, T., Muto, S., and Miyachi, S. 1986. Biosynthesis and intracellular processing of carbonic anhydrase in *Chlamydomonas reinhardtii. Eur. J. Biochem.* **158**: 443–50.

24. Yagawa, Y., Aizawa, K., Yang, S.-Y., and Miyachi, S. 1986. Release of carbonic anhydrase from the cell surface of *Chlamydomonas reinhardtii* by trypsin. *Plant Cell. Physiol.* **27**: 215–21.

25. Yagawa, Y., Muto, S., and Miyachi, S. 1988. Carbonic anhydrase of *Chlamydomonas reinhardtii*: Cellular distribution and subunit constitution in two strains. *Plant Cell Physiol.* **29**: 185–8.

26. Yang, S.-Y., Tsuzuki, M., and Miyachi, S. 1985. Carbonic anhydrase of *Chlamydomonas*: Purification and studies on its induction using antiserum against *Chlamydomonas* carbonic anhydrase. *Plant Cell. Physiol.* **26**: 25–34.

V
THE ENVIRONMENT

15

Photosynthesis, CaCO₃ Deposition, Coccolithophorids, and the Global Carbon Cycle

C. STEVEN SIKES and VICTORIA J. FABRY

Photosynthetic, CaCO₃-depositing organisms are widespread in nature, particularly in the marine environments. They potentially have significant effects on the global carbon cycle. The coccolithophorids—unicellular marine algae having CaCO₃ cell coverings—are perhaps the most abundant of such organisms in terms of mass and numbers, with *Emiliania huxleyi* the most commonly occuring. In this organism, photosynthesis and calcification are linked through complementary reactions that depend on HCO_3^- utilization from the sea. The overall photosynthetic properties of *E. huxleyi* are intermediate between C_3 and C_4 metabolism, owing probably to the way the cells handle dissolved inorganic carbon. There is no clear evidence as yet for an inorganic carbon concentrating mechanism in *E. huxleyi* such as has been hypothesized to operate in other algae. This chapter examines the potential for inorganic carbon fixation in *E. huxleyi* to act as a sink for atmospheric CO_2.

There are many organisms that are both photosynthetic and CaCO₃ depositing. The great majority of these creatures live in marine environments, although there are some prominent examples among the freshwater algae. In the oceans, the principal groups of photosynthetic, calcifying organisms are (a) reef-building corals that are replete with symbiotic algae; (b) calcareous algae often associated with coral reef communities; (c) planktonic unicellular foraminifera that have symbiotic algae; and (d) coccolithophorids, the group that may produce the greatest amount of CaCO₃ on a global basis (31, 73). Coccolithophorids are unicellular, marine algae that form a sphere of ornate CaCO₃ structures, the coccoliths, as a cell covering. The coccoliths are formed intracellularly within Golgi vesicles; then, through processes that are poorly understood, they are extruded to the cell surface, where they interlock to form the coccosphere. Among the coccolithophorids, *E. huxleyi* (Fig. 15.1) is the most abundant and widely distributed.

The purpose of this chapter is to review the general aspects of the relationship between photosynthesis and calcification, using *E. huxleyi* as a model system. The possibilities for interactions between photosynthetically linked calcification and atmospheric CO_2 are discussed.

GLOBAL OCCURRENCE OF COCCOLITHOPHORIDS

Coccolithophorids are distributed throughout the world's oceans in both coastal and open ocen environments. The species assemblages vary and frequently are closely associated with water masses (44, 49). Coccolithophorids live in the upper 200 m or less of the water column, with typical densities ranging from 10^2 to 10^5 cells $\cdot$ L^{-1} (49–51, 56). Episodic blooms of coccolithophorids (Fig. 15.2) have been observed in surface waters of several oceanic and coastal regions (1, 3, 4, 6, 11, 29, 46). These blooms often consist of a dominant species such as *E. huxleyi* at densities greater than 10^6 cells $\cdot$ L^{-1}.

Although the global rates of photosynthesis on land and in marine (and freshwater) environments are known imprecisely, and the relative contributions of particular types of cells are even more difficult to know, estimates suggest that photosynthesis on land is roughly equivalent to that in the sea (23, 28, 66, 75). The principal primary producers in the oceans traditionally have been thought to be the diatoms, the dinoflagellates, and the coccolithophorids, each with

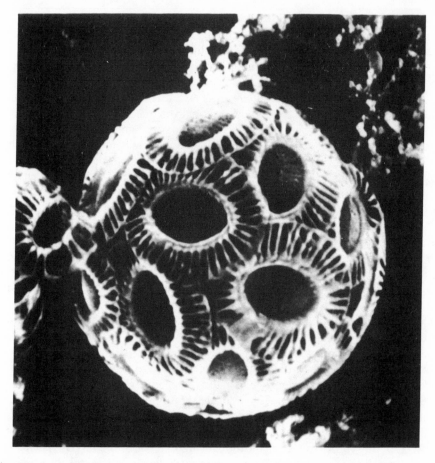

Fig. 15.1. Scanning electron micrograph of *Emiliania huxleyi*. The cell is about 10 µm in diameter. The cell covering (coccosphere) is made of about 15–30 CaCO$_3$ coccoliths.

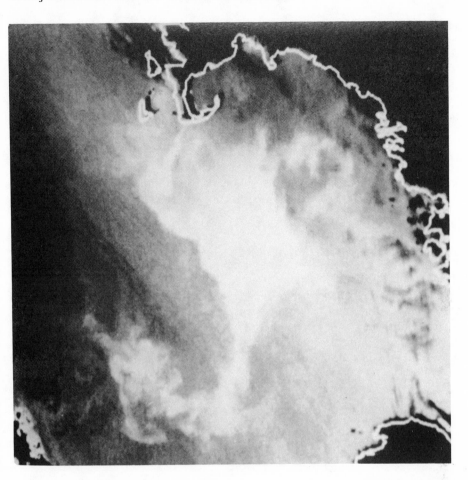

Fig. 15.2. Satellite image of a latter stage of a bloom of *E. huxleyi* in the Gulf of Maine in July 1988 (1). The white areas of the photograph result from the backscattering of light mainly by free-floating CaCO₃ coccoliths, as verified by concurrent ship studies. The coccoliths are shed from the surface of cells, particularly as the cells age.

Table 15.1. CaCO₃ Fluxes Collected in a Sediment Trap at 3200 m in the Sargasso Sea.

	Coccolithophorids	Foraminifera	Pteropods and Other Sources
Flux (mg CaCO₃·m⁻²·d⁻¹)	11	6	5
Percentage of total CaCO₃ flux	50	27	23

Note: Values are averaged over a 14-month period.
Sources: Devser and Ross (20) and Fabry and Deuser (unpublished).

approximately equal contributions (65). More recent research has revealed that picoplankton ($<1.0\,\mu m$) can be an important component of phytoplankton biomass, particularly in the open ocean (25, 26, 33, 40, 71). Although the contribution of coccolithophorids to global photosynthesis has not been quantified, their abundance and wide distribution suggest that it is significant.

Because coccolithophorids also fix carbon into calcareous coccoliths, they are dually important in the carbon cycle. In large areas of the oceans, coccolithophorids are the major producers of $CaCO_3$. For example, sediment trap measurements of sinking particulate matter reveal that coccolithophorids account for half of the total $CaCO_3$ flux to the deep Sargasso Sea (Table 15.1) (20, 21). Coccoliths are a principal component of calcareous sediments, constituting up to 30% of recent sediments (45) and up to 60% of fossil carbonates (10).

RELATIONSHIP BETWEEN PHOTOSYNTHESIS AND CaCO₃ DEPOSITION

Algal calcification and the link between photosynthesis and $CaCO_3$ deposition have been reviewed often (16, 38, 53, 54), and there are also excellent recent reviews (8, 9). The account and reactions given by Tolbert (Chapter 1, this volume) are indeed representative of the majority of the evidence. For the coccolithophorids, as shown in Figure 15.3, the inorganic reactions of $CaCO_3$ deposition and photosynthesis include both a direct uptake of external CO_2 as a substrate for

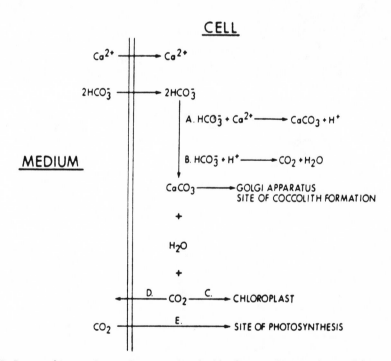

Fig. 15.3. Proposed inorganic reactions associated with photosynthesis and coccolith formation in cocclithophorids (53).

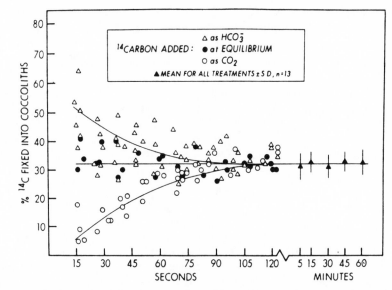

Fig. 15.4. Results from use of the "isotopic disequilibrium" approach to measuring carbon fixation by *E. huxleyi* (53). The values were corrected for results from experiments with a noncalcifying strain of the organism in which there was no coccolith fraction. As explained in the text, HCO_3^- appears to be the form of carbon supplied for coccolith formation, whereas CO_2 is the substrate of photosynthesis.

photosynthesis and an influx of HCO_3^- from the sea that supplies both carbonate deposition and CO_2 for photosynthesis through complementary reactions. In general, the amount of photosynthesis is greater than the amount of coccolith formation, with a ratio often between 1.0 and 2.0, with values as high as 9 reported (9).

The carbon sources for photosynthesis and coccolith formation and the net inorganic reaction of deposition are important in determining the extent to which carbon fixation by coccolithophorids might be a sink for atmospheric CO_2. Unambiguous identification of CO_2, HCO_3^-, or CO_3^{2-} as the carbon sources for photosynthesis and calcification is difficult because of the inherent complexity of the dissolved inorganic carbon (DIC) system. The DIC equilibrium essentially consists of five simultaneously interacting variables: H^+, CO_2, HCO_3^-, CO_3^{2-}, and OH^-. Although complex, the DIC system is well defined, and it has been possible to elucidate the carbon sources through use of the properties of the system.

For example, the relative contributions of CO_2 and HCO_3^- (and CO_3^{2-}) were measured via the isotopic disequilibrium approach (15). This method takes advantage of the relatively slow reversible hydration and dehydration reactions:

$$CO_2 + H_2O \leftrightarrow H_2CO_3 \leftrightarrow HCO_3^- + H^+$$

which have half-lives of 15–20 s at 20°C. As shown in Figure 15.4, when $[^{14}C]DIC$ was supplied as $H^{14}CO_3^-$, the initial relative fixation into coccoliths by *E. huxleyi* was enhanced. When supplied as $^{14}CO_2$, the initial relative fixation as photosynthate was enhanced (57). This type of experiment was repeated using both calcifying and noncalcifying strains of *E. huxleyi*, and included control measurements in which [^{14}C]DIC was supplied at equilibrium levels of CO_2 and HCO_3^-. The results clearly suggested that CO_2 is the substrate of photosynthesis in these cells, but that HCO_3^- is the form of DIC that supplies coccolith formation (61).

This approach was supplemented with a set of experiments in which pH, alkalinity, and ^{14}C incorporation by *E. huxleyi* into both coccoliths and photosynthetic products were monitored in closed vessels (61). In this method, CO_2 fluxes were linked to pH changes and HCO_3^- fluxes to alkalinity changes because, under the chosen conditions, CO_2 fluxes affect pH but not alkalinity whereas HCO_3^- fluxes affected alkalinity but not pH. *Alkalinity* is defined as the acid-consuming capacity of a liquid, which, with respect to DIC, is equal to:

$$[HCO_3^-] + 2[CO_3^{2-}] + [OH^-] - [H^+]$$

The calculated amounts of photosynthesis directly attributable to CO_2 influx and coccolith formation attributable to HCO_3^- influx were compared to direct measurements by [^{14}C]DIC incorporation. The difference between the [^{14}C]DIC measurements of photosynthesis and the calculated photosynthesis attributable to CO_2 influx was assigned as photosynthesis that was supplied by HCO_3^- influx. Overall, the results from this method were consistent with the previously described approach utilizing isotopic disequilibria.

The assignment of CO_2 as the source for photosynthesis and HCO_3^- for coccolith formation, which in turn generated additional CO_2 for photosynthesis, was also supported by $\delta^{13}C$ measurements of coccoliths and photosynthate of *E. huxleyi*. The photosynthate was enriched in ^{12}C versus ^{13}C (low $\delta^{13}C$ values), and the carbon in coccoliths approached an equilibrium distribution (60). The photosynthetic discrimination in favor of ^{12}C was attributed to the hydration of CO_2, which itself favors ^{12}C (19, 72), and also a discrimination by Rubisco for $^{12}CO_2$ (7).

The key to the action of the mechanism of Figure 15.3 is the HCO_3^- influx. Given the flow of HCO_3^- into the cells, the rest of the inorganic reaction sequence is essentially predetermined by the nature of DIC equilibrium, particularly in a cell that is both photosynthesizing and forming coccoliths.

It is important to note that mechanisms of calcification by algae and plants are highly variable, ranging from relatively amorphous extracellular deposits to the highly sculpted and rigidly controlled intracellular formation of coccoliths. The net reaction of Figure 15.3 may not apply exactly in other systems such as Characean (freshwater, filamentous algae) calcification, recent studies of which suggested that bicarbonate, calcium, and proton fluxes produce internal CO_2 that may be the source for external $CaCO_3$ deposition (42, 43). However, the reactions of Figure 15.3 are supported by much evidence, in addition to that discussed herein, and they represent a general case.

PROPERTIES OF PHOTOSYNTHESIS IN *E. HUXLEYI*

In spite of their abundance and global occurrence, the coccolithophorids have not received much attention by students of photosynthetic mechanisms. The properties

Table 15.2. Some Properties of Photosynthesis in Coccolithophorids and C_3 and C_4 Plants.

	C_3 Plants	C_4 Plants	Coccolithophorids
Primary enzyme of carbon fixation	Rubisco	Phosphoenolpyruvate carboxylase	?(Rubisco)
First fixation products	3-PGA	Oxaloacetate, malate, aspartate	?(3-PGA)
CO_2 compensation point (ppm)	30 to 70	0–10	?($\sim$0)
O_2 sensitivity (0–21%)	Yes	No	?
Photorespiration	Yes	No	?
Optimum temperature	15–25°C	30–47°	18°C
Isotopic rate $\delta^{13}C$	−22 to −34	−11 to −19	−18
HCO_3^- active	No	Yes	Yes

of photosynthesis in *E. huxleyi* are summarized, and comparisons to C_3 and C_4 metabolism are given in Table 15.2.

The activities of the carbon fixing enzymes have been measured for *Coccolithus pelagicus* (24), but not for *E. huxleyi*. Both Rubisco and PEP carboxylase were active in *C. pelagicus*, with Rubisco about eight-fold higher for log phase cells. Presumably, Rubisco is the primary enzyme of carbon fixation in *E. huxleyi* and other coccolithophorids as well, but this needs to be addressed. The fact that CO_2 is the substrate of photosynthesis in *E. huxleyi* suggests that Rubisco almost certainly is predominent in this organism, as CO_2 is the substrate for Rubisco (15) and HCO_3^- for PEP carboxylase (17). It follows that 3-phosphoglycerate (3-PGA) is the first fixation product, but again this should be confirmed. Carbon fixation in *E. huxleyi* presumably exhibits O_2 sensitivity and photorespiration, both of which are inherent in Rubisco activity.

The CO_2 compensation point has not been established in *E. huxleyi* or other coccolithophorids. However, carbon fixation as uptake of [^{14}C]DIC has been measured at a level of external DIC (0.5 mM, pH 8.0), which is equivalent to 0.005 mM CO_2, or less than 1 ppm (52, 64). If this turns out to be close to the compensation point, then *E, huxleyi* is capable of net photosynthesis at very low external CO_2, as are C_4 plants.

Photosynthesis in *E. huxleyi* also is similar to that in C_4 plants, in that the cells can assimilate HCO_3^- directly from the medium, but this similarity is largely a reflection of the calcification mechanism rather than the action of PEP carboxylase. The $\delta^{13}C$ values of whole cells of *E. huxleyi* also fall into the C_4 range, but again this convergence is a result of the presence of the coccoliths and not of isotopic discrimination attributable to HCO_3^- utilization via PEP carboxylase.

In summary, although some outcomes still await direct measurement, the overall tabulation of photosynthetic properties of *E. huxleyi*, is intermediate between established C_3 and C_4 parameters. Such a picture, while seen in other algae as well as in coccolithophorids, is generally a reflection of the way the cells handle DIC rather than the operation of a pathway other than the C_3 photosynthetic carbon cycle (2, 41).

IS THERE A DIC PUMP IN COCCOLITHOPHORIDS?

Much evidence has accumulated to suggest the presence of CO_2- and HCO_3^--concentrating mechanisms in algae. A cornerstone of this concept is the observation that some algae contain much higher concentrations of DIC within their cells than are found in the ambient medium (2, 70), as measured following a rapid filtration of the cells through silicone oil. The exact mechanism of accumulation of DIC is unknown; it may involve action of a CO_2 pump, although such a pump would be difficult to envision because of the high mobility of CO_2 across membranes.

The existence of a HCO_3^- pump also is often invoked, sometimes as a HCO_3^-/ATPase that may involve simultaneous balancing fluxes of H^+ or OH^- (9, 34, 47, 55, 63, 70, 77). A central feature of the DIC concentrating process is the action of carbonic anhydrase (CA). This enzyme has been observed to increase in cells at low levels of CO_2 (ambient atmosphere), but is repressed at higher (1–5% CO_2 in air) concentrations (70). It functions to maintain, instantaneously, CO_2 within cells and at their surfaces at equilibrium levels, thus minimizing diffusion limitations of carbon fixation (27). Because it affects only the rate of hydration and dehydration of CO_2, CA does not influence the DIC equilibrium itself, and therefore, "pumps" of some types are required to maintain the elevated internal DIC levels. Calcification studies have shown that CA acts to maintain DIC levels at equilibrium values in the region of $CaCO_3$ deposition, which may facilitate removal of the CO_2 that accompanies the deposition of CO_3^{2-} (58, 76).

The odd thing is that *E. huxleyi*, like certain other marine microalgae (48), has essentially no measureable CA activity (57). These results are consistent with the results of the "isotopic disequilibrium" experiments that were used to establish the carbon sources for photosynthesis and calcification in the organism. That is, if there had been significant CA activity in or around the cells, the slow hydration and dehydration of CO_2 on which the experiments were based would have become nearly instantaneous, yielding equilibrium results in all cases. The other principal feature of a DIC concentrating system in algae—a HCO_3^- pump—has not been studied directly in coccolithophorids. There is, however, some evidence of ATPase activity associated with the membranes of coccolith vesicles (37).

Is there a DIC pump in coccolithophorids? As we have seen, there are some interesting puzzles associated with this question. A final problem is that, on the one hand, *E. huxleyi* is capable of photosynthesizing at very low levels of external CO_2 and is one of the most successful photosynthetic cells. On the other hand, as shown in Figure 15.5, carbon fixation in the organism is not saturated until supplied with very high levels of external DIC (>10 mM)—much higher than occur in the sea, and much higher than are saturating for other algae. This raises the possibility that growth of *E. huxleyi* may be carbon rate-limited in some situatons. Contemplating this situation, Steemann Nielsen (65) wrote,

---

→

Fig. 15.5. Dissolved inorganic carbon saturation of photosynthesis in *E. huxleyi* and *C. reinhardtii*. In both sets of experiments, photosynthesis was measured at light saturation (around 200 $\mu E \cdot m^{-2} \cdot s^{-2}$) using log-phase cells. The experiments with *E. huxleyi* were originally done by Paasche (52), at different pH values as another approach to investigating the inorganic carbon sources. (Redrawn from Refs. 60 and 4.)

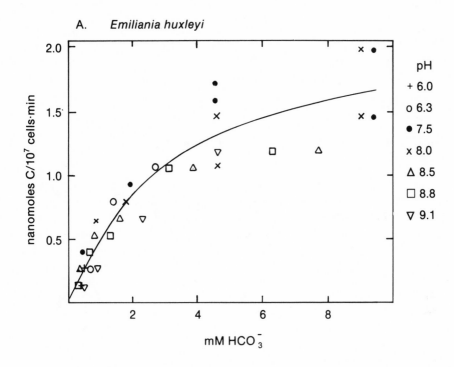

A.   *Emiliania huxleyi*

pH
+ 6.0
o 6.3
● 7.5
x 8.0
△ 8.5
□ 8.8
▽ 9.1

nanomoles C/10⁷ cells·min

mM HCO₃⁻

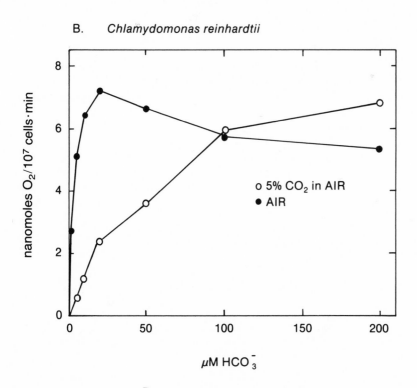

B.   *Chlamydomonas reinhardtii*

nanomoles O₂/10⁷ cells·min

o 5% CO₂ in AIR
● AIR

μM HCO₃⁻

"Coccolithophorids may be an old and very special physiological group of algae lacking a mechanism, which in other taxonomic groups of algae provides for the concentration of CO_2 before the proper carboxylation process in the chloroplasts." This is an intriguing statement that merits further investigation.

CAN CARBON FIXATION IN THE COCCOLITHOPHORIDS AFFECT ATMOSPHERIC CO_2?

As major primary produces in the oceans, like all photosynthetic cells whether on land or in the sea, coccolithophorids remove CO_2 from the environment. The result is that the concentration of CO_2 in surface waters is lowered and influx of atmospheric CO_2 into the oceans is promoted. Of course, much of the photosynthetic CO_2 may be returned to the sea via respiration, dampening any net influx that may occur. What is more interesting is consideration of the effect that $CaCO_3$ formation as coccoliths may have an DIC dynamics in the sea, the subsequent effects on atmospheric CO_2 influx, and the removal of both photosynthetic carbon and $CaCO_3$ deposits from the surface waters to the depths and the sediments.

As shown in Figure 15.6, biogenic $CaCO_3$ produced in the euphotic zone has several possible fates. For example, it may sink to the sea floor and accumulate in sediments. If these sediments are at depths where waters are saturated with respect to $CaCO_3$, the $CaCO_3$ is essentially removed from the system for long periods of time. Alternatively, if the $CaCO_3$ sinks below the depth of the $CaCO_3$ saturation horizon, it may dissolve, thereby transporting DIC from surface layers to deeper waters. The depth of the calcite saturation horizon varies with location, but waters typically become undersaturated with respect to calcite at 4000–5000 m in the Atlantic Ocean and 500–3000 m in the Pacific Ocean (14, 39).

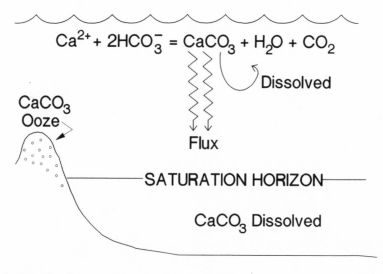

Fig. 15.6. Generalized inorganic $CaCO_3$ cycle in the sea.

Some types of biologically produced $CaCO_3$ are more resistant to dissolution than are nonbiogenic carbonates (12, 32, 35), owing to the protective action of organic matrix molecules that occur within and around biominerals (74). Presumably, matrix molecules bind to sites of crystal growth or dissolution, and can retard either process (13, 66, 67). This binding thus affects the depth at which some biological carbonates dissolve. The phenomenon is well known and has been considered in estimates of the saturation horizon.

In the case of the coccolithophorids, the $CaCO_3$ is in the form of coccoliths that envelop a photosynthetic cell. Therefore, when coccolithophorids sink, both $CaCO_3$ and photosynthetic carbon are exported out of surface waters. E. huxleyi routinely sheds coccoliths, and these sink as well (29, 59). The photosynthetic carbon of coccolithophorids may be recycled through respiration in the upper ocean by the cells themselves, or cells may be consumed by zooplankton or other predators. In passage through the guts of some zooplankton, coccolithophorids (with the coccoliths largely intact) become packaged into fecal pellets, which sink rapidly and accelerate the downward transport of coccolithophorid carbonate (11, 30, 60, 62). The extent to which coccoliths may be dissolved through predation, and thereby recycled in the upper ocean, is unknown.

The equation of Figure 15.6 shows that simple inorganic $CaCO_3$ deposition may actually increase the level of free CO_2 in the water and therefore would not be a sink for atmospheric CO_2. This reasoning is also applied sometimes in biological contexts (36, 68). However, there are a number of reasons why this equation is an oversimplification of photosynthetically enhanced calcification and therefore should be applied with caution.

First, CO_2 generated intracellularly as a by-product of coccolith formation is fixed photosynthetically. The return of photosynthate to the sea as respiratory CO_2 is more likely to occur at depth rather than in surface waters. Furthermore, much of this photosynthate may become fossilized—for example, the North Sea oil, which derives from deposits of coccolithophorids (22).

Second, even in the case of simple inorganic $CaCO_3$ precipitation, which more precisely follows the following equation:

$$CO_2 + H_2O \leftrightarrow H_2CO_3 \leftrightarrow H^+ + HCO_3^- \leftrightarrow H^+ + CO_3^{2-} + Ca^{2+} \leftrightarrow CaCO_3$$

The removal of carbonate does not follow a 1:1 correspondence with the production of CO_2. As seen in this equation, removal of CO_3^{2-} sets up a contradictory situation of pulling the equilibrium to the right by mass action which produces protons, lowers pH, and in turn acts to drive the reaction in the reverse direction to the left, again by mass action. This is one way to envision how buffers such as DIC act: The opposing reactions act to moderate each other. Because the exact outcome of a reaction like this is often not intuitive, we rely on the equilibrium equations for calculation of the result. For example, suppose seawater at pH 8.20, 2.20 mM DIC is in a closed vessel and is undergoing an inorganic precipitatioon of $CaCO_3$ that results in a decrease in pH from 8.2 to 8.0. The effects on the DIC situation during the course of this reaction are shown in Table 15.3.

Several interesting features are observed in this reaction. Although the amount of CO_3^{2-} in solution decreases substantially from 0.235 mM to 0.148 mM, the

Table 15.3. The Effect of Precipitation in a Close Seawater System on Aspects of the Dissolved Inorganic Carbon (DIC) Equilibrium.

pH	C_T	a_2	CO_3^{2-}	a_1	HCO_3^-	a_0	CO_2	CO_3^{2-} ppt.	Increase in CO_2	CO_3^{2-} ppt./CO_2 Increase
8.20	2.200	0.1070	0.235	0.889	1.956	0.0040	0.0088	—	—	—
8.15	2.169	0.0960	0.208	0.899	1.950	0.0050	0.0108	0.031	0.0020	15.5
8.10	2.146	0.0865	0.185	0.908	1.949	0.0055	0.0118	0.054	0.0030	18.0
8.05	2.125	0.0780	0.165	0.916	1.947	0.0060	0.0127	0.075	0.0039	19.2
8.00	2.107	0.0705	0.148	0.923	1.945	0.0065	0.0137	0.093	0.0049	19.0

Note: a_0, a_1, and a_2 = the ionization fractions of CO_2, HCO_3^-, and CO_3^{2-} respectively. C_T = total DIC. $C_T a_0$, $C_T a_1$, and $C_T a_2$ = the concentrations of CO_2, HCO_3^-, and CO_3^{2-}, respectively. All DIC values are listed as millimolar concentrations. For convenience, CO_2 herein refers to dissolved CO_2 plus H_2CO_3, although H_2CO_3 is a very minor component. The amount of CO_3^{2-} precipitated = C_T at pH 8.2 minus C_T at the lower pH values. The increase in CO_2 = the difference between CO_2 at pH 8.2, and at the lower pH values.

$$a_0 = (1 + K_1/[H^+] + K_1 K_2/[H^+]^2)^{-1}$$

$$a_1 = ([H^+]/K_1 + 1 + K_2/[H^+])^{-1}$$

$$a_2 = ([H^+]^2/K_1 K_2 + [H^+]/K_2 + 1)^{-1}$$

$$K_1 = 1.288 \times 10^{-6} = [H^+][CO_3^-]/[CO_2]$$

$$K_2 = 7.586 \times 10^{-10} = [H^+][CO_3^{2-}]/[HCO_3^-]$$

$$C_T = ([acyl] - [H^+] + [OH^-])/(a_1 + 2a_0)$$

$$Acy = acidity = 2[CO_2] + [HCO_3^-] + [H^+] - [OH^-]$$

To determine the amount of CO_3^{2-} added to or removed from the solution to produce a specific change in pH, calculate C_T for the new pH value.

Source: Equations are from Stumm and Morgan (66), adapted from Deffeyes (18).

amount of HCO_3^- is relatively constant from 1.956 mM to 1.945 mM, with CO_2 increasing only from 0.0088 to 0.0137 mM.

If this reaction occurred near the surface of the sea, we could assume that the CO_2 concentration would be at equilibrium with the atmosphere. Given an atmospheric pCO_2 of $10^{-3.45}$ atm, the equilibrium concentration of dissolved CO_2 can be calculated from:

$$CO_2 \text{ dissolved} = K_H p(CO_2) \quad \text{where } K_H = \text{Henry's law constant}$$
$$= 4.8 \times 10^{-2} \text{ M atm}^{-1} (3.5 \times 10^{-4} \text{ atm})$$
$$= 1.68 \times 10^{-5} \text{ M or } 0.0168 \text{ mM}$$

Comparisons of this value with the values of Table 15.3 for increases in CO_2 during precipitation (0.002–0.0049 mM) shows that some CO_2 outgassing could occur in such an experiment, but the amount would be low compared to the amount of CO_3^{2-} deposited.

Actually, the values chosen for demonstration here are quite high as compared to events that are likely to occur in nature. For example, the bloom of *E. huxleyi* that occurred along the northwest European continental shelf in 1982 resulted in cell densities as high as $8.5 \times 10^6 \cdot L^{-1}$ with an integrated production of $CaCO_3$ of

40 g·m^{-2} for the upper 60 m (29). This corresponds to 0.4 mole CaCO$_3$·6 × 10^4 L, or 6.67 μM CaCO$_3$ precipitation. Even as an inorganic precipitation, this would generate very little dissolved CO$_2$ in the external medium. However, the main point is that any CO$_2$ produced during the process of calcification by *E. huxleyi* is immediately fixed photosynthetically. Moreover, during carbon fixation by photosynthetic, calcifying organisms, the pH of the medium goes up, not down, due to net removal of CO$_2$ (57), which would tend to draw CO$_2$ into the medium from the atmosphere.

The available evidence suggests that much of the carbon fixed by coccolithophorids is transported out of the upper ocean. Therefore, photosynthetically enhanced cocclith formation is a sink for oceanic DIC that may draw down atmospheric CO$_2$ as well, owing to the way in which the cells handle DIC and its fate after fixation. The size of this sink is unknown, however. Many questions remain concerning the mechanisms of photosynthesis and calcification in coccolithophorids and the quantitative significance of these cells in the global carbon cycle.

ACKNOWLEDGMENTS

This work was supported by National Science Foundation Grants DMB-8916407 to S. Sikes and OCE-8716589 to W. Deuser in support of V. J. Fabry. This is contribution number 7488 of the Woods Hole Oceanographic Institution.

REFERENCES

1. Ackleson, S., Balch, W. M., and Holligan, P. M. 1988. White waters of the Gulf of Maine. *Oceanog. Mag.* **1**(1): 18–22.

2. Badger, M. A., Kaplan, A., and Beery, J. 1978. A mechanism for concentrating CO$_2$ in *Chalmydomonas reinhardtii* and *Anabaena variabilis* and its role in photosynthetic CO$_2$ fixation. *Carnegie Inst. Washington Year Book* **77**: 261–61.

3. Balch, W. M., Holligan, P. M., Ackleson, S. G., and Voss, K. J. 1991. Biological and optical properties of mesoscale coccolithophore blooms in the Gulf of Maine. *Limnol. Oceanogr.* **36**: 629–43.

4. Berge, G. 1962. Discoloration of the sea due to the *Coccolithus huxleyi* "bloom." *Sarsla* **6**: 27–40.

5. Berry, J., Boynton, J., Kaplan, A., and Badger, M. 1976. Growth and photosynthesis of *Chalmydomonas reinhardtii* as a function of CO$_2$ concentration. *Carnegie Inst. Washington Year Book* **75**: 423–32.

6. Birkenes, E. and Braarud, T. 1952. Phytoplankton in the Oslo Fjord during a "*Coccolithus huxleyi*-summer." *Avhandl Norske Vidanskaps Akad Oslo.* **27**: 1–23.

7. Black, C. C. 1973. Photosynthetic carbon fixation in relation to net CO$_2$ uptake. *Annu. Rev. Plant Physiol.* **24**: 253–86.

8. Borowitzka, M. A. 1987. Calcification in algae: Mechanisms and the role of metabolism. *CRC Crit. Rev. Plant Sci.* **6**: 1–45.

9. Borowitzka, M. A. 1989. Carbonate calcification in algae initiation and control. In *Biomineralization: Chemical and Biochemical Perspectives*, ed. S. Mann, J. Webb, and R. J. P. Williams, pp. 63–94. VCH Publishers, Weinheim, FRG.

10. Bramlette, M. N. 1958. Significance of coccolithophorids in calcium-carbonate deposition. *Bull. Geol. Soc. Am.* **69**: 121–6.

11. Cadee, G. C. 1985. Macroaggregates of *Emiliania huxleyi* in sediment traps. *Mar. Ecol. Prog. Ser.* **24**: 193–6.

12. Chave, K. E. 1965. Carbonates: Association with organic matter in surface seawater. *Science* **148**: 1723–4.

13. Chave, K. E., and Suess, E. 1970. Calcium carbonate saturation in seawater: Effects of dissolved organic matter. *Limnol. Oceanogr.* **15**: 633–7.

14. Chen, C. T. A., Feely, R. A., and Gendron, J. F. 1988. Lysocline, calcium carbonate compensation depth, and calcareous sediments in the North Pacific Ocean. *Pacific Sci.* **42**(3–4): 237–52.

15. Cooper, T. G., Filmer, D., Wishnick, M., and Lane, M. D. 1969. The active species of CO_2 utilized by ribulose diphosphate carboxylase. *J. Biol. Chem.* **244**: 1081–3.

16. Darley, W. M. 1974. Silicification and calcification. In *Algal Physiology and Biochemistry*, ed. W. D. Stewart, pp. 655–75. Blackwell, Boston.

17. Davies, D. D. 1979. The central role of phosphoenolpyruvate in plant metabolism. *Annu. Rev. Plant Physiol.* **30**: 131–58.

18. Deffeyes, K. S. 1965. Carbonate equilibria: A graphic and algebraic approach. *Limnol. Oceanogr.* **10**: 412–26.

19. Deuser, W. G., and Degens, E. T. 1967. Carbon isotope fractionation in the system CO_2 (gas)–CO_2 (aqueous)–HCO_3 (aqueous). *Nature* **215**: 1033–5.

20. Deuser, W. G., and Ross, E. H. 1989. Seasonally abundant planktonic foraminifera of the Sargasso Sea: Succession, deepwater fluxed, isotopic compositions and paleoceanographic implications. *J. Foraminiferal Res.* **19**(4): 268–93.

21. Fabry, V. J., and Deuser, W. G. 1991. Aragonite and magnesian calcite fluxes to the deep Sargasso Sea. *Deep-Sea Res.* **33**: 713–28.

22. Gallois, R. W. 1976. Coccolith blooms in the Kimmeridge clay and origin of North Sea oil. *Nature* **259**: 473–5.

23. Garrels, R. M., and Perry, E. A., Jr. 1974. Cycling of carbon, sulfur, and oxygen through geologic time. In *The Sea*, ed. E. D. Goldberg, Vol. 5, pp. 303–36. Wiley-Interscience, New York.

24. Glover, H. E., and Morris, I. 1970. Photosynthetic carboxylating enzymes in marine phytoplankton. *Limnol. Oceanogr.* **24**: 510–19.

25. Glover, H. E., Campbell, L., and Prezelin, B. B. 1986. Contribution of *Synechococcus* spp. to size-fractioned primary productivity in three water masses in the Northwest Atlantic Ocean. *Mar. Biol.* **91**: 193–203.

26. Glover, H. E., Prezelin, B. B., Campbell, L., and Wyman, M. 1988. Pico- and ultraplankton Sargasso Sea communities: Variability and comparative distributions of *Synechoccus* spp. and algae. *Mar. Ecol. Prog. Ser.* **49**: 127–39.

27. Gutknecht, J., Bisson, M. A., and Tosteson, D. C. 1977. Diffusion of carbon dioxide through lipid bilayer membranes. *J. Gen. Physiol.* **69**: 779–94.

28. Holland, H. D. 1978. *The Chemistry of the Atmosphere and Oceans*, pp. 351. Wiley-Interscience, New York.

29. Holligan, P. M., Viollier, M., Harbour, D. S., Camus, P., and Champagne-Philippe, M. 1983. Satellite and ship studies of coccolithophore production along a continental shelf edge. *Nature* **304**: 339–42.

30. Honjo, S. 1976. Coccoliths: Production, transportation and sedimentation. *Marine Micropaleontol.* **1**: 65–79.

31. Honjo, S. 1977. Biogenic carbonate particles in the ocean; do they dissolve in the water column? In *The Fate of Fossil Fuel CO_2 in the Oceans*, ed. N. R. Andersen and A. Malahoff, pp. 269–94. Plenum Press, New York.

32. Honjo, S., and Erez, J. 1978. Dissolution rates of calcium carbonate in the deep ocean: An in-situ experiment in the North Atlantic Ocean. *Earth Planet. Sci. Lett.* **40**: 287–300.

33. Johnson, P. W. J., McN. Sieburth, J. 1979. Chrococcoid cyanobacteria in the sea: A ubiquitous and diverse phototrophic biomass. *Limnol. Oceanogr.* **24**: 928–35.

34. Kaplan, A., Zenvirth, D., Reinhold, L., and Berry, J. A. 1982. Involvement of a primary electrogenic pump in the mechanism for HCO_3 uptake by the cyanobacterium *Anabaena* variabilis. *Plant Physiol.* **69**: 978–82.

35. Keir, R. S. 1980. The dissolution kinetics of biogenic calcium carbonates in seawater. *Geochim. Cosmochim. Acta* **44**: 241–52.

36. Kinsey, D. W., and Hopley, D. 1991. The significance of coral reefs as global carbon sinks—response to Greenhouse. *Paleogeogr. Palaeoclimatol. Palaeoecol.* **89**: 363–77.

37. Klaveness, D. 1976. *Emiliania huxleyi* (Lohmann) Hay and Mohler. 3. Mineral deposition and the origin of the matrix during coccolith formation. *Protistologica* **12**: 217–24.

38. Klaveness, D., and Paasche, E. 1979. Physiology of coccolithopnorids. *Biochem. Physiol. Protozoa* **1**: 191–213.

39. Li, W. K. W., Subba Rao, D. V., Harrison, W. G., Smith, J. C., Cullen, J. J., Irwin, B , and Platt, T. 1983. Autotrophic picoplankton in the tropical ocean. *Science* **219**: 292–5.

40. Li, Y. H., Takahashi, T., and Broecker, W. S. 1969. Degree of saturation of $CaCO_3$ in the oceans. *J. Geophys. Res.* **74**(23): 5507–25.

41. Lucas, W. J., and Berry, J. A. 1985. Inorganic carbon transport in aquatic photosynthetic organisms. *Physiol. Plant* **65**: 539–43.

42. McConnaughey, T. 1991. Calcification in *Chara corallina*: CO_2 hydroylation generates protons for bicarbonate assimilation. *Limnol. Oceanogr.* **36**: 619–28.

43. McConnaughey, T. A., and Falk, R. H. 1991. Calcium–proton exchange during algal calcification. *Biol. Bull.* **180**: 185–95.

44. McIntyre, A., and Be, A. W. H. 1967. Modern coccolithophoridae of the Atlantic Ocean I. Placoliths and crytoliths. *Deep-Sea Res.* **14**: 561–97.

45. McIntyre, A., and McIntyre, R. 1971. Coccolith concentrations and differential solution in oceanic sediments. In *The Micropaleontology of Oceans*, ed. B. M. Funnell and W. R. Riedel, pp. 253–61. Cambridge University Press, Cambridge.

46. Milliman, J. D. 1980. Coccolithophorid production and sedimentation, Rockall Bank. *Deep-Sea Res.* **27**: 959–63.

47. Moroney, J. V., Husic, H. D., and Tolbert, N. E. 1985. Effect of carbonic anhydrase inhibitors on inorganic carbon accumulation by *Chlamydomonas reinhardtii*. *Plant Physiol.* **79**: 177–83.

48. Munoz, J., and Merrett, M. J. 1989. Inorganic-carbon transport in some marine eukaryotic microalgae. *Planta* **178**: 450–5.

49. Okada, H., and Honjo, S. 1973. The distribution of oceanic coccolithophorids in the Pacific. *Deep-Sea Res.* **20**: 355–74.

50. Okada, H., and McIntyre, A. 1977. Modern coccolithophores of the Pacific and North Atlantic Oceans. *Micropaleontology* **23**: 1–55.

51. Okada, H., and McIntyre, A. 1979. Seasonal distribution of modern coccolithophores in the western N. Atlantic Ocean. *Mar. Biol.* **54**: 319–28.

52. Paasche, E. 1964. A tracer study of the inorganic carbon uptake during coccolith formation and photosynthesis in the coccolithophorid *Coccolithus huxleyi*. *Physiol. Plant* **3** (Suppl.): 1–82.

53. Pautard, F. G. 1970. Calcification in unicellular organisms. In *Calcification: Ceulluar and Molecular Aspects*, ed. H. Shraer, pp. 105–201. Appleton-Century-Crofts, New York.

54. Raven, J. A. 1970. Exogenous inorganic carbon sources in plant photosynthesis. *Biol. Rev.* **45**: 167–221.

55. Raven, J. A., and Osborne, B. A. 1985. Uptake of CO_2 by aquatic vegetation. *Plant Cell Environ.* **8**: 417–25.

56. Reid, F. M. H. 1980. Coccolithophorids of the North Pacific Central Gyre with notes on their vertical and seasonal distribution. *Micropaleontology* **26**: 151–76.

57. Sikes, C. S., and Wheeler, A. P. 1982. Carbonic anhydrase and carbon fixation in coccolithophorids. *J. Phycol.* **18**: 423–6.

58. Sikes, C. S., and Wheeler, A. P. 1983. A systematic approach to some fundamental questions of carbonate calcification. In *Biomineralization and Biological Metal Accumulation*, ed. P. Westbbroek, and E. W. DeJong, pp. 285–9. D. Reidel Publishing, Leiden.

59. Sikes, C. S., and Wilbur, K. M. 1980. Calcification by coccolithophorids: Effects of pH and Sr. *J. Phycol.* **16**: 433–6.

60. Sikes, C. S., and Wilbur, K. M. 1982. Functions of coccolith formation. *Limnol. Oceanogr.* **27**: 18–26.

61. Sikes, C. S., Roer, R. D., and Wilbur, K. M. 1980. Photosynthesis and coccolith formation: Inorganic carbon sources and net inorganic reaction of deposition. *Limnol. Oceanogr.* **25**: 248–61.

62. Silver, M. W , and Bruland, K. W. 1981. Differential feeding and fecal pellet composition of salps and pteropods, and the possible origin of the deep-water flora and olive-green "cells." *Mar. Biol.* **62**: 263–73.

63. Spalding, M. H., Spreitzer, R. J., and Ogren, W. L. (1983) Reduced inorganic carbon transport in a CO_2-requiring mutant of *Chlamydomonas reinhardtii*. *Plant Physiol.* **73**: 273–6.

64. Steemann Nielsen, E. 1966. The uptake of free CO_2 and HCO_3^- during photosynthesis of plankton algae with special reference to the coccolithophorid *Coccolithus huxleyi*. *Physiol. Plant* **19**: 232–40.

65. Steemann Nielsen, E. 1975. *Marine Photosynthesis, with Special Emphasis on the Ecological Aspects.* Elsevier, New York.

66. Stumm, W., and Morgan, J. J. 1981. *Aquatic Chemistry: An Introduction Emphasizing Chemical Equilibria in Natural Waters*, Wiley Interscience, New York.

67. Suess, E. 1970. Interaction of organic compounds with calcium carbonage. I. Association phenomena and geochemical implications. *Geochem. Acta* **34**: 157–68.

68. Taylor, A. H., Watson, A. J., Ainsworth, M., Robertson, J. E., and Turner, D. R. 1991. A modelling investigation of the role of phytoplankton in the balance of carbon at the surface of the North Atlantic. *Glob. Biogeochem. Cycles* **5**: 151–71.

69. Tsuzuki, M., Miyachi, S., and Berry, J. A. Intracellular accumulation of inorganic carbon and its active species taken up by *Chlorella vulgaris* 11 h. In *Inorganic Carbon Uptake by Aquatic Photosynthetic Organisms*, ed. W. J. Lucas and J. A. Berry, pp. 53–66. Am. Soc. Pl. Physiol. Symp. Ser., Rockville, Md.

70. Tsuzuki, M., and Miyachi, S. 1989. The function of carbonic anhydrase in aquatic photosynthesis. *Aq. Bot.* **34**: 85–104.

71. Waterbury, J. B., Watson, S. W., Guillard, R. R., and Brand, L. E. 1979. Widespread occurrence of a unicellular marine planktonic cyanobacteria. *Nature* **277**: 293–4.

72. Wendt, I. 1968. Fractionationn of carbon isotopes and its temperature dependence in the system CO_2–gas CO_2 in solution and HCO_3^-–CO_2 in solution. *Earth Planet Sci. Lett.* **4**: 64–8.

73. Westbroek, P., Young, J. R., and Linschooten, K. 1989. Coccolith production (biomineralization) in the marine alga *Emiliania huxleyi*. *J. Protozool.* **36**: 368–73.

74. Wheeler, A. P., and Sikes, C. S. 1989. Matrix–crystal interactions in $CaCO_3$ biomineralization. In *Biomineralization: Chemical and Biochemical Perspectives*, ed. S. Mann, J. Webb and R. J. P. Williams, pp. 95–131. VCH Publishers, Weinheim.

75. Whittaker, R. H , and Likens, G. E. 1973. Carbon in the biota. In *Carbon and the Biosphere*, ed. G. M. Woodwell and E. V. Pecan, pp. 281–302. U.S. Atomic Energy Commission.

76. Wilbur, K. M., and Saleuddin, A. S. M. 1983. Shell formation. In *The Mollusca*, ed. A. S. M. Saleuddin and K. M. Wilbur, Vol. 4, pp. 235–87. Academic Press, New York.

Zenvirth, D., Volokita, M., and Kaplan, A. 1984. Evidence against H^+–HCO_3^- symport as the mechanism for HCO_3^- transport in the cyanobacterium *Anabaena variabilis*. *J. Memb. Biol.* **79**: 271–4.

16

The Compensation Point: Can a Physiological Concept Be Applied to Global Cycles of Carbon and Oxygen?

JOSEPH A. BERRY, G. JAMES COLLATZ, ROBERT D. GUY, AND MARILYN D. FOGEL

Recent evidence that the CO_2 concentration of the atmosphere is increasing (21), along with evidence from ice cores that relates past changes in climate with changes in the concentration of CO_2 (1), has stimulated a great deal of interest in the mechanisms that determine the global CO_2 balance. This turns out to be an extremely complex problem because it involves a hierarchy of processes including not only the obvious metabolic processes of photosynthesis and respiration but also a whole range of associated biological, chemical, and geological processes that control the fraction of the geologically active pools of carbon and nutrients in circulation and, ultimately, constrain the size of the biosphere. In fact, many studies of the carbon cycle tend to focus on the physical processes, and, much to the bewilderment of most biologists, tend to ignore or attach little control to the biological processes. For example, Kastings et al. (20), in discussing the constraints on CO_2 and O_2 concentration in the atmosphere, acknowledge that "living organisms play an important role in the exchange of CO_2 with the atmosphere," but go on to state that these activities will have little impact on the long-term steady state established by the tectonic cycling of carbon. Broecker (6), speculating on the basis for presuming lower CO_2 concentrations during glacial as opposed to interglacial periods (intervals of thousands of years), argues, "The atmospheric CO_2 content on this time scale must be slave to the ocean's chemistry." These statements have merit, and should serve notice to biologists that the problem is more complicated than it might at first appear.

Nevertheless, we do have good reason to suspect that over the short term, the steady-state concentrations of CO_2 and O_2 in the atmosphere must be at their respective compensation points, where there is stoichiometric balance between photosynthesis and respiration. Furthermore, physiological and biochemical studies have provided a good deal of information on the kinetics of these processes and their sensitivity to factors such as the CO_2 concentration and temperature. In this chapter, we review some of the basic physiological and biochemical concepts that apply at the compensation point, attempt to convince the reader that these have some relevance for understanding the control of atmospheric CO_2 concentration, and illustrate how this information might be applied to interpret and predict the responses of complex systems.

THE CO_2 COMPENSATION POINT OF C_3 PHOTOSYNTHESIS

When a leaf of a C_3 plant is illuminated in a closed space, it either takes up or evolves CO_2 (depending upon the initial concentration of CO_2) until the CO_2 concentration reaches a stable steady-state value ($\Gamma = 45$ µbar CO_2 at 25°C and 210 mbar O_2) at which the rate of net CO_2 assimilation (A_n) is zero. The value of Γ is independent of the intensity of illumination (above a threshold intensity), strongly dependent upon $[O_2]$ and temperature, and, under any given condition, similar for leaves of all C_3-plants. The value of Γ for a single leaf is obviously quite different from the apparent CO_2 compensation point of the global system (near 280 ppm in preindustrial times). Nevertheless, it is useful to consider the leaf system in some detail since it provides a good model system to develop a quantitative basis for relating the value of Γ to the kinetics of enzymes and the structure of the metabolic pathways involved in uptake and production of CO_2 and O_2.

The stability of Γ from leaf to leaf, under a given condition, and its remarkable dependence on temperature and O_2, presented quite a puzzle in the "old days," since these factors indicated that the rates of photosynthetic CO_2 uptake and respiratory CO_2 release were linked. Yet these were apparently controlled by separate metabolic pathways. Tregunna and co-workers (27) postulated that the CO_2 release governing the CO_2 compensation point of leaves emanated from a special type of respiration—*photorespiration*—which was distinct from normal mitochondrial respiration; and they postulated that the rates of photorespiration and gross CO_2 uptake were linked. The basis for this linkage became clear with the discovery that Rubisco catalyzes the first step in both metabolic sequences (4, 5). It is now generally accepted that the properties of Rubisco that contribute to its capacity to catalyze (or to its inability to avoid catalyzing) both the carboxylation and oxygenation of RuBP, also directly allow the existence of a kinetically defined compensation point.

The structure of metabolic pathways linked to the reactions catalyzed by Rubisco is also needed to define the compensation point. The products formed from carboxylation and oxygenation of RuBP are metabolized in intact leaf cells by the photosynthetic carbon reduction (PCR) cycle and photorespiratory carbon oxidation (PCO) cycle; the biochemistry and compartmentation of the latter were largely worked out by Tolbert (14). A scheme showing the linkage between these two pathways at the CO_2 compensation point is given in Figure 16.1. The PCO cycle accepts phosphoglycolate from the oxygenation of RuBP and produces CO_2, NH_3, and 3-phosphoglycerate (3-PGA). The CO_2 mixes rapidly with other pools of CO_2 in the leaf; the NH_3 is quantitatively reassimilated, and the 3-PGA may enter the PCR cycle. The PCR cycle is considered to accept 3-PGA regardless of its origin, and its product is ribulose-1,5-biphosphate (RuBP). The stoichiometry

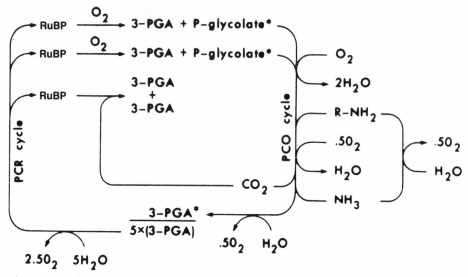

Fig. 16.1. A scheme relating photorespiratory O_2 uptake and photosynthetic O_2 production to the linked operation of the photosynthetic carbon reduction cycle (PCR cycle) and the photorespiratory carbon oxidation cycle (PCO cycle) at the CO_2 compensation point.

of the cycle varies with the v_c/v_o ratio, where v_c is the rate of carboxylation and v_o the rate of oxygenation of RuBP. At the CO_2 compensation point, the reactions are balanced such that there is one CO_2 released by the PCO cycle for each one fixed in carboxylation of RuBP. Figure 16.1 also shows that there is net balance in the uptake and production of O_2. O_2 uptake occurs in the synthesis of phosphoglycolate, in the glycolate oxidase reaction, and in the glycine decarboxylase reaction of leaf mitochondria; and O_2 is produced in photosystem II to provide NADPH or reduce ferredoxin used in the reassimilation of NH_3 and reactions of the PCR and PCO cycles. The sum is 3.5 mole of O_2 per mole CO_2. The net balance between O_2 uptake and production at the CO_2 compensation point, and the expected stoichiometric balance between the fluxes were verified in experiments using $^{18}O_2$ (3).

From the above scheme we may deduce that the compensation should occur at any combination of CO_2 and O_2 concentrations that give $v_c/v_o = 0.5$. Laing et al. (22) derived an equation relating the ratio of the rates of carboxylation (v_c) and oxygenation (v_o) of RuBP to the concentration of the substrates CO_2 and O_2:

$$\frac{v_c}{v_o} = \frac{V_c/K_c}{V_o/K_o} \cdot \frac{[CO_2]}{[O_2]} = \tau \cdot \frac{[CO_2]}{[O_2]} \tag{16.1}$$

where v_c and v_o are the rates of the carboxylation and oxygenation; K_c, K_o, V_c, and V_o are the corresponding K_m and V_{max} terms for the carboxylase and oxygenase functions of Rubisco, respectively; and τ is a constant used to abbreviate the ratio of kinetic constants. Substituting $v_c/v_o = 0.5$ and solving for $[CO_2]$, we obtain, $\Gamma = 0.5[O_2^2]/\tau$. Laing et al. (24) evaluated the kinetic constants that make up τ with purified Rubisco in vitro over a range of temperatures, and showed that the

Table 16.1. Values of Kinetic Parameters for Rubisco of C_3 Higher Plants

Symbol	Value	Units	Q_{10}	Description
K_c	300^a	$\mu mol \cdot mol^{-1}$	2.1	Michaelis constant for CO_2
K_o	300^a	$mmol \cdot mol^{-1}$	1.2	Inhibition constant for O_2
τ	2600		0.57	CO_2/O_2 specificity ratio

^{a} Assuming an atmospheric pressure $= 10^5$ Pa.

observed dependence of Γ values on $[O_2]$ and temperatures could be approximated by use of this model. Jordan and Ogren (15) developed a more accurate procedure for determining τ based upon a simultaneous assay of the v_c/v_o ratio. Furthermore, they showed that τ is not significantly affected by the concentration of RuBP, the pH of the medium, or the activation state of Rubisco, all of which may vary in vivo (18). Table 16.1 presents a compilation of the current best estimates of the kinetic parameters of Rubisco. These kinetic constants have generally been measured in terms of the molar concentrations of CO_2 and O_2 dissolved in the action medium (28). For the purpose of this analysis, the chemical activity is specified here in terms of the partial pressure of O_2 or CO_2 in the gas phase at equilibrium with the reaction medium (28). These constants are strongly temperature dependent, and the temperature coefficients (Table 16.1) reflect temperature-dependent changes in both the solubility of the gases (Henry's constants) and the reaction kinetics.

THE PHYSIOLOGICAL COMPENSATION POINT

The biochemical structure of the metabolic pathways and the kinetic properties of Rubisco provide a basis to estimate the "theoretical" value of the compensation point. As noted, this should be when $v_c/v_o = 0.5$. However, as always happens, reactions are more complicated in a real C_3 plant than on paper. The most obvious complication is the simultaneous flux of CO_2 release associated with dark respiration. This complication is related to the fact that the Γ of intact leaves is typically about 45 μbar at normal ambient O_2 levels and 25°C, but when we use the value of τ (Table 16.1) and Eq. 16.1, we obtain a slightly lower estimate of Γ: 40.4 μbar.

From the equation for net CO_2 exchange (11),

$$A_n = \left[v_c \left(1 - 0.5 \frac{v_o}{v_c} \right) \right] - R \tag{16.2}$$

it may be deduced that $v_o/v_c = 2$ under $[CO_2]$ and $[O_2]$ conditions where $A_n = -R$, the rate of normal respiration continuing in the light. As first proposed by Laisk (23), the "true" compensation point of photosynthesis is not at Γ where $A_n = 0$, but at a slightly lower CO_2 concentration, where $A_n = -R$. He defined this point as Γ_*. The difference between Γ and Γ_*, although only a few microbars (μbar), is significant given the precision of the measurements of τ by Jordan and

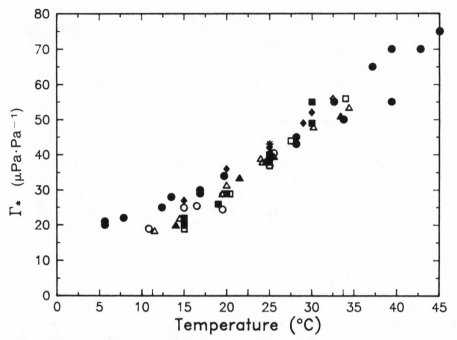

Fig. 16.2. The "true" CO_2 compensation point, Γ^*, as influenced by temperature at 209 mbar O_2. (Note that $\tau = [O_2]/2\Gamma^*$.) Data from spinach Rubisco assayed in vitro (●) (18) are compared with in vivo gas exchange measurements with spinach (■) (7). Data for eight other species are shown.

Ogren (18), and the contribution of respiration becomes more pervasive at larger scales of organization.

Brooks and Farquhar (7) developed gas exchange procedures to measure Γ^* for leaves. The basic idea is that net CO_2 exchange should be totally independent of the light intensity only at the true value of Γ^*, and in fact, careful studies show a small dependence of Γ on light (Fig. 16.2). The CO_2 concentration where A_n is independent of light Γ^*, can be estimated by constructing curves of the response of A_n to intercellular CO_2 at different light intensities. The point of interest is at the intersection of these curves, as illustrated in Figure 16.2.

The method of Brooks and Farquhar (7) is essentially an in vivo assay for τ, and it is of interest to compare values obtained with this in vivo assay to values obtained in vitro. Figure 16.3 is a plot of Γ^* at 21% O_2 for several species of C_3 plants measured over a range of temperature. Data obtained in vitro by Jordan and Ogren (15–18) are plotted as the equivalent Γ^* together with data obtained by gas exchange. These measurements agree very well, given the difficulty of cross-calibrating among the assay systems. The specificity of Rubisco for CO_2 is apparently very similar among several C_3 plants from different taxa.

Other kinetic information, concerning the dependence of gross CO_2 uptake ($J_p = v_c - 0.5v_o$) on CO_2 concentration over the range of Γ^* to Γ and the rate of dark respiration, is required to predict the actual physiological CO_2 compensation point—which is of more direct interest to us here. A photosynthetic model provides

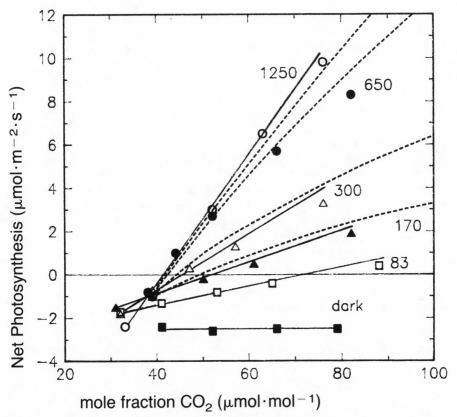

Fig. 16.3. Net CO_2 exchange of an intact attached leaf of *Xanthium strumarium* as a function of the intercellular partial pressure of CO_2. The different curves were obtained with the same leaf at different light intensities (numbers to the right of the curves are in units of $\mu mol \cdot m^{-2} \cdot s^{-1}$). The curves at 1250, 650, and 300 intersect at 38 μbar. The flux of net CO_2 exchange at Γ^* is the rate of respiration in the light. Note that the rate of respirations is lower in the light than in darkness.

a convenient way to assemble this information. Besides providing the data points, Figure 16.3 also shows simulations (dashed lines) obtained from a model using only environmental input variables. A description of the model [based on Farquhar et al. (11)] and more complete tests of its accuracy are presented by Collatz et al. (9).

COMPENSATION POINTS OF ALGAE AND C₄ PLANTS

It has long been known that C_4 plants have a much lower Γ value than C_3 plants and that this value is not sensitive to O_2 (12); similar observations were made with algae (8). It is now well established that both photosynthetic systems are based on CO_2 fixation by Rubisco, and both possess fully functional PCR and PCO cycles. The low CO_2 compensation point is related to the presence of

"CO_2-concentrating mechanisms." An elegant paper by Peisker and Bauwe (25) points out that the "local" CO_2 compensation point inside the bundle sheath cells of a C_4 plant at its CO_2 compensation point is likely to be about 45 µbar—similar to that reached by a C_3 plant in air. The "local" CO_2 concentration is higher than the actual Γ of the leaf because CO_2 transported into the bundle sheath cells (coupled to fixation into C_4 acids) has difficulty leaking out through the thick cell walls. In the case of algae, a CO_2-concentrating mechanism is based on ion transport (24), and unlike C_4 plants, algae may apparently adjust the activity of their CO_2-concentrating mechanism according to the availability of CO_2. In abundant CO_2 they may lack any evidence of reduced CO_2 compensation points, but they can adjust to maintain growth at very low CO_2 concentrations. For example, the green alga *Chlamydomonas reinhardtii* can grow in culture bubbled with 92% O_2 and 33 ppm CO_2 (about $0.1 \times \Gamma_*$), at a rate only slightly lower than it has when bubbled with 5% CO_2 (2). It is interesting to speculate that the global CO_2 could be maintained a much lower value if either of these photosynthetic systems came to dominate the biosphere.

ISOTOPIC COMPENSATION POINTS

A diagram of the global oxygen cycle is shown in Figure 16.4. Many of the reactions of this cycle are linked to the carbon cycle by reactions we have already discussed, but it is also important to note that O_2 is coupled to the redox state of other elements—principally sulfur, iron, and nitrogen. There are very large reservoirs of some of these elements in rocks of the crust, and the atmosphere is not strictly a closed system on geological time scales (20).

One very interesting observation concerning the oxygen cycle is the presence of a large steady-state difference between the isotope ratio of oxygen in atmospheric O_2 and that in its source, H_2O. The $\delta^{18}O_2$ equals 23.5‰ (i.e., the $^{18}O/^{16}O$ ratio of O_2 is 1.0235 times that of standard mean ocean water). This is known as the *Dole effect*, in honor of its discoverer, Malcolm Dole (10). The Dole effect is the result of isotopic discrimination in the reactions that recycle oxygen in the global biogeochemical oxygen cycle; this is, in effect, an *isotopic compensation point*.

Recent studies by Guy et al. (13) have examined the isotopic compensation point in so-called *microcosm* experiments. In these experiments, mesophyll cells of the asparagus fern were suspended in He-sparged buffer containing approximately 250 µM HCO_3^- in a 400-ml vessel fitts with an O_2 electrode. The vessel was closed and illuminated, and the O_2 concentration within it was increased, eventually reaching about air levels. Great care was taken to prevent mixing of the O_2 within the vessel with that of the air, and the system was kept illuminated for different intervals of time before a sample of the medium was withdrawn for analysis of the isotope ratio of the O_2 and water. As shown in Figure 16.5, the isotope ratio changed with time; initially it was close to that of the water in the vessel, and with time the $\delta^{18}O$ value became displaced by about 22‰ (e.g., $R_{O_2} = 1.022 \times R_{H_2O}$ where R is the $^{18}O/^{16}O$ ratio). When steady-state isotopic balance is established, $d[^{18}O]/dt = 0$, and we may write that

THE DOLE EFFECT:

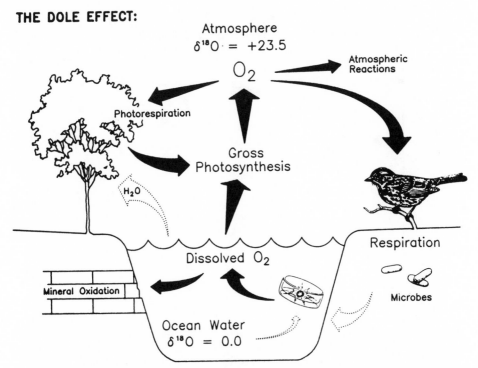

Fig. 16.4. A scheme showing the geochemical cycling of oxygen. O_2 is produced from water during photosynthesis and is consumed in several biological and chemical reactions. The bold, solid arrows represent the transfer of the oxygen as O_2; and the light, dotted arrows represent transfer as H_2O. The observations that the O_2 of the atmosphere is enriched in ^{18}O ($\delta^{18}O = 23.5$; $R_{O_2}/R_{H_2O} = 1.0235$) is referred to as the "dole effect."

$$\delta^{18}O_{O_2} = \delta^{18}O_{H_2O} + D_{PSII} + D_{Resp} \qquad (16.3)$$

where the D terms are the apparent isotopic discrimination [$D = 1 - (^{18}k/^{16}k)$] in the photosystem II, D_{PSII}, and in oxygen uptake, D_{Resp}. The discrimination occurring in various biochemical reactions that could contribute to the steady-state isotope displacement was examined in vitro, and results of these studies are summarized in Figure 16.6. No fractionation was detected in O_2 production from water, whereas substantial fractionation was observed in reactions contributing to photorespiration. Referring to Figure 16.1, we see that Rubisco oxygenase should account for $2/3.5 = 0.57$; glycolate oxidase, $1/3.5 = 0.28$; and cytochrome oxidase, $0.5/3.5 = 0.15$ of the total oxygen taken up in this system at the compensation point. The weighted average $D_{Resp} \simeq 22\permil$, and substituting in Eq. 16.3 yields a $\Delta\delta^{18}O_{O_2}$, very close to the actually observed displacement (Fig. 16.5).

It is particularly significant that the isotope ratio of the microcosm experiments comes very close to that in the planetary atmosphere (Fig. 16.4), and that, at least in the microcosm, we can explain the displacement by the observed discrimination in respiration and photorespiration. There are some complicating factors in the global system that cannot be fully discussed here. Nevertheless, these results

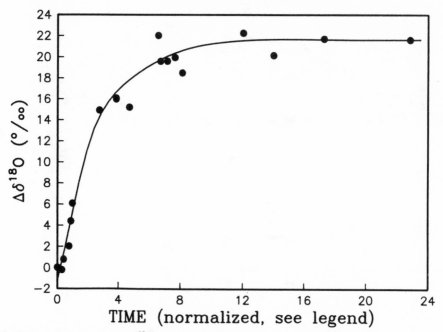

Fig. 16.5. Divergence in the $\delta^{18}O$ of dissolved O_2 from the source water in *Asparagus* cell microcosm experiments. Data from six separate experiments are presented. The time axis is normalized to the $t_{1/2}$ (time required to reach 1/2 the final O_2 concentration in each experiment). This is approximately the turnover time of the O_2 pool.

indicate that properties of biochemical reactions appear to be relevant to the global oxygen cycle. Drawing on this insight, we now return to consider the cycling of CO_2 in large-scale systems.

THE ECOLOGICAL CO_2 COMPENSATION POINT

When we brought respiration into the treatment of the CO_2 compensations of C_3 plants, some of the strict kinetic control of Γ was lost. We now have that term, R (Eq. 16.2), which may vary with time and scale. However, if the focus is expanded to the scale of ecosystems and we consider long periods of time, then the quantity of biomass should be constant and the R term is constrained; gross CO_2 uptake must equal respiration ($J_p = R$). We need not concern ourselves here with the way that an ecosystem arrives at this steady-state condition. It is obviously not a strict stoichiometric requirement of the metabolic systems; rather, it is imposed by analogous allometric requirements of the biogeochemical cycles operating within the ecosystem. It seems reasonable to assume that these structural properties have long time constants that will not interact strongly with short-term physiological responses of the system.

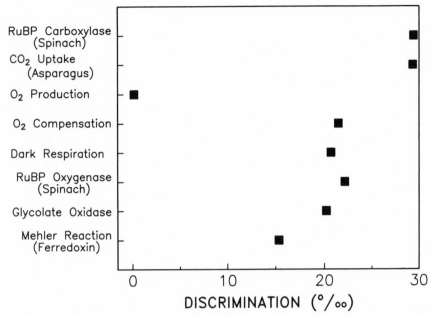

Fig. 16.6. O_2 discrimination, D (‰), by different processes associated with O_2 metabolism (13).

If we had an ecosystem that we could isolate (i.e., put a bell-jar over it), we could conduct some experiments to examine metabolic control of the ecosystem Γ. For example, what would happen to the CO_2 compensation point if we lowered the oxygen concentration within the bell jar to 50% of the current value? We know that Γ_* would be reduced proportionally, but what would happen to Γ, given that most of the respiration is heterotrophic? This type of problem can be approached with the photosynthesis model. Figure 16.7 illustrates a graphical solution. We start by constructing a curve for the response of gross photosynthesis to CO_2 in normal O_2 (solid line). We assume that the original system was at steady state in the present atmosphere (here we use the preindustrial level of 280 μbar CO_2). Also, respiration is relatively independent of CO_2 (see Fig. 16.2), and thus we can represent it as a horizontal line (dashed line) intersecting the photosynthesis curve at the original Γ (Fig. 16.7). If we now calculate another photosynthesis response curve using the same parameters in our photosynthesis model, changing only the O_2, we obtain the dotted line shown in Figure 16.7 which intersects the respiration line at a lower CO_2 concentration, the new compensation point. This analysis assumes that the uptake of CO_2 required to reach the new value of Γ does not alter the structure of the ecosystem (i.e., the volume of the bell jar is not large). Of course, it is not really necessary to use a model to solve this problem. The model is composed of kinetic expressions, and if these are simple enough, we can solve for the new value of Γ, Γ'. For example, the response of gross photosynthetic fixation of CO_2 (J_p) to CO_2 is given by

$$J_p = V_m \frac{c_i - \Gamma_*}{c_i + K_{ap}} \qquad (16.4)$$

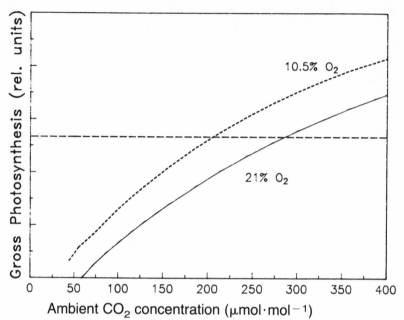

Fig. 16.7. A graphical solution for the new CO_2 concentration that would yield a steady-state balance of an ecosystem at reduced O_2 (105 mbar). A model of photosynthesis was used to construct response curves of gross photosynthesis to CO_2 using identical parameter values except that $[O_2] = 105$ mbar (*dotted line*) and $[O_2] = 209$ mbar (*solid line*). It is assumed that respiration equals J_p at 280 μbar in normal O_2, and that respiration is not affected by O_2. The intersection of the dashed line with the dotted line is the new CO_2 compensation concentration.

when Rubisco is rate-limiting, and

$$J_p = \alpha I \frac{c_i - \Gamma^*}{c_i + 2\Gamma^*} \tag{16.5}$$

when light is limiting; c_i is the intercellular CO_2 [taken here as 0.7 × the ambient CO_2 (c) to allow for diffusion through stomata and cells walls], V_m is the maximum catalytic capacity of Rubisco, and K_{ap} is the apparent Michaelis–Menten constant for CO_2 [$K_{ap} = K_m(1 + O_2/K_i)$; see Table 16.1]. We do not know whether light or Rubisco is limiting, and it may be different in different places. The model makes a choice for us. We can be open-minded and examine both possibilities. If we take light as limiting and assume that value of $I\alpha$ (Eq. 16.5) is constant, then we may write that

$$\frac{c - \Gamma^*}{c + 2\Gamma^*} = \frac{c' - \Gamma'^*}{c' + 2\Gamma'^*}$$

where the prime refers to the value in altered O_2, and all except c' are known. Solving for c':

$$c' = c \frac{\Gamma'^*}{\Gamma^*} \tag{16.6}$$

Similarly, if we take light as saturating and V_m as constant, we may obtain an equation

$$c' = \frac{K'_{ap}(c - \Gamma^*) + \Gamma'^*(c + K_{ap})}{\Gamma^* + K_{ap}} \tag{16.7}$$

Using the values of K_{ap} and Γ^* in 21% or 10.5% O_2, we estimate that the new ecosystem Γ will be between 140 ppm CO_2 (when light is limiting) and 207 ppm CO_2 (when Rubisco is limiting). Using a similar approach, we can estimate that replacing the Rubisco of all C_3 plants with Rubisco from *Rhodispirillum rubrum* [$K_{ap} = 5$ mbar $\Gamma^* = 400$ µbar; (16)] would increase the apparent ecosystem CO_2 compensation point concentration to somewhere in the range of 1740–1960 ppm. These calculations serve to illustrate that the present atmospheric CO_2 concentration is sensitive to the O_2 concentration and to the kinetic properties of Rubisco and those enzymes that govern respiration and decomposition.

It is interesting that temperature affects the kinetic properties of both photosynthesis and respiration, as this is the parameter most likely to vary over short time frames in nature. Using a model that includes both effects, we estimate thst $\Delta C/\Delta T$ ranges from 18 to 55 ppm $\cdot {}^\circ C^{-1}$, depending upon whether we assume photosynthesis is limited by Rubisco or light, respectively. This is a very large temperature sensitivity, and it suggests that the global CO_2 concentration might be expected to vary with changes in temperature.

TEMPERATURE AND THE GLOBAL CO_2 COMPENSATION POINT

The above example leads us to suspect that the global compensation concentration might be rather sensitive to temperature, and there is evidence that temperature and Γ have covaried during the wax and wane of the ice ages. Evidence for this comes from analysis of ice cores. Small volumes of air become entrapped as accumulated snow is compacted to form glacial ice, and a core of ice approximately 2 km in depth and spanning the past 160,000 years was obtained from a drilling project at the Russian station at Vostok, Antarctica (19). This ice contains a continuous record of the regional temperature at the time of the snowfall (indicated by the H/D ratio of H_2O in the ice), and the trapped air gives the global CO_2 concentrations at approximately the same time. Figure 16.8 plots CO_2 concentration versus the difference in temperature (taking the present temperature as a reference). These data were obtained by digitizing from published curves of CO_2 and ΔT as a function of time (1). This plot shows that changes in the apparent global CO_2 compensation point and temperature have been correlated over this time interval, which spans two major glacial epochs.

If the present analysis is correct, then it follows that a long-term drop in global temperature could cause a drop in CO_2, which in turn (through the greenhouse effect) could result in a positive feedback on the temperature when a decrease in temperature is imposed—for example, by changes in the orbital geometry. Similarly, positive feedback should occur when an increase in CO_2 is imposed on the system. These response characteristics would tend to destabilize the global temperature. We should caution, however, that the strength of these positive

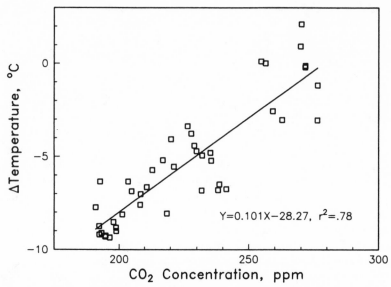

Fig. 16.8. Co-variation of temperature and atmospheric CO_2 concentration over the past 160,000 years as recorded in an ice core from Vostok, Antarctica. The CO_2 concentration is obtained by analyzing air bubbles trapped in ice from different depths (that formed at different times), and the Δ temperature is inferred from analysis of the isotopic composition of the water of that ice. These data were obtained from plots of the original measurements published by Barnola et al. (1) and Jouzel et al. (19). The slope of the regression line indicates that the apparent CO_2 compensation point of the global system has changed by about 10 ppm $\cdot$ $^0C^{-1}$ as the climate changed.

feedback loops would depend on the extent of damping and time delays associated with the large reservoir of CO_2 in the atmosphere and surface ocean. The present analysis does not take into account the fact that a change of atmospheric CO_2 concentration requires substantial changes in the biomass of the biosphere. Other analyses of the carbon cycle (26) have focused primarily on the changes in pool size. The present analysis focuses on the kinetics of reactions that communicate between pools. It would seem valuable to combine these approaches.

ACKNOWLEDGMENTS

This is CIW-DPB Publication No. 1078.

REFERENCES

1. Barnola, J. M., Raynaud, D., Korotkevich, Y. S., and Lorius, C. 2987. Vostok ice core provides 160,000-year record of atmospheric CO_2. *Nature* **329**: 408–14.

2. Berry, J. A., Boynton, J., Kaplan, A., and Badger, M. R. 1976. Growth and photosynthesis of *Chlamydomonas reinhardtii* as a function of CO_2 concentration. *Carnegie Inst. Year Book* **75**: 423–32.

3. Berry, J. A., Osmond, C. B., and Lorimer, G. H. 1978. Fixation of O_2 during photorespiration. *Plant Physiol.* **62**: 954–67.

4. Bowes, G., Ogren, W. L., and Hageman, R. H. 1971. Phosphoglycolate production catalyzed by ribulose diphosphate carboxylase. *Biochem. Biophys. Res. Commun.* **45**: 716–22.

5. Bowes, G., and Ogren, W. L. 1972. Oxygen inhibition and other properties of soybean ribulose 1,5-disphosphate carboxylase. *J. Biol. Chem.* **247**: 2171–6.

6. Broecker, W. S. 1982. Glacial to interglacial changes in ocean chemistry. *Prog. Oceanogr.* **11**: 151–97.

7. Brooks, A., and Farquhar, G. D. 1985. Effects of temperature on the CO_2/O_2 specificity of ribulose-1,5-biphosphate carboxylase/oxygenase and the rate of respiration in the light. *Planta* **165**: 397–406.

8. Brown, D. L., and Tregunna, E. B. 1967. Inhibition of respiration during photosynthesis by some algae. *Can. J. Bot.* **45**: 1135–43.

9. Collatz, G. J., Ball, J. T., Grivet, C., and Berry, J. A. 1991. Regulation of stomatal conductance and transpiration: A physiological model of canopy processes. *Agric. Forest Meterol.* **54**: 107–36.

10. Dole, M. 1935. The relative atomic weight of oxygen in water and air. *J. Am. Chem. Soc.* **57**: 2731.

11. Farquhar, G. D., von Caemmerer, S., and Berry, J. A. 1980. A biochemical model of photosynthetic CO_2 fixation in leaves of C_3 species. *Planta* **149**: 78–90.

12. Forrester, M. L., Krotkov, G., and Nelson, C. D. 1966. The effect of oxygen on photosynthesis photorespiration and respiration in detached leaves. II. Corn and other monocotyledons. *Plant Physiol.* **41**: 428–531.

13. Guy, R. D., Fogel, M. F., Berry, J. A., and Hoering, T. H. 1986. Isotope fractionation during oxygen production and consumption by plants. In *Progress in Photosynthesis Research*, ed. J. Biggins, Vol. III, pp. 597–600. Martinus Nijhoff, Dordrecht.

14. Husic, D. W., Husic, H. D., and Tolbert, N. E. 1987. The oxidative photosynthetic carbon cycle or C_2 cycle. In *Critical Reviews in Plant Science*, Vol. 5, pp. 45–100. CRC Press, Boca Raton, Fla.

15. Jordan, D. B., and Ogren, W. L. 1981. A sensitive assay procedure for simultaneous determination of ribulose-1,5-biphosphate carboxylase and oxygenase activities. *Plant Physiol.* **67**: 237–45.

16. Jordan, D. B., and Ogren, W. L. 1981. Species variation in the specificity of ribulose bisphosphate carboxylase/oxygenase. *Nature* **291**: 513–15.

17. Jordan, D. B., and Ogren, W. L. 1983. Species variation in kinetics properties of ribulose 1,5-bisphosphate carboxylase/oxygenase. *Arch. Biochem. Biophys.* **227**: 425–33.

18. Jordan, D. B., and Ogren, W. L. 1984. The CO_2/O_2 specificity of ribulose-1,5-bisphosphate carboxylase/oxygenase. *Planta* **161**: 308–13.

19. Jouzel, J., Lorius, C., Petit, J. R., Genthor, C., Barkov, N. I., Kotlyakov, V. M., and Petrov, V. M. 1987. Vostoke ice core: A continuous isotope temperature record over the last climate cycle (160,000 years). *Nature* **329**: 403–8.

20. Kastings, J. F., Toon, O. B., and Pollock, J. B. 1988. How climate evolved on the terrestrial planets. *Sci. Am.* **XX**: 90–97.

21. Keeling, C. D. 1973. Industrial production of carbon dioxide from fossil fuels and limestone. *Tellus* **25**: 174–98.

22. Laing, W. A., Ogren, W. L., and Hageman, R. H. 1974. Regulation of soybean net photosynthetic CO_2 fixation by the interaction of CO_2, O_2, and ribulose-1,5-bisphosphate carboxylase. *Plant Physiol.* **54**: 678–85.

23. Laisk, A. 1977. *Kinetics of Photosynthesis and Photorespiration in C_3 Plants.* Nauka, Moscow.

24. Lucas, W. J., and Berry, J. A., eds. 1985. *Inorganic Carbon Uptake by Aquatic Photosynthetic Organisms,* p. 493. Am. Soc. Plant Physiologists, Rockville, Md.

25. Peisker, M., and Bauwe, H. 1984. Modelling carbon metabolism in C_3–C_4 intermediate species. I. CO_2 compensation concentration and its O_2 dependence. *Photosynthetica* **18**: 9–19.

26. Sundquist, E. T. 1985. Geological perspective on carbon dioxide and the carbon cycle. In *The Carbon Cycle and Atmospheric CO_2: Natural Variation Archaen to Present,* ed. E. T. Sundquist and W. S. Broecker. Geophys. Monogr. Serv. No. 32. American Geophysical Union, Washington, D.C.

27. Tregunna, E. B., Krotkov, G., and Nelson, C. D. 1966. Effect of oxygen on the rate of photorespiration in detached tobacco leaves. *Physiol. Plant.* **19**: 723–33.

28. Woodrow, I. E., and Berry, J. A. 1988. Enzymatic regulation of photosynthetic CO_2 fixation in C_3 plants. *Annu. Rev. Plant Physiol. Plant Mol. Biol.* **39**: 533–94.

17

From Corn Shucks to Global Greenhouse: Stable Isotopes as Integrators of Photosynthetic Metabolism from Tissue to Planetary Scale

BARRY OSMOND, DAN YAKIR, LARRY GILES, and
JOHN MORRISON

Photosynthetic metabolism is both the cause and concern of the problems discussed in this volume. Ironically, it is the combustion of about a billion years worth of fossil photosynthate in the course of about 200 years that appears to be responsible for rapid global climatic change at the turn of the twentieth century. This acceleration of normal planetary respiration by 5 million fold potentially provides a feed-forward stimulus to photosynthesis, the speed and magnitude of which may be greater than at any other time in the Earth's history. Humans, as the dominant mammals, have short-circuited the normal processes of respiration and sequestration of photosynthates on Earth to the extent that "the scale of human activity on the surface of the Earth is now comparable with the scale of still poorly understood external influences which regulate the environment" (6).

Living plants are the principal natural transducers of water vapor and chemical elements in the atmosphere–biosphere–geosphere continuum that makes the surface of this planet habitable. In addition to sustaining these fluxes of gases, water vapor, and chemical elements, plant processes also respond markedly to changes in concentration of the component gases in the atmosphere and nutrients in the geosphere. That is, plant processes may also serve as indicators of change in a changing global environment.

It is difficult to measure the fluxes of water vapor and chemicals through the thin green veil of the biosphere on the global scale. The reservoirs of the atmosphere and the soil and sea are huge, compared with the reservoirs of the biosphere; and changes in these pool sizes are small and slow, compared with biospheric changes. Until very recently, the biosphere was considered to be passively respondent to the state of the Earth as driven by solar energy flux, and the atmospheric, oceanic, and soil variables. The Ottawa conference of the International Council of Scientific Unions (ICSU) in 1984, which led to the

International Geosphere Biosphere Program, was distinguished by the chauvinisms of atmospheric, oceanic, and information scientists, all of which were directed toward the preoccupation of biologists with the detailed facts of life. Since then, a number of research teams from these disciplines have made the observation that *plants matter*; that *plants are indeed the heart and lungs of the living planet*. Researchers in photosynthesis need to grasp at this dawning awareness of the importance of what they take for granted, and to contribute new insights on the larger scale of things.

Fortunately, plant metabolic processes are most discriminating with respect to the transduction of water vapor and chemicals, and are especially selective about the naturally occurring isotopes of the elements of life (34). The isotopic discrimination of photosynthetic metabolism has permitted significant integration of these processes into ecological and agricultural contexts since the 1970s. We have been extraordinarily successful in leaping up and down a scale of about 12 powers of 10 in space and time from the molecule to the individual plant (28).

Until very recently, atmospheric chemists were reluctant to accept a significant role for the plant biosphere in global exchange processes. Plant effects in climate models were relegated to vegetation-dependent changes in albedo, with gross uncertainties due to cloud formation over transpiring vegetation. The appreciation of the role of the plant biosphere is changing as a result of careful CO_2 analyses of isotopic and CO_2-concentration changes in the global atmosphere, which are partly driven by photosynthesis. The isotopic signatures left in carbon of atmospheric CO_2 by isotopic discrimination due to the Earth's most abundant protein (Rubisco), confirm that photosynthetic research has much to contribute to global CO_2 exchange research. Unexpectedly, the isotopic signature of oxygen in atmospheric CO_2 has given us another strong indicator for the participation of plants in global carbon fluxes.

To be heard in the current debate on changing climates, photosynthetic research has to expand its scale of thinking another six or more powers of ten in space and time, to the scale of the global biosphere. In this chapter, we outline some ways in which the isotopic discriminations of photosynthetic metabolism may help achieve the necessary integration of plant processes into climate change research.

OVERVIEW

Plant biologists owe their opportunities in this field to the observations of an atmospheric scientist in a meteorological laboratory and to her dabblings in archeology. As a result of a chance conversation with R. H. Burris, a distinguished plant biochemist at the University of Wisconsin and mentor of Ed Tolbert, what seemed to be a technical difficulty was transformed into biological insight. These serendipitous exchanges allowed Bender (2) to correlate errors in the ^{14}C dating of artifacts from plants with the C_4 photosynthetic pathway, which has markedly less discrimination against ^{13}C (less negative $\delta^{13}C$ values when referenced against a limestone standard). Some years earlier, Vogel in South Africa, and Craig in the United States had noted some less negative $\delta^{13}C$ values in unknown grass species. Craig (4) attributed this unusual value to the utilization of limestone, rather than

CO_2, as photosynthetic substrate—an interpretation not inappropriate for an earth scientist.

The $\delta^{13}C$ value of plant material and its association with specific pathways of photosynthesis have given us many intrinsically interesting applications in photosynthetic research, such as the evaluation of the induction of CAM (46). However, some of the most spectacular applications have been in environmental and ecophysiological studies, many of which have been reviewed recently by Rundel et al. (33). Analysis of $\delta^{13}C$ values has become a routine interpretive tool in archeology and an indicator with wide ramifications in paleoclimatology. For example, at about the time *Sivapithecus* (a relative of the tree-dwelling orangutangs) disappeared from the fossil record of the Potowar Plateau of Pakistan (7.4 to 7.0 million years ago), the $\delta^{13}C$ of paleosols changed abruptly to those characteristic of C_4 grasslands (31). These data are important as an early record of replacement of C_3 woodlands by early C_4 grasslands, but more important, perhaps, as indicators of the onset of the Asian monsoon climate.

The distinct isotopic fractionations associated with the different photosynthetic carboxylations were established by the in vitro studies of Whelan et al. (43), and have been confirmed by subsequent studies with purified enzymes under catalytically and analytically optimal conditions (26, 32). The formal analysis of isotopic fractionation during CO_2 exchange (35, 41), and direct determination of the in vivo fractionation associated with the 4-C carboxyl of malate during CAM (25), opened the way for a general treatment of diffusional and biochemical components of carbon isotope fractionation in photosynthesis (9, 24).

Although there are many indications that photosynthetic reductions give rise to large discriminations against deuterium (^2H, or D), mechanistic understanding has not progressed much since the 1980s (50, 51). Almost certainly, the primary fractionation involves water dissociation and NADPH photoreduction (7), but there is clear evidence of additional fractionation associated with specific metabolisms (36, 38, 49). Oxygen in plant organic matter is derived from CO_2. However, there is isotopic exchange of these oxygen atoms with those in water during photosynthesis (5). One hypothesis suggests that the enrichment of ^{18}O in photosynthetic products is due to a carbonyl hydration reaction involving isotopic exchange between leaf water and early photosynthetic products such as triose phosphates (38, 39). A great deal of basic biochemical research is still needed in order to bring studies concerned with both hydrogen and oxygen stable isotopes into any reasonably clear picture.

In particular, much remains to be done to understand the baseline, or source effects, underlying these processes—that is, the hydrogen and oxygen isotopic composition of H_2O in the metabolic compartments of leaves. The daily course of enrichment of these isotopes in leaf water is still poorly understood (43, 49). Most workers (11, 20) agree that the leaf is unlikely to be a simple, one-compartment evaporating pan insofar as changes in the isotopic composition of water is concerned. Yakir et al. (47), have directly demonstrated this, using physical methods, and found evidence of at least three compartments of different isotopic composition. Only the isotopic composition of the water pool directly involved in photosynthesis is relevant to the oxygen and hydrogen isotopic signals preserved in plant organic matter. Moreover, only the water directly involved in photosynthesis is responsible for the $\delta^{18}O$ value of atmospheric oxygen, all of which originated from the water-splitting reaction of photosynthesis. Yakir and colleagues (46, 50) have recently identified this water compartment directly, by means of the $\delta^{18}O$ value of photorespiratory CO_2 and photosynthetic O_2.

FROM CORN SHUCKS...

The general model of carbon isotopic fractionation in photosynthesis (9) has led to a wave of applications based on analyses of the isotopic composition of plant materials that are used to infer conditions of past photosynthetic CO_2 exchange. These range from breeding for water use efficiency in crops (10, 18), to implications for thermal hydraulic aspects of paleoclimatology (12, 21). Progress has depended on unfettered access to modern, automated mass spectrometers of greatly enhanced precision. By these means, the scale of evaluation of photosynthetic processes have been expanded through about 12 powers of ten in space and time. The problem has moved from one of technique to one of imagination, at least for carbon.

For the most part, single-isotope analyses are used, but there is immense scope for a multiple-isotope approach. For example, as mentioned above, it is important to distinguish between source effects and subsequent biochemical isotope effects. Because water is the source for both the oxygen and hydrogen isotopic signals in plants, a tight correlation between changes in the isotopic composition of the two elements indicates source effects, whereas the breakdown of this correlation indicates additional biochemical isotope effects that can be very different for the two elements, and that vary with environmental conditions (48).

As another illustration of the potential of the multiple-isotope approach, we can consider an evaluation of autotrophy in the humble hypsophyll of corn (51). The corn cob is an uncertain structure insofar as C_4 photosynthesis is concerned. Early infrared gas analyses (IRGA)-based measurements show astonishingly high rates of photosynthesis in the outer husks (17), whereas $^{14}CO_2$ studies suggest rates 5–650 times lower than in leaves (36). Anatomists noted that the corn shuck does not have normal Kranz anatomy (1); and immunolocalization of Rubisco, malic enzyme, and pyruvate phosphate dikinase (19) confirms the confused state of functional anatomy in this tissue. The rate of photosynthesis in air of the outer husk is a small fraction of that in leaves (Table 17.1) and is inhibited by O_2 (19), indicating that some direct C_3 photosynthetic CO_2 fixation occurs.

The existence of C_3 metabolism in the shuck is strongly supported by the variability observed in its $\delta^{13}C$ value (Fig. 17.1). Green parts of the outer shucks have considerably more negative $\delta^{13}C$ values than do the pale green to yellow parts. This suggests the presence of some C_3 metabolism which contributes organic carbon having a $\delta^{13}C$ of about $-27‰$ to a predominantly C_4 tissue that would otherwise have a $\delta^{13}C$ value of about $-11‰$. However, on the basis of the carbon isotope data alone, we cannot distinguish between local autotrophic C_4 fixation and C_4 carbon from sucrose translocated from other leaves.

In order to partition local autotrophic and imported heterotrophic C_4 components, we have to resort to a second isotopic analysis, that of hydrogen.

Table 17.1. Photosynthetic Response by *Zea mays* to Oxygen

Tissue	Photosynthetic response ($\mu mol\ CO_2 \cdot m^{-2} \cdot s^{-1}$)	
	21% O_2	2% O_2
Leaf	18.0	18.4
Hyposophyll		
Bright green	0.71	1.03
Green	0.45	0.74
Pale green	0.02	0.15

Note: Leaves and hyposophylls were measured in 350 μbar CO_2 at 1400 μmol photons·m^{-2}·s^{-1} PFD.

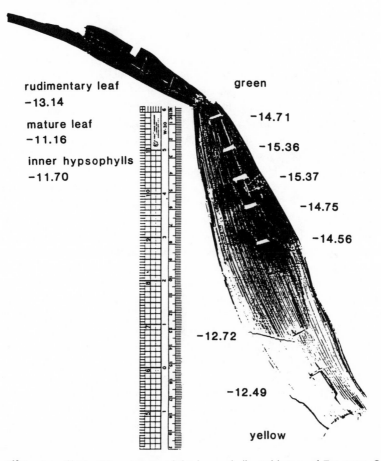

Fig. 17.1. $\delta^{13}C$ values (‰) in different parts of the hypsophylls and leaves of *Zea mays*. Cobs and leaves were harvested from cv Pioneer 33 grown outdoors in summer at Raleigh, NC; similar results were obtained with random samplings of materials from supermarkets at other seasons. Measurements were made on fresh tissue (1–2 mg), following combustion in a Carlo Erba NA1500, interfaced through an automated trapping system to a SIRA Series II isotope ratio mass spectrometer.

Analyses were made of the $\delta^{13}C$ and δD of cellulose fractions from shucks, cobs, and leaves. The δD values of cellulose are very different, depending on hetero-trophic and autotrophic origins (48), and on the δD value of water in the tissues. In order to remove the effect of variation in the δD of tissue water, we used only the difference in δD values in the mass balance calculations to estimate the proportion of shuck cellulose from local, autotrophic metabolism (51). Taking the isotopic composition of both elements into account, we concluded that only 16% of the autotrophic cellulose carbon in the shuck is derived from direct C_3 photosynthesis; about 22% is derived from sucrose imported from other leaves, and the balance (62%) from local C_4 photosynthesis. Clearly, the humble hypsophyll is a laboratory in itself, one that is valuable for further elucidation of these complex relationships.

... TO PHOTOSYNTHETIC MICROCOSMS ...

The discriminations against heavier isotopes during carboxylation, respiration, and transpiration, which lead to changes in the isotopic composition of plant material, leave resultant isotopic signatures in the atmospheres over the leaves which can be measured directly. By these means Evans et al. (8) measured a resistance to CO_2 exchange in wheat leaves that could be ascribed neither to stomatal diffusion nor to Rubisco, and which they attributed to liquid-phase resistance of the cell walls. Subsequent studies using on-line techniques by von Caemmerer and Hubick (41) help to sort out some of the component carboxylations in different classes of C_3–C_4 intermediate species. Griffiths et al. (15) used these methods to identify transitions in the different carboxylation phases of CAM.

O'Leary et al. (27) studied these effects using a leaf in a closed gas volume. This system represents a microcosm of the plant biosphere of the planet on a vastly smaller scale in which responses are accelerated. If soybean leaflets are enclosed in a bell jar and air is circulated in the closed system, CO_2 concentration is drawn down rather rapidly to the CO_2 compensation point set by photo-respiratory conditions. The $\delta^{13}C$ value of the air also changes rapidly, becoming enriched (less negative) owing to the discrimination of Rubisco. A very different picture is obtained using C_4 plants, with a slower rate of enrichment due to the smaller discrimination by PEP carboxylase. In a number of other studies, Rooney (32) described the CO_2 exchanges of different plants under different conditions, in a similar closed chamber. Of particular interest were her comparison of day and night CO_2 fixation in CAM, and the C_3/C_4 components of the mixed phases of CAM. All showed predictable changes in the isotopic composition of the atmosphere around the leaves in response to the different carboxylation processes in these plants.

There are now two independent microcosm studies, using carbon (32) and oxygen (16) stable isotopes, which demonstrate that the isotopic composition of the atmosphere is at, or very near, steady-state conditions predominantly imposed by plants. In the first, a stable $\delta^{13}C$ value of about $-7.5‰$ was established in the CO_2 of air above the plants in the microcosm. In the second, a stable $\delta^{13}C$ value of $+22‰$ was established in the CO_2 of air above a solution in which plant cells were suspended. These values, almost identical to those of the planetary atmosphere, may simply reflect the isotopic steady state balance between photosynthesis and photorespiration (see also Berry et al., Chapter 16, this volume).

...TO THE GLOBAL GREENHOUSE

The $\delta^{13}C$ value of atmospheric carbon is becoming inexorably more negative as the CO_2 concentration rises, due to the input of CO_2 from combustion of fossil C_3 photosynthates. Mook et al. (23) found a seasonably adjusted trend of $-0.02‰$ per ppm increase in CO_2 concentration which, in the last 200 years amounted to about 80 ppm or $-1.6‰$. This trend is precisely tracked in the annual rings of bristlecone pines in the White Mountains of Nevada and California, which were about $-1.6‰$ more negative in 1983 than in 1783 (21).

Seasonal changes, which cannot be resolved in the above analysis, show an oscillation of about 10 ppm in atmospheric CO_2 in the Northern Hemisphere. In addition, every spring and summer the atmosphere becomes about $0.8‰$ more enriched in $\delta^{13}C$, as Rubisco discriminates against $^{13}CO_2$ during the global photosynthetic draw-down of CO_2 concentration (Fig. 17.2). Every fall and winter, the atmospheric CO_2 becomes depleted in ^{13}C as respiration of photosynthetic products in the biosphere returns carbon depleted by Rubisco during the previous growing season. Fossil fuel expenditure adds a net $1–1.5$ ppm each year, all of it derived from fossil photosynthate which is only slightly more negative in $\delta^{13}C$ value, and which cannot be distinguished globally.

It is indicative of the traditions of global change research that small long-erm trends have been emphasized while 10–40 times larger predictable annual oscillations have been ignored. These atmospheric changes imposed by plants demonstrate that the faster response times of plants and microorganisms overwhelm the slow mixing of global pools. This is indisputable evidence for the role of photosynthesis in global atmospheric CO_2 regulation.

There is no argument that the ocean and other nonbiotic pools contain vastly greater amounts of carbon than the terrestrial biosphere, but for practical purposes this is irrelevant when plants can impose such rapid changes on the atmosphere. These plant activities can provide a feedback mechanism that will determine the rate, and perhaps even the direction, of long-term changes. For example, as Berry demonstrates in Chapter 16, an increase in atmospheric CO_2 that brings about a rise in temperature will affect photosynthesis and respiration, differentially enhancing the latter to a greater extent. The result will be a rapid increase in the biospheric compensation point for CO_2, and amplification of the slow, long-term trend.

Early comprehensive models (29) confirm that photosynthesis in the Northern Hemisphere biosphere is a principal factor in these changes of CO_2 concentration and isotopic composition (Fig. 17.3). These models also identify key uncertainties associated with marine photosynthesis at high southern latitudes. Tans et al. (39) calculate that there should be a terrestrial sink for carbon at temperature latitudes that balances 38–64% of the annual fossil carbon input. We are prepared to accept

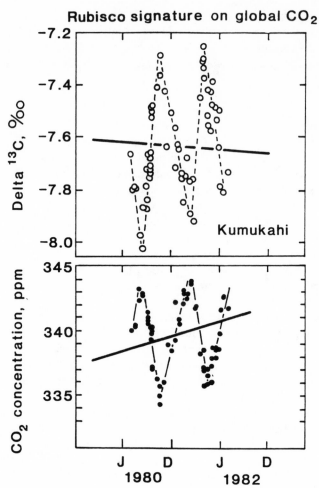

Fig. 17.2. Seasonal changes in atmospheric CO_2 concentration, and $\delta^{13}C$ value of atmospheric CO_2 at Kumukahi, Hawaii. The data show the isotopic signature which links the summer draw-down of CO_2 to isotope discrimination by Rubisco. [Redrawn from Mook et al. (23).]

that this sink is land plant photosynthesis, mostly of the C_3 variety. We also endorse the pleas of Tans et al. (39) for more high-precision, large-scale measurements of atmospheric CO_2 concentration and the $\delta^{13}C$ value of CO_2, as a means to better assess these possibilities.

This significant role of terrestrial plant photosynthesis on the global scale has been emphasized by another set of large-scale observations. Friedl et al. (14) reported, and Francey and Tans (13) independently confirmed, that the $\delta^{18}O$ value of atmospheric CO_2 reflected not the $\delta^{18}O$ of oceanic and cloud water with which it was previously thought to exchange, but water in plant leaves, which is enriched in ^{18}O as a result of transpiration. The latter authors calculated that each year, approximately 200 G tonnes of carbon in atmospheric CO_2 are exposed

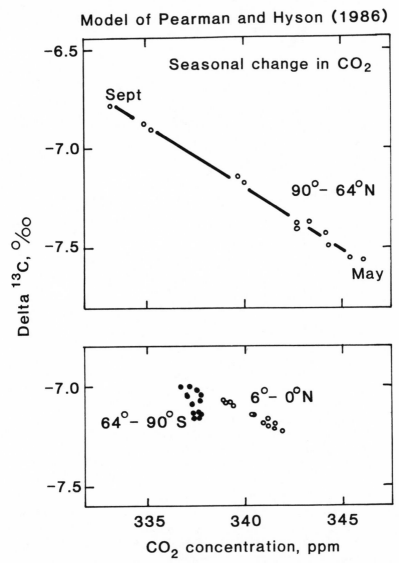

Fig. 17.3. Simulation of the coordinated changes in CO_2 concentration and $\delta^{13}C$ value of CO_2 in the atmosphere using a 3 reservoir model. [Redrawn from Pearman and Hyson (29).] The model conforms to observations in northern latitudes, and shows the expected damping at equatorial and southern latitudes, as expected from oceanic interactions and the paucity of biospheric processes, respectively.

to leaf water in which carbonic anhydrase facilitates oxygen isotope exchange between H_2O and CO_2. Figure 17.4 shows that the maximum value for $\delta^{18}O$ in atmospheric CO_2 at one Northern Hemisphere site occurs well before the maximum value for $\delta^{13}C$, as would be expected from massive CO_2 exchange with ^{18}O-enriched leaf water that follows stomatal opening with the greening of the terrestrial biosphere during spring and early summer.

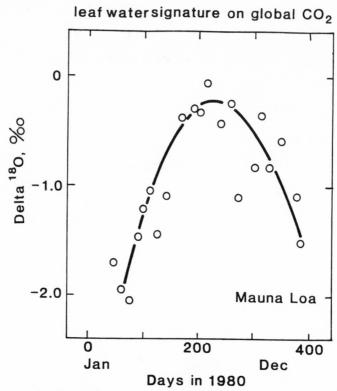

Fig. 17.4. Seasonal change in $\delta^{18}O$ value of atmospheric CO_2 at Mauna Loa, Hawaii. The data show a seasonal trend and isotopic signature consistent with the equilibration of atmospheric CO_2 with leaf water in the Northern Hemisphere, facilitated by carbonic anhydrase. [Data of Mook et al. (23); interpretations proposed by Freidli et al. (14), Francey and Tans(13).]

These observations opened up a big picture for carbonic anhydrase in photosynthesis and prompted a new series of experiments on the way it facilitates isotopic exchange of ^{18}O with water in the metabolic compartment. Its potential roles in facilitating CO_2 exchange in C_3 plants have been modeled with the usual perspicacity by Cowan (3), and Graham Farquhar is acknowledged to have anticipated the potential global scale of this process (13). Yet we are still poorly prepared to put this insight into context, as judged by our rudimentary knowledge of the dynamics of leaf water δD and $\delta^{18}O$ values when the stomata are open. Although Yakir et al. (47) have suggested the existence of three compartments for leaf water with different $\delta^{18}O$ values, we do not know how their specific isotopic compositions are determined (46). Until recently, we knew little about which compartment has the carbonic anhydrase and, therefore, which one is likely to impart the $\delta^{18}O$ signal to atmospheric CO_2.

Nevertheless, our physiological insights into leaf water compartmentation already have global implications. Francey and Tans (13) commented that "we find from global model calculations (based on exchange with leaf water) that the $\delta^{18}O$ of CO_2 would be more enriched than what is observed by 1–2‰. The discrepancy

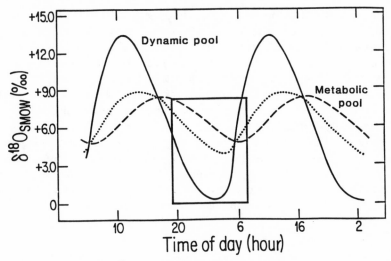

Fig. 17.5. Model of average $\delta^{18}O$ value of leaf water ($\cdots\cdots$) throughout the diurnal cycle based on data from cotton (49). The model includes the estimated pool sizes of transpiring water (dynamic pool) and cellular water (metabolic pool) from Yakir et al. (47). The $\delta^{18}O$ value of leaf water modeled by Francey and Tans (13) only corresponds to that of the dynamic pool. This figure shows that the assumptions of these authors may have been displaced several ‰ if CO_2 exchanges with leaf water in which carbonic anhydrase is located, i.e. in the metabolic pool.

would be greater if the relative humidity is lower during the time stomates are open." To explain this discrepancy, they speculated on another nonbiotic CO_2 exchange mechanism. However, our data suggest that an alternative mechanism may be unnecessary. Figure 17.5 shows the diurnal fluctuations in $\delta^{18}O$ of leaf water and its two major components, modeled on observations with cotton plants. Because Francey and Tans (13) considered leaf water as an open evaporating pan, their calculations are based only on the dynamic (transpiring) pool shown in Figure 17.5. It can be readily seen that both the fixed total leaf water and the more isolated metabolic pool, in which carbonic anhydrase-facilitated isotopic exchange may occur, are 2–4‰ less enriched than the dynamic pool.

Direct proof of these speculations has been obtained recently by Yakir et al. (46, 50). The photorespiratory CO_2 of leaves held at the CO_2 compensation point under conditions of controlled water flux and steady-state water compartmentation, shows the isotopic composition of water in the metabolic compartment of leaves. These physiological and biochemical insights provide a simple explanation for the discrepancy identified by Francey and Tans (13), and further understanding of the ^{18}O signal in atmospheric CO_2. Moreover, these results provide the key needed to disclose the most elusive isotopic process, the $\delta^{18}O$ mass balance of the atmosphere.* Such mass balance calculations, based on the isotopic signature

---

* Note added in proof. See the recent paper by Farquhar, G. D., Lloyd, J., Taylor, J. A., Flanagan, L. B., Sylvestson, J. P., Husick, K. T., Wong, S. C., and Ehleringer, J. P. 1993. Vegetation effects on the isotope composition of oxygen in the atmosphere. *Nature* **363**: 439–43.

of chloroplast water, are analogous to those done for the corn shuck described above, and may provide estimates of the gross primary productivity of the biosphere. Certainly they will provide higher spatial and temporal resolution than was possible previously—resolution that is needed to improve the current, extremely coarse global models (50). Clearly, much more research into leaf photosynthetic CO_2/H_2O exchange properties is needed to bring them into a global context, before this particular isotopic signature can be fully evaluated.

It is especially fitting, in the context of this symposium and volume, that photorespiratory CO_2 should hold so much promise for understanding photosynthetic impacts on the global atmosphere (50). Our understanding of the origin of photorespiratory CO_2 from glycine in the photorespiratory oxidation cycle owes much to the personal research efforts of Ed Tolbert. In this context, may we be permitted a few laconic, and some not so serious, observations by way of conclusions?

CONCLUSIONS

When Ed Tolbert left the Calvin laboratory, the photosynthetic cycle was a beautiful thing on paper, but it could not function very well in the present atmosphere. It was known to be driven by Rubisco, but what a miserable catalyst it was then. If left to themselves in the light with air levels of CO_2, laboratory-grown algae or laboratory-prepared chloroplasts tended to dump all of the carbon of the photosynthetic carbon reduction (PCR) cycle as glycolate (Fig. 17.6A). From the point of view of efficiency of solar energy utilization, this seemed a rather pointless expenditure of energy. So far as CO_2 assimilation in air was concerned, the early photosynthetic bicycle could only move backwards.

Kortschak and Karpilov discovered that at least in sugar cane and maize, the cycle was operated through an entirely different axle. Early enthusiasm for this new design was such that, for some years, Hatch and Slack thought that they could dispense with Rubisco altogether. But this was not so. Björkman and Gauhl, and Berry, Downton, and Tregunna demonstrated the need for persistent grinding to extract all of the chloroplasts in saltbush leaves; thus the location of the elusive Rubisco in bundle sheath cells of C_4 plants was revealed. The discovery of obligingly separable mesophyll and bundle sheath cells in crabgrass enabled Black and Edwards to demonstrate how this modification of the photosynthetic cycle could really move into gear. With the C_4 cycle elevating internal CO_2, the even more indifferent Rubisco (with a higher K_m for CO_2) in these plants nevertheless could move the photosynthetic cycle in a forward direction (Fig. 17.6B).

Recognizing that the world belongs to C_3 plants, Ed Tolbert and his colleagues spearheaded an entirely new design—the integrated photosynthetic bicycle. With Ogren and Bowes, throughout the 1970s, they elucidated the full contrarotatory predilections of Rubisco. They also recognized the engineering opportunities presented by mitochondria and peroxisomes as gears that could convert the inevitabilities of Rubisco function in air, into recovery of CO_2 and NH_3 in the photorespiratory carbon oxidation (PCO) cycle. These minor players are indispensable to the forward movement of the two big wheels (Fig. 17.6C). Proof

Evolution of the photosynthetic bicycle

circa 1960
Your basic Bassham–Benson–Calvin cycle

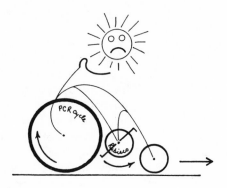

circa 1970
The Kortschak Hatch–Slack modification

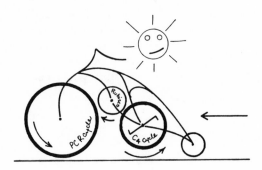

circa 1980
The integrated Tolbert bicycle

Fig. 17.6. Evolution of the photosynthetic bicycle, as described in the text. [Redrawn from many idle moments in which the opinions expressed were not necessarily those of the sponsors; CBO is entirely responsible for the representation of this figure).

of the concept was provided by Ogren and Somerville, who established that if one sought to rationalize the design by eliminating one enzyme at a time, the bicycle would not work at all! It is an elegant bicycle, and there are few vehicles in plant biology so comprehensively adequate to explain photosynthetic responses to atmospheric CO_2 and O_2 concentration, from molecular mechanisms through to environmental effects on the whole plant.

At this meeting Ed Tolbert has given us the challenge of contributing to an understanding of the biggest cycle of them all, the biogeochemical cycle. We have argued that stable isotopic evidence has established the role of land plant photosynthesis as a key transducer in global cycles of carbon, oxygen, and water vapor. The physiology and biochemistry of photosynthesis in plant leaves give rise to isotopic signals that are key indicators of the fluxes involved. As researchers in photosynthesis we are much more experienced in working from the leaf level down to the molecular level. We have the capacity to generate ever increasingly detailed accounts of the fundamental processes that sustain plants and, hence, all other life processes on earth. However, we have argued here that it is important to reverse some of the reductionist tendencies in photosynthetic research if we are to make credible contributions to global change research. This will require us to reassess many issues, such as the isotopic composition of water that participates in photosynthetic O_2 evolution in different plants, the locations and activities of carbonic anhydrase, and the isotopic fractionations associated with reductive and oxidative events. In the nature of things, the big issues that often seem so vague will force us to reconsider fundamental mechanisms, about which we often become so glib. We thank Ed Tolbert for this opportunity to come down to earth, and up for fresh air.

ACKNOWLEDGMENTS

We are grateful for the stimulus to expand our horizons that we have enjoyed in collaborations with Dr. Joseph Berry, and for his original, unpublishable speculations on the evolution of photosynthetic bicycles over many years. Our research using stable isotopes is supported in part by DE-FG05-89ER14005 from the Department of Energy.

REFERENCES

1. Antonielli, M., and Venanzi, G. 1979. Structural properties of the rachis and hypsophyll of the maize ear. *Plant Sci. Lett.* **15**: 302–4.

2. Bender, M. M. 1968. Mass spectrometric studies of carbon-13 variation in corn and other grasses. *Radiocarbon* **10**: 468–72.

3. Cowan, I. R. 1986. Economics of carbon fixation in higher plants. In *On the Economy of Plant Form and Function*, ed. T. J. Givnish, pp. 133–70. Cambridge Univ. Press, Cambridge.

4. Craig, H. 1953. The geochemistry of stable carbon isotopes. *Geochim. Cosmochim. Acta* **3**: 53–92.

5. DeNiro, M. J., and Epstein, S. 1979. Relationship between oxygen isotope ratios of terrestrial plant cellulose, carbon dioxide and water. *Science* **204**: 51–3.

6. [Editorial] 1986. *Nature* **323**: 189.

7. Estep, M. F., and Hoering, T. C. 1981. Stable hydrogen isotope fractionation during autotrophic and mixotrophic growth of microalgae. *Plant Physiol.* **67**: 474–7.

8. Evans, J. R., Sharkey, T. D., Berry, J. A., and Farquhar, G. D. 1986. Carbon-isotope discrimination measured concurrently with gas-exchange to investigate CO_2 diffusion in leaves of higher plants. *Aust. J. Plant. Physiol.* **13**: 281–92.

9. Farquhar, G. D., O'Leary, M. H., and Berry, J. A. 1982. On the relationship between carbon isotope discrimination and the intercellular carbon dioxide concentration in leaves. *Aust. J. Plant Physiol.* **9**: 121–37.

10. Farquhar, G. D., and Richards, R. A. 1984. Isotopic composition of plant carbon correlates with water use efficiency of wheat genotypes. *Aust. J. Plant Physiol.* **11**: 539–52.

11. Farris, F., and Strain, B. R. 1978. The effects of water stress on leaf $H_2^{18}O$ enrichment. *Rad. Environ. Biophys.* **15**: 167–202.

12. Francey, R. J., and Farquhar, G. D. 1982. An explanation of C-13/C-12 variations in tree rings. *Nature* **297**: 28–31.

13. Francey, R. J., and Tans, P. P. 1987. Latitudinal variation in oxygen 18 in atmospheric CO_2. *Nature* **327**: 495–7.

14. Friedl, H., Sigenthaler, U., Rauber, D., and Oeschger, H. 1987. Measurements of concentration, $^{13}C/^{12}C$ and $^{18}O/^{16}O$ ratios of tropospheric carbon dioxide over Switzerland. *Tellus* **39B**: 80–8.

15. Griffiths, H., Boardmeadow, M. S. J., Borland, A. M., and Hetherington, C. S. 1990. Short-term changes in carbon-isotope discrimination identify transitions between C_3 and C_4 carboxylations during Crassulacean acid metabolism. *Planta* **181**: 604–10.

16. Guy, R. D., Fogel, M. F., Berry, J. A., and Hoering, T. H. 1986. Isotope fractionation during oxygen production and consumption by plants. In *Progress in Photosynthesis Research*, ed. J. Biggins, Vol. III, pp. 597–600. Marthinus Nijhoff, Dordrecht.

17. Hesketh, J. D., and Musgrave, R. B. 1962. Photosynthesis under field conditions IV. Light studies with individual corn leaves. *Crop Sci.* **2**: 311–15.

18. Hubick, K. T., Shorter, R., and Farquhar, G. D. 1989. Heritability and genotype × environment interactions of carbon isotope discrimination and transpiration efficiency in peanut (*Arachis hypogea* L.). *Aust. J. Plant Physiol.* **15**: 799–814.

19. Langdale, J. A., Zelitch, I., Miller, E., and Nelson, T. 1988. Cell position and light influence C_4 versus C_3 photosynthetic gene expression in maize. *EMBO J.* **7**: 3643–51.

20. Leaney, F. W., Osmond, C. B., Allison, G. B., and Ziegler, H. 1985. Hydrogen isotope composition of leaf water in C_3 and C_4 plants: Its relationship to the hydrogen isotope composition of dry matter. *Planta* **164**: 215–20.

21. Long, A., Leavitt, S., and Chen, S. 1987. Carbon-13/carbon-12 variations in bristlecone pine over the past 500 years and their relation to climate and global atmospheric CO_2. In *Proc. Int. Symp. Ecol. Aspects of Tree Ring Anal.*, compiled by G. C. Jacoby and J. W. Hornbeck, pp. 485–93. CONF8608144. U.S. Department of Energy, Washington, D.C.

22. Masle, J., and Farquhar, G. D. 1988. Effects of soil strength in relation to water use efficiency and growth to carbon isotope discrimination in wheat seedlings. *Plant Physiol.* **86**: 32–8.

23. Mook, W. G., Koopmans, M., Carter, A. F., and Keeling, C. D. 1983. Seasonal latitudinal and secular variations in the abundance and isotope ratios of atmospheric carbon dioxide 1. Results from land stations. *J. Geophys. Res.* **88**: 10915–33.

24. O'Leary, M. H. 1981. Carbon isotope fractionation in higher plants. *Phytochemistry* **20**: 553–7.

25. O'Leary, M. H., and Osmond, C. B. 1980. Diffusional contribution to carbon isotope fractionation during dark CO_2 fixation in CAM plants. *Plant Physiol.* **66**: 931–4.

26. O'Leary, M. H., Reife, J. E., and Slater, J. D. 1981. Kinetic and isotope effect studies of maize phosphoenol pyruvate carboxylase. *Biochemistry* **20**: 7038–41.

27. O'Leary, M. H., Treichel, I., and Rooney, M. 1986. Short term measurement of carbon isotope fractionation in plants. *Plant Physiol.* **80**: 578–82.

28. Osmond, C. B. 1989. Photosynthesis from the molecule to the biosphere: A challenge for integration. In *Photosynthesis*, ed. W. R. Briggs, pp. 5–17. A. R. Liss, New York.

29. Pearman, G. I., and Hyson, P. 1986. Global transport and interreservoir exchange of carbon dioxide with particular reference to stable isotope distributions. *J. Atmosph. Chem.* **4**: 81–124.

30. Quade, J., Cerling, T. E., and Bowman, J. R. 1989. Development of Asian monsoon revealed by marked ecological shift during the latest Miocene in northern Pakistan. *Nature* **342**: 163–6.

31. Roeske, C. A., and O'Leary, M. H. 1984. Carbon isotope effects on the enzyme-catalyzed carboxylation of ribulose bisphosphate. *Biochemistry* **23**: 6275–84.

32. Rooney, M. A. 1988. *Short-Term Carbon Isotope Fractionation in Plants*. Ph.D. Thesis, University of Wisconsin, Madison.

33. Rundel, P. W., Ehleringer, J. R., and Nagy, K. A. eds. 1988. *Stable Isotopes in Ecological Research*. Ecological Studies, Vol. 68. Springer Verlag, Heidelberg.

34. Schmidt, H.-L., and Winkler, F. J. 1979. Einige Ursachen der Variations breite von ^{13}C- Werten bei C_3- und C_4- Pflanzen. *Ber. Deutsch. Bot. Ges.* **92**: 185–91.

35. Smith, B. N., and Epstein, S. 1970. Biochemistry of the isotopes of hydrogen and carbon in salt marsh biota. *Plant Physiol.* **46**: 738–42.

36. Soldatini, G. F., Antonielli, M., Venanzi, G., and Lupattelli, M. 1982. A comparison of the metabolism of the ear and accompanying tissues in *Zea mays* L. *Z. Pflanzenphysiol.* **108**: 1–8.

37. Sternberg L da, S. L. 1988. Oxygen and hydrogen isotope ratios in plant cellulose: Mechanisms and applications. In *Stable Isotopes in Ecological Research*. Ecological Studies, eds. P. W. Rundel, J. R. Ehleringer and K. A. Nagy. Springer-Verlag, Heidelberg, Vol. 68, 124–41.

38. Sternberg L da, S. L., DeNiro, M. J., and Savidge, R. A. 1986. Oxygen isotope exchange between metabolites and water during biochemical reactions leading to cellulose synthesis. *Plant Physiol.* **82**: 423–7.

39. Tans, P. P., Fung, Y., and Takahashi, T. 1990. Observational constraints on the global atmospheric CO_2 budget. *Science* **247**: 1431–8.

40. Vogel, J. C. 1980. Fractionation of the Carbon Isotopes During Photosynthesis. Sitz. Ber. Heidelberg Akad. Wiss. Math. Naturwiss. Kl., pp. 113–35.

41. von Caemmerer, S., and Hubick, K. T. 1989. Short-term carbon-isotope discrimination in C_3–C_4 intermediate species. *Planta* **718**: 475–81.

42. Walker, C. D., Leaney, F. W., Dighton, J. C., and Allison, G. B. 1989. The influence of transpiration on the equilibration of leaf water with atmospheric water vapour. *Plant Cell Environ.* **12**: 221–34.

43. Whelan, T., Sackett, W. M., and Benedict, C. R. (1973). Enzymic fractionation of carbon isotopes by phosphoenolpyruvate carboxylase from C_4 plants. *Plant Physiol.* **51**: 1051–4.

44. Winkler, F. J., Wirth, E., Latzko, E., Schmidt, H.-L., Hoppe, W., and Winner, P. 1978. Einfluss von wachstumsbedingungen und Entwicklung auf $\delta^{13}C$-Werte in verschiedenen Organen und Inhaltsstoffen von Weizen, Hafer und Maiz. *Z. Pflanzenphysiol.* **87**: 255–63.

45. Winter, K., Lüttge, U., Winter, E., and Troughton, J. H. 1978. Seasonal shift from C_3 photosynthesis to Crassulacean acid metabolism in *Mesembryanthemum crystallinum* growing in its natural environment. *Oecologia* **34**: 225–37.

46. Yakir, D. 1992. Water compartmentation in plant tissue. In Water and Life, ed. G. N. Somero, C. B. Osmond, and L. S. Bolis, Springer-Verlag, Heidelberg, Chap. 12, pp. 205–22.

47. Yakir, D., DeNiro, M. J., and Rundel, P. W. 1989. Isotopic inhomogeneity of leaf water: Evidence and implications for the use of isotopic signals transduced by plants. *Geochim. Cosmochim. Acta* **53**: 2769–73.

48. Yakir, D., and DeNiro, M. J. 1990. Oxygen and hydrogen isotope fractionation during metabolism in *Lemna gibba* L. *Plant Physiol.* **93**: 325–32.

49. Yakir, D., DeNiro, M. J., and Gat, J. R. 1990. Natural deuterium and oxygen-18 enrichment in leaf water of cotton plants grown under wet and dry conditions: Evidence for water compartmentation and its dynamics. *Plant Cell Environ.* **13**: 49–56.

50. Yakir, D., Berry, J. A., Giles, L., and Osmond, C. B. 1993. The ^{18}O of water in the metabolic compartment of transpiring leaves. In: *Stable Isotopes and Plant-Water Relations.* eds. J. R. Ehleringer, A. E. Hall, and G. D. Farquhar, pp. 529–40. Academic Press, San Diego, Calif.

51. Yakir, D., Osmond, C. B., and Giles, L. 1992. Autotrophy in maize husk leaves. *Plant Physiol.* **97**: 1196–8.

52. Ziegler, H. 1988. Hydrogen isotope fractionation in plant tissues, ed. P. W. Rundel et al., pp. 105–23.

53. Ziegler, H., Osmond, C. B., Stickler, W., and Trimborn, P. 1976. Hydrogen isotope discrimination in higher plants: Correlations with photosynthetic pathway and environment. *Planta* **128**: 85–92.

Index